The Life and Science of Léon Foucault

The Man Who Proved the Earth Rotates

In 1851 a young French physicist erected a giant pendulum in the heart of Paris and showed astonished spectators that the Earth was turning beneath their feet. Pendulum mania swept the learned and everyday worlds and Léon Foucault's name became synonymous with his famous pendulum. The demonstration continues to captivate a century and a half later.

The history and interpretation of the pendulum experiment are described in terms suitable for the general reader in this abundantly illustrated biography of Foucault. His contributions to science went well beyond the pendulum, however: most notably to the gyroscope, to decisive laboratory measurements of the speed of light, and to the invention of the telescope in its modern form. Foucault was a talented early photographer. Gifted with his hands, he valued precision and loved clockwork. He worked in optics and electricity. Though steadfast in friendship, he could be stubborn and blunt. His collaboration with Hippolyte Fizeau ended in rift while the frankness of the newspaper articles that he wrote provoked hostilities that hindered his acceptance by academic peers. Telescope making and dreams of wealth from an industrial governor were curtailed by an agonizing early death. The blend of pure and applied in Foucault's work and the ordeals he suffered make him an intriguing case study as one of the last amateur scientists at a time when science was becoming institutionalized. The book assumes some familiarity with simple scientific terms, but no detailed knowledge of physics is required.

This biography offers a fascinating read about an unconventional scientific pioneer whose independent spirit led to acclaimed and unexpected discoveries but whose horizon was limited by his disdain for abstraction.

WILLIAM TOBIN, M.A., Ph.D., F.R.A.S., was born in Manchester, England, in 1953 and attended Stockport Grammar School. He read Natural Sciences at Emmanuel College, Cambridge and took his doctorate in astronomy at the University of Wisconsin at Madison. Since then he has worked at the University of St Andrews, the Laboratoire d'Astronomie Spatiale in Marseilles, the Marseilles Observatory (where Foucault's largest telescope is preserved and Tobin's own interest in Foucault was sparked) and the Université de Provence. Since 1987 he has been at the University of Canterbury in Christchurch, New Zealand, where he has been Director of the Mount John University Observatory and is currently a part-time senior lecturer in astronomy, splitting his time between New Zealand and France. He has spent sabbatical leave at the Institut d'Astrophysique de Paris. Research interests besides Foucault centre on eclipsing binary stars in the Magellanic Clouds and the comet-like objects surrounding the deep southern star Beta Pictoris. He is a Fellow of the Royal Astronomical Society and a member of the International Astronomical Union. Among his distinctions are the New Zealand Institute of Physics Journalism Award (1993), the Arthur Beer Memorial Prize for the best article in the journal *Vistas in Astronomy* (1994), and the Mechaelis Memorial Prize administered by the University of Otago for contributions to astronomy (1997).

LÉON FOUCAULT in 1867 (aged forty-seven).

Despite the scorn of this throng of drudges who believe that they are building something when they only dress stones, and notwithstanding their claim that genius is no more than the passive contemplation of facts, imagination will always be one of the most powerful levers of the intelligence; it is what gives intelligence wings, it is what lifts intelligence up and allows it to soar above mere observation.

In the *Journal des Débats*, 1845 November 14

The Life and Science of
LÉON FOUCAULT

The Man Who Proved the Earth Rotates

William Tobin

Senior Lecturer in Physics and Astronomy
University of Canterbury

PUBLISHED BY THE PRESS SYNDICATE OF THE UNIVERSITY OF CAMBRIDGE
The Pitt Building, Trumpington Street, Cambridge, United Kingdom

CAMBRIDGE UNIVERSITY PRESS
The Edinburgh Building, Cambridge CB2 2RU, UK
40 West 20th Street, New York, NY 10011–4211, USA
477 Williamstown Road, Port Melbourne, VIC 3207, Australia
Ruiz de Alarcón 13, 28014 Madrid, Spain
Dock House, The Waterfront, Cape Town 8001, South Africa

http://www.cambridge.org

© William John Tobin 2003

This book is in copyright. Subject to statutory exception
and to the provisions of relevant collective licensing agreements,
no reproduction of any part may take place without
the written permission of Cambridge University Press.

First published 2003

Printed in the United Kingdom at the University Press, Cambridge

Typeface Times 10/12pt.   *System* LATEX 2_ε [TB]

A catalogue record for this book is available from the British Library

Library of Congress Cataloguing in Publication data

Tobin, W. (William)
The life and science of Léon Foucault: the man who proved the earth rotates / William Tobin.
    p.   cm.
Includes bibliographical references and index.
ISBN 0 521 80855 3
1. Foucault, Léon, 1819–1868. 2. Pendulum. 3. Physicists – France – Biography.
I. Title.
QC16.F626 T63 2003
530′.092—dc21   2003043577

ISBN 0 521 80855 3 hardback

French adaptation: *Léon Foucault. Le miroir et le pendule.* EDP-Sciences, Les Ulis, France, 2002 (J. Lequeux, translator).

The publisher has used its best endeavours to ensure that the URLs for external websites referred to in this book are correct and active at the time of going to press. However, the publisher has no responsibility for the websites and can make no guarantee that a site will remain live or that the content is or will remain appropriate.

Contents

Foreword		vii
Preface		ix
Acknowledgments		xiii
1	Introduction	1
2	Early years	11
3	The metallic eye: photography	21
4	The 'delicious pastime' applied to science	37
5	The beautiful science of optics	57
6	Order, precision and clarity: reporter for the *Journal des Débats*	79
7	Mixed luck	95
8	The speed of light. I. Demise of the corpuscular theory	117
9	The rotation of the Earth: pendulum and gyroscope	133
10	Biding time	173
11	The Observatory physicist	183
12	Perfecting the telescope	199
13	The speed of light. II. The size of the solar system	227
14	Recognition	235
15	Control: the quest for fortune	247

16 Unfinished projects	**263**
17 Commentary	**279**
Colour plates	**285**
Appendices	
A Maps and chronology	293
B Extracts from the *Journal des Débats*	296
C Photographs and instruments	305
D Building a Foucault pendulum	307
Notes	**311**
Index	**333**

Foreword

The work before you is a means to help an English-speaking audience better appreciate the complex intellectual origins of modern physics and, in particular, of astrophysics. Léon Foucault and his contemporaries provided many of the strategies that evolved into the practice of first empirical and ultimately rational observation in modern astrophysics.

The roots of astrophysics can be traced back to many sources. In its greatest generality, it began when the Universe was considered physical, subject to physical law. Thus Galileo, Kepler and Newton gave us the Newtonian clockwork universe, which was elaborated as celestial mechanics. Its continued refinement brought Lagrange, Laplace, Gauss and Le Verrier to fame. The other side of astrophysics, the side most associated with the name today – not the dynamics of planetary and stellar systems, but the study of the form and substance of those bodies comprising the systems, and the courses of their lives – remained far more speculative. Still, John Mitchell, Henry Cavendish and, especially, William Herschel explored matter theory, its cosmical ramifications, and speculated on the solar constitution. The observational tools for inspection and analysis that would turn speculation into scientific exploration only began to emerge during the time Foucault was active.

During most of Foucault's short life, of course, the term astrophysics was not yet in use. It was hardly codified, or even recognized as a new discipline. The term was first coined in German in the mid-1860s and the first popular reference in English came in 1869 when the Greenwich astronomer Edwin Dunkin called someone who performed telescopic observations of the spectra of celestial bodies and compared them to the spectra of laboratory substances an 'astro-physicist'.[1] By then Foucault was already dead by a year. The definition was elaborated in the following decades of the century, reaching a near-modern state only in the late 1880s when Charles Augustus Young wrote that astronomical physics, or astro-physics 'treats of the physical characteristics of the heavenly bodies, their brightness and spectroscopic peculiarities, their temperature and radiation, the nature and condition of their atmospheres and surfaces, all the phenomena which indicate or depend on their physical condition'.[2] So although astrophysics would not be a word known or used by Foucault, in many ways, his life and times coincided with an era when the nascent tools of astrophysics were being honed to the point where they became usable, and finally did establish a wholly new astronomy, a new way to look at and interpret the universe.

If Foucault is not remembered today as a dominant figure in the history of modern physics and astrophysics, it is only because our histories do not yet fully elucidate the complex root structure of the system that created those fields. Most writers and chroniclers of the history of astronomy and astrophysics in the late nineteenth and most of the twentieth centuries concentrated their attention and praise on the application of the spectroscope by Kirchhoff and his followers, and the spectacular elaboration of spectrum analysis to many problems in astronomy over the following decades. Slowly, some more contemporary writers in the late twentieth century began to realize that the roots to astrophysics were more complex, engaging broad swaths of physics and chemistry including electricity and magnetism, physical optics and thermodynamics. And even more recently, some of those writers began to emphasize the importance of getting at the changing nature of problem formation in astrophysics, as it migrated from a highly empirical science to a rational enterprise shaped by modern physical theory. These are all positive historiographical steps, to be sure, but still elusive is a fully adequate elucidation of the interdependence of the science with the creation and evolution of its observational tools. This is where understanding Foucault's life and times will help.

As you read through the following chapters, keep in mind all the wonders of modern astronomical technology that were touched upon by Foucault. Don't worry or fret over the assignment of credit or who did what first – concentrate on Foucault's independence of mind and the tireless search for certainty through precision, and above all the prodigious concentration of intellectual energy and enthusiasm that led him to explore so many tools and techniques that today define the practice of modern astrophysics.

William Tobin's goal is to use biography as an organizing tool to illuminate the specific areas of science and technology explored by Foucault. There is no better framework to use to explore a life and its times than biography. Focusing on a person is the best way to carry the reader through what otherwise might be rather difficult literary terrain. By organizing descriptive and explanatory narrative around a person's life, Tobin has provided a rational structure that every reader will identify and appreciate and follow with certainty. So even though you will often find yourself on a branch off the biographical trunk of the book that Tobin has crafted, you will know exactly where you are and how you got there, exploring the scientific principles that underlie Foucault's work: encountering a life in science, with all its right and wrong turns, risks, successes and failures, uncertainties and convictions.

David DeVorkin

Preface

Why Léon Foucault?

> His life presents nothing worth telling except the discoveries he made.

This judgment by the editor of the *Recueil* (or *Collected Works*) published after Foucault's death may seem unpropitious for a biographer, but more than a century later his life as well as his science is instructive and full of fascination.

Foucault's name is known to millions because of his eponymous pendulum, which through a gradual drift of its swing plane reveals the slow, daily rotation of the Earth. The rotation of a merry-go-round gives a sharp, sideways kick to a child jumping from one wooden horse to another, but the apparent absence of any similar dynamical effects resulting from the Earth's rotation was an objection levied against the heliocentric solar systems of both Copernicus and the ancients. Since Copernicus' time, and more particularly since Galileo's, savants had been dropping weights from towers or down wells and firing cannon balls vertically upwards in a vain search for evidence of the terrestrial rotation in subtle deviations of the flight paths of these projectiles (Fig. P.1). It was Foucault, with apparatus no more complex than a weighty ball swinging on a wire, who finally produced the long-sought proof in 1851. His pendulum experiment caused a sensation then and continues to be repeated to this day.

Fig. P.1. Where will the vertical cannon-shot fall – or will it even fall at all? See Chapter 9.

Less well known is that in the following year, 1852, Foucault constructed another apparatus, based on a rotating torus, which also demonstrates the Earth's rotation. He named this device the gyroscope, and mechanical gyroscopes were of immense practical importance in navigation for most of the twentieth century. Earlier, he had driven the final nail into the coffin of Descartes' and Newton's corpuscular theory of light when, with a fast-spinning, fingernail-sized mirror, he had shown that light travels faster in air than in water. Subsequently, he adapted this experiment to obtain the first accurate laboratory measurement of the speed of light in absolute terms (i.e. in metres per second). This confirmed a prediction made by his boss, the autocratic Urbain Le Verrier, director of the Paris Observatory, and shrank the estimated size of the solar system to essentially the now-accepted value. While working at the Paris Observatory, Foucault also devised the optical testing techniques which permitted the fabrication of large, perfectly shaped optical surfaces. The palm-sized telescope lenses and mirrors, which were all that was possible with hit-or-miss polishing, gave way to the man-sized and larger mirrors of modern reflecting telescopes.

The pendulum and gyroscope, the differential and absolute speeds of light, and authorship of the telescope in its modern form are Foucault's five great achievements. They have immortalized his name and provide more than sufficient reason to study his life. But Foucault made many other contributions to nineteenth-century science. He was one of the first to apply the daguerreotype to science. He invented an automatic arc light. He worked in optics, mechanics and electricity. For fifteen years he wrote reports of the weekly meetings of the French Academy of Sciences for an influential Paris newspaper, the *Journal des Débats*, where his frankness made enemies. In the final years before his early death, aged forty-eight, he tried to make his fortune with mechanical governors. These additional achievements, the blend of pure and applied in his work, and his status as an outsider at a time when science was becoming institutionalized are further reasons for the modern reader to be interested in Léon Foucault.

Who should read this book, and how is it arranged?

Science is often presented through the discoveries of its greatest practitioners with supposed crucial experiments and eureka moments. Foucault was no Newton or Einstein and his science is less fundamental. His genius is correspondingly more accessible, and although in some ways a tragic character, he is a hero to inspire the young as well as to charm and intrigue everyone.

Foucault was a scientist, and more particularly a physicist, though in English these terms were little-used during the nineteenth century. (The word *physicists* 'where four sibilant consonants fizz like a squib', was being deplored in 1843, while *man of science* was considered preferable to *scientist* as

Preface

late as the 1910s.[1]) One cannot write honestly about a scientist without explaining his or her work, at least qualitatively. In part, this is therefore a book of popular physics, written for the canonical *Scientific American* reader who is familiar with elementary technical terms and concepts such as sine (even if the exact definition has been forgotten), density, velocity and refraction, but who is not necessarily acquainted with topics such as the aberration of starlight or the optical layout of a telescope. The main text provides a taste of both Foucault's physics and what was required to be a scientist in the middle of the nineteenth century. It also offers an accessible case study of how, in science, one idea leads to another. I hope it provides an entertaining as well as an instructive read.

Appendices follow. Appendix A aids the reader with maps of France and Paris, and a chronology of Foucault's life. Additional extracts from Foucault's newspaper articles are given in Appendix B; they may also interest students of journalism. Appendix C describes where Foucault photographs, instruments and replicas can be found. Readers who wish to build a Foucault pendulum should consult Appendix D, the only part of this book which presumes prior knowledge of physics. The small, superscript numerals that pepper the text refer to notes and references for those such as historians of science who wish to check sources. Persons' years of birth and death are given systematically in the index. For dates, I adopt the astronomical format Year Month Day.

Foucault was no mathematician. His science is classical, and is often easy to quantify. In it, teachers of introductory physics will find beautiful examples to enliven their classes. To keep the book short, Foucault's lesser experiments are treated in shallower depth, but the interested reader can consult Foucault's original papers, most of which are available on-line through the GALLICA project of the French National Library or in a recent reprint of the *Recueil*.[2] Readers with deeper knowledge of physics and astronomy may be interested by the specialized articles that I wrote while researching this book.[3]

Women

Foucault slots between the two great women icons of nineteenth-century French science, the mathematician Sophie Germain (1776–1831), and the physicist Marie Curie (1867–1934). The few women who appear in this biography are relatives, empresses or anaesthetized. I hope no reader will think this scarcity of women indicates how science should be advanced in the twenty-first century. Women have much to contribute both to scientific discovery and in humanizing the presently tarnished social face of physics. I am encouraged by the relatively large proportion of women in French physics today, which puts English-speaking and especially Germanic and Nordic countries to shame. This book reports what was a man's world. As a counterbalance, I am pleased to dedicate it to women everywhere.

Fig. P.2. An unreliable source: the science popularizer Louis Figuier (*d*.1894), who like Foucault was born in 1819 and abandoned medicine for science.

Sources

In science, ideas stand or fall when confronted with experiment, observation or calculation. The reconstruction of a life, however, relies on the vagaries of the evidence that has survived. I can emit a hypothesis: for example, that Foucault did not marry because of fears of mental instability; but without documentary evidence this must remain speculative. Nor is documentary evidence the same thing as proof, since documents may mislead or be misunderstood. I have always preferred original manuscripts and papers to the writings of derivative authors such as the popularizer Louis Figuier (Fig. P.2), who worried little about factual accuracy provided the story-line was good; but many of the details of Foucault's life have perforce been gleaned from secondary sources, or inferred from circumstantial evidence. I discuss the difficulties of biography-writing at greater length in the final chapter.

Details about Foucault's family and style of life were furnished by notarial papers and the records of the *état civil*, or births, marriages and deaths. Files in the National Archives and elsewhere were crucial concerning Foucault's employment. But there are lacunae; the older Parisian *état civil*, for example, burned during the Paris Commune in 1871, and though Foucault was born in Paris, I have been unable to determine exactly where.

The evidence for Foucault's science is much more reliable and comes primarily from papers printed in the scientific journals of the day. Foucault's articles in the *Journal des Débats* illuminate human aspects of what he did, as do reports in general-readership magazines such as *Cosmos*. My scouring of archives revealed that sadly few manuscript papers in Foucault's characteristically tidy hand have survived. Some four thousand scientific manuscripts were inventoried after his death. A detailed index of them survives in the care of the Foucault family (Fig. P.3), and a selected few were reproduced in the *Recueil* published in 1878; but the papers themselves, which would answer so many questions, are tantalizingly missing, destroyed, according to family legend, in a fire. We may regret these lost details, but Foucault himself would probably not have been too concerned:

> ...while acknowledging the interest which attaches to the discovery of the manuscript papers of famous men, one cannot, however, expect to derive evidence from them comparable to that which comes from printed sources when it is a question of establishing the rights to priority of a scientific discovery.[5]

Fig. P.3. Inventory of Foucault's lost scientific papers.[4]
(Family collection)

Acknowledgments

First, I should like to thank my wife Laurence, her mother Gisèle Bon and her aunt Noëlle Saunier for their encouragement, aid and understanding during the decade and a half that Léon Foucault has occupied my life. I should also like to thank my father, John O'H. Tobin, FRCP, and his colleagues for advice concerning medical matters treated in this biography.

Second, my thanks go to members of Foucault's family who without exception received me hospitably and openly made available the few items and documents that have survived in their care: Bruno, Claude and Colette Chaumet, Jacques and William Foucault, André, Cecile and Philippe Gutzwiller, Daniel and Dominique Prest, and Alain and Sylvie Sourrieu.

Third, the research and writing of this book involved the libraries and other facilities of numerous institutions and societies and their personnel. I should particularly like to mention the help provided by the late Marie-Julie Meynent (Observatoire de Marseille), Josette Alexandre (Observatoire de Paris) and the staff of the Physical Sciences Library (University of Canterbury). My warm appreciation goes to them as well as to the personnel, often anonymous in addition to those stated, of the following organizations: the Académie nationale de médecine (Michelle Lenoir), the Archives de l'Académie des sciences (G. Darrieus, M. J. Mine, C. Demeulanaere-Douyère), the Archives du Bureau des longitudes (Y. de Kergrohen), the Archives nationales, the Archives de la Ville de Paris, the Bibliothèque nationale de France (Sylvie Aubenas, Bernard Marbot), the Centre d'histoire des sciences et techniques (Dominique De Place, the late Jacques Payen, André Guillerme), the Collège de France (Mme Roussell, Marcel Froissart), the Conservatoire national des arts et métiers (Edith Delroche), the École de médecine (M. Rivet), the École nationale des ponts et chaussées (Florence Doux), the École polytechnique (Claudine Billoux), George Eastman House (Joe Struble, Barbara Galasso, Janice Madhu), the Institut d'Astrophysique de Paris, the Institut de France (Françoise Quinton), the Institut national de la propriété industrielle, the libraries of the Universities of Auckland, Cambridge, Canterbury, Otago, Oxford, Victoria (Wellington) and Wisconsin, the Médiathèque d'histoire des sciences at the Cité des Sciences et de l'Industrie, the Musée français de la photographie (André Fage), the Musée Nicéphore Niepce (P. Jay), the Musée national des techniques (Jeanne Bruno, Frédérique Desvergnes, Anne-Catherine Hauglustaine, Thierry Lalande), the Museum national d'histoire naturelle (Françoise Serre), the Patent Office London, the Observatories of Bonn, Marseilles and Paris (Frédérique Auffret, Laurence Bobis, Dominique Monseigny), the Royal Society (Mary Sampson), the Sorbonne (Mme Magnaudet), the Science Museum (the late Jon Darius, Kevin Johnson, Rhiannan Sullivan, A. Vincent), the Science Reference Library, the Smithsonian Institution (Steven Turner), the Société française de photographie (Michel Poivert,

Katia Busch), Trinity College Dublin (Jane Maxwell), the Wellcome Institute (W. M. Schupbach), and the University of Canterbury's photographic service (Barbara Cottrell, Merilyn Hooper, Duncan Shaw-Brown) and printery (Ken Spall). I also wish to thank André Bertrand, René Blanchet, Jacques Bon, Mylène and Maurice Bon, F. F. Bonnart, Jacques Boulon, Dave Burba, Alain Bussard, Jean-Pierre Clavier, Stéphane Deligeorges, Lucio Fregonese, Philippe Garner (of Sotheby's), Manuel vaz Guedes, J. Goumard, Jacques Guilbert, Alden Hyashi (*Scientific American*), Marie-Thérèse and André Jammes, Roberto Mantovani, Loïc Métrope, Anthony Michaelis, Philippe Pajot, Bernard Rattoni, Archie Roy, Roberto Semanzato, Carolyn Shuster-Fournier, Hervé Theis, Christian Warolin, Odile Welfelé-Capy, Pascale and Jean-Marc Yersin-Bonnard and Phil Yock.

Fourth, many academic colleagues and friends have provided invaluable guidance and criticism. I am especially grateful to James Lequeux for careful reading of the manuscript and its adaptation for the French edition, as well as Paul Acloque, Mike Bradstock, John Campbell, Jim Caplan, Philip Catton, Rod Claridge, Marie Connolly, Georges Courtès, Dick Crane, Suzanne Débarbat, Alison Downard, Kelly Duncan, Ken Entwistle, the late Jacques Foiret, David Gallagin, Yves Georgelin, Alan Gilmore, Owen Gingerich, Joan Gladwyn, Alison Griffith, John Hearnshaw, Alice Houston, Jürg Hönger, Graeme Kershaw, Pam Kilmartin, Julie King, Gerard Lemaître, Matt Lloyd, Guy Mathez, David Miller, Simon Mitton, Alison Morrison-Low, the late Garry Nankivell, Sarah Nichols, Margaret Patterson, Norman A. Phillips, Brian Pippard, John Pritchard, Christopher Rose-Innes, Lewis Ryder, Geoff Stedman, Sarah Wheaton and David Wiltshire. I thank Hubert Curien and David DeVorkin for the forewords to the French and English editions.

Fifth, my work on Foucault has been facilitated through grants, prizes and salaries and I thank the Université de Provence, the Observatoire de Marseille, the University of Canterbury, the Institut d'Astrophysique de Paris, *Vistas in Astronomy,* the Mechaelis Memorial Trust (administered by the University of Otago), the Marsden Fund of New Zealand, which also provided a grant towards black-and-white illustration costs, and the Planet Earth Fund set up under the will of the late George Eiby, which funded the colour photographs. I also thank the photograph libraries which either waived or reduced their reproduction fees.

Finally, I should like to apologize to anyone I have omitted and to thank you, the reader, for opening this book. I hope you enjoy it.

Chapter 1

Introduction

Henri Sainte-Claire Deville leaned back in his heavy academician's chair, sick with emotion. It was a grim Monday afternoon, a grey March day, so different from the carefree warmth of his childhood days as a colonial in the Caribbean.[1] There he sat, amid the wood-panelled splendour of the elegantly proportioned chamber where he and his fellow members of the exalted Académie des Sciences were holding their regular weekly meeting. The damp vapours rising from the River Seine flowing silently a hundred metres away, and the sharp bite of the watery west wind would normally have found a chilly echo in the uncertain light of the flickering candles, which the Académie, in its stolid perversity, continued to use to illuminate its deliberations, even though this was the year 1868. But the 49-year-old chemist was hot. The chamber was packed. The public had read the newspapers and had turned up in force. People were standing in the aisles. Sainte-Claire Deville held no illusions: they had come to enjoy a cock fight, and he was one of the cocks. It was only a small comfort to know that the spectators were mostly on his side, as were the newspapers and magazines. In an epoch when any opinion might be wrapped in a cocoon of polite words, there had been no misunderstanding the full meaning when *Les Mondes* had reported of the previous week's meeting that:

> In a solemn and sorrowful voice, M. Henri Sainte-Claire Deville read out an eloquent protest made in the name of M. Léon Foucault and his friends. It was an outcry against the admittedly indirect accusations from which M. Le Verrier had felt himself unable to spare [M. Foucault].[2]

There had been a religious silence then. The public benches had been empty, and, unaware of what was in store, the odious Le Verrier had been absent. This week, the Imperial Senator and Director of the Paris Observatory *was* present, and as soon as the Permanent Secretary had finished reading out the correspondence, he had jumped up and been given the floor. Le Verrier had just boomed out that Sainte-Claire Deville's intervention had said 'nothing new, was unnecessary', adding, preposterously, that it was 'absolutely unjust'.

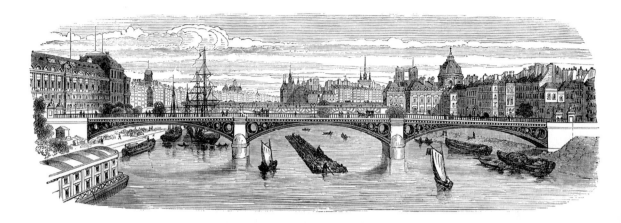

Fig. 1.1. The prominent dome to the right of this 1860s engraving of Paris is part of the Institut National, which included the Académie des Sciences as one of its constituent academies. The Institut marked the northern limit of the Latin Quarter on the left bank of the River Seine. The bridge in the foreground is the Pont du Carousel; behind it is the Pont des Arts. The imposing building on the left is part of the Palais des Tuileries. The rectangular, box-like boat moored in the left foreground is a floating wash house.

'What does he know?' thought Sainte-Claire Deville, his anger flaring. 'Foucault – my dear, good, *quaint* friend Foucault – is hardly cold in his grave after a prolonged and difficult death. *I* was his confidant, *I* know what troubles he suffered at the hands of the unspeakable Director of the infernal Observatory, but here is the dictator himself claiming they always enjoyed the most cordial relations.'

Le Verrier had prepared his remarks on paper, and it was obvious he was approaching their end. Eyes everywhere were turning to Sainte-Claire Deville. He swung his weight forward in his seat, readying himself to get up. 'Can I contain myself?' he wondered, 'I must stick to my written text.' The candles flickered. The busts and statues looked down impassively. 'M. Sainte-Claire Deville has the floor,' the Chairman announced.

* * *

'No man is an island,' sang the poet. We cannot hope to understand who this Foucault was, the significance of his achievements, and why his friends were so incensed, without also understanding his world and its preoccupations. This world was mid-nineteenth century France and will be foreign to many readers, so let us set the scene with some historical background.

The *Ancien régime*

Let us begin in the seventeenth century with Louis XIV, the *roi soleil*, or Sun King, whose court was characterized by lavish entertainments and sumptuous pleasures. Behind the throne of this absolute and initially debauched monarch was an active and intelligent finance minister, Jean-Baptiste Colbert (1619–83). Colbert saw that the basis of power was wealth, and that the basis of wealth was production. He endeavoured to advance national and regal wealth

The French Revolution

Fig. 1.2. Paris Observatory in the 1840s, viewed from the north. When completed in 1672 it was outside Paris and enjoyed an open horizon. The architect was Claude Perrault, brother of the fairy-tale teller; his building is more grand than practical.

with sound, forward-looking accounting, and via the promotion of industry, transport and trade. Among the numerous institutions founded through his influence were the Académie Royale des Sciences, or Royal Academy of Sciences, in 1666,[3] and the Observatoire de Paris, or Paris Observatory, in 1667.[4] The Academy was founded with the practical aim of applying science to industry in order to improve manufacturing and increase exports. No utilitarian goal constrained the Observatory, however, where the astronomers were free to study whatever they wished.

The Academy initially found lodgings near the Louvre, but the Observatory needed purpose-built quarters (Fig. 1.2). The Italian-born Jean-Dominique Cassini stamped his mark on the early Observatory.[5]

The French Revolution

The causes of the French Revolution in 1789 have been debated at length. The absolutism of the Bourbon kings and the inflexible grip of the nobility on its privileges had produced widespread discontent amongst the middle classes who were the mainspring of economic production. The eighteenth-century enlightenment advanced the supremacy of reason against the inequality and injustice of established authority and institutions. The scorn of superstition and passionate belief in the benefits of science and reason were most notably expressed in the seventeen volumes of the *Encyclopédie*, or Encyclopedia, published between 1751 and 1765; its attempted suppression epitomized the rigidity of the ruling classes. In 1789 May, social unrest combined with urgent financial problems to persuade Louis XVI to summon the *États généraux*, or States-General. This advisory body of the church, the nobility, and city corporations (in effect, the middle classes) was traditionally summoned to rubber-stamp controversial royal policy, but had not met for over 150 years. The King and nobles lost control; and the *États généraux* transformed itself into the Assemblée Nationale Constituante, or National Constitutional Assembly, intent on reform. The fortress-prison of the Bastille in eastern Paris, emblematic of the old order and its injustices, was stormed on July 14.

The National Assembly tried to create a constitutional monarchy, but war with Austria and the King's attempted flight and efforts to mobilize foreign military support in 1792 led to more radical policies. A Republic was declared and the King was guillotined, but Revolutionary government was unstable as different factions vied for power. By the standards of modern terror, the French Revolution was comparatively bloodless, but scientists executed included the chemist Antoine Lavoisier and several astronomers (Fig. 1.3).

For the Revolution, the Académie Royale des Sciences was a suspect institution and was abolished in 1793 along with all other learned academies, literary societies, universities and medical schools. Suppression was short lived, however, and two years later the Académie des Sciences was resurrected as one of five academies forming the newly created Institut National, or National Institute (now the Institut de France, visible in Fig. 1.1). Other academies included the Académie Française, devoted to literature, and the Académie des Beaux-Arts, dedicated to fine art. Amongst institutions created by the Revolution, the so-called *grandes écoles* (literally, major schools) were set up to provide and develop technical services and knowledge, as well as to train the administrators, scientists and engineers that the new order required. These institutions included the École Normale Supérieure, set up to educate an intellectual élite who would go on to propagate technical knowledge and the spirit of enlightenment as secondary and tertiary teachers throughout France; the École des Mines, for mining; and the Conservatoire des Arts et Métiers, established to act as a repository of scientific and industrial devices, to encourage technical innovation, and to spread scientific, technical and industrial know-how. There were also some older institutions. The École des Ponts et Chaussées (bridges and carriageways), set up to train civil engineers, dated from 1747. But the pick of the Revolutionary institutions was the École Polytechnique, founded for instruction in science and engineering, and which during the first half of the nineteenth century was a world leader in scientific research, most notably in mathematics and mechanics. Entry to the *grandes écoles* was by competitive examination and provided a route for young men of intelligence, but little wealth, to rise in the new, meritocratic France.

Desiring to 'make our shipping flower' and to develop navigation and trade, the Revolution founded a Bureau des Longitudes (Board of Longitude). The Bureau was to be in charge of the Republic's astronomy, a discipline which the Revolution felt had already produced so many benefits: it had chased

Fig. 1.3. Jean Sylvain Bailly, astronomer and Mayor of Paris, guillotined on the Champ de Mars in 1793. Other astronomers amongst the Revolution's twelve thousand or so victims were J. B. G. Bochart de Saron and Johan Wilhelm Wallot.[6]

away superstition, and it had provided theoretical foundations for geography and navigation, even though the means of determining longitude at sea were still imperfect. The Bureau was equipped with a staff and a budget and given charge of the Paris and other observatories, such as the naval ones at Brest and Toulon.[7]

The Revolution introduced a uniform system of weights and measures to promote trade, but the metric system was slow to take root. In his newspaper articles, Foucault often used pints, feet, leagues, pounds and toises, and his pendulums were engineered in inches and lines.

Napoléon I

The political disorder of the Revolution ended in 1799 with a *coup d'état* by Napoléon Bonaparte and conspirators. A Corsican by birth, Bonaparte had won prestige by leading a brilliant military campaign in Italy with decisive victories over the Austrians, and from an expedition aimed to strike at Britain's wealth by occupying Egypt and closing the route to India. Although the Egyptian expedition had been thwarted when Nelson sank the French fleet, it had had the merit of audacity. Bonaparte's new constitution was apparently liberal, but in fact gave him the bulk of power as First Consul, elected for ten years. During the next few years he set about his long-lasting reform of the legal system, the church, education and the administration of the French state. In 1804, with the pope reduced to the rank of a spectator, he crowned himself Emperor of the French. Already at war with Britain, he set about dominating continental Europe. Initial success was followed by ruin, and in 1814 he abdicated. He returned after a short exile in Elba, but defeat at the battle of Waterloo ended his reign after only one hundred days.

Fig. 1.4. Pierre Simon Laplace (1749–1827), the pre-eminent French physicist at the end of the eighteenth and beginning of the nineteenth centuries, but a man of ungenerous character.

Napoléon had intellectual aspirations and hoped that France would be the seat of all future science. While in Cairo he founded a local Institut d'Egypte, modelled on the Institut National. He famously said that:

> To divide the night between a pretty woman and a starry sky, and to spend the day working on one's observations and calculations, seems to me to be heaven on earth. Of all the sciences, astronomy is the one which has been most useful to reason and commerce. Astronomy particularly has need of long-distance communications and the Republic of Learning...[8]

The Marquis de Laplace (Fig. 1.4), pre-eminent at the time in French physics, sent the first two volumes of his *Mécanique céleste*, or Celestial Mechanics, to Napoléon, who with wit responded, 'The first six months which I can spare will be employed in reading it.'[9] Napoléon re-established the universities, with the doctorate as their highest degree. The university in Paris was the Sorbonne, with origins dating back to the thirteenth century.

The Restoration

The Bourbon restoration brought Louis XVI's brothers to the throne: first Louis XVIII; and then Charles X in 1824. Foucault was born in 1819, midway through Louis XVIII's reign. Charles X's claim to rule by divine right and his suppression of liberties provoked the 'July Revolution' in 1830 and his abdication after five days of bitter street fighting in Paris. His elected successor, Louis-Philippe, possessed astonishing republican, royalist *and* liberal credentials. Foucault lived his formative years and early adulthood under this 'citizen king's' rule. But during this reign political corruption, judicial malpractice and a restricted parliamentary franchise united liberals and extremists in calls for reform. In 1848 February, riots caused Louis-Philippe to flee to England. One legacy of his reign, however, was the re-establishment of the Collège de France. Unlike the universities or *grandes écoles*, the Collège de France offered and offers no degrees or diplomas. The professors taught as they saw fit in classes open to all.

François Arago

At this point it is appropriate to introduce François Arago (Fig. 1.5), who was Director of the Paris Observatory from 1843 until his death in 1853. He was born in 1786 at Estagel in the Roussillon, adjacent to the Spanish border, and

Fig. 1.5. François Arago (1786–1853): astronomer, physicist, politician, Permanent Secretary of the Académie des Sciences and Director of the Paris Observatory.

François Arago

Fig. 1.6. Was Foucault among the audience represented in this engraving of one of Arago's public astronomy lectures? The purpose-built lecture theatre at the Observatory could hold 800 people.

spent his adolescent years in Perpignan, where his father was treasurer of the mint. He entered the École Polytechnique with ambitions for a military career. Soon, however, his staunchly republican views collided with political authority. Bonaparte as First Consul was preparing his transformation to Emperor. Arago was among those at the École who refused to sign a petition urging this change. Bonaparte resolved to dismiss these student republicans, but on seeing a list of their names and marks, with Arago first in his year, he sighed, 'One can't send down the top student. Oh, if only he'd been at the bottom.'[10]

Arago's mathematical prowess was such that within two years he was offered a post at the Paris Observatory to work on the refraction of light by the Earth's atmosphere with Jean-Baptiste Biot (1774–1862). Triangulation of the Paris meridian was required, supposedly for the practical implementation of the metric-system metre as one ten-millionth of the distance from pole to equator, but really for investigation of the Earth's flattening.[11] The meridian had already been surveyed from Dunkirk to Barcelona. Early in 1806, Arago and Biot left for Spain to extend the triangulation to the Balearic Islands. The heroic enterprise was completed, but Arago's return to France was obstructed by French military involvement in Spain in 1808. After several escapes and imprisonments, attempted poisoning, capture by pirates, and two passages through Algiers, Arago regained French shores in 1809 July.[12] Two months later, he was elected to the Académie des Sciences,[13] more in recognition of the romance of his exploits than for any fundamental contribution to science. In the 1810s his scientific work was mostly in optics and in later

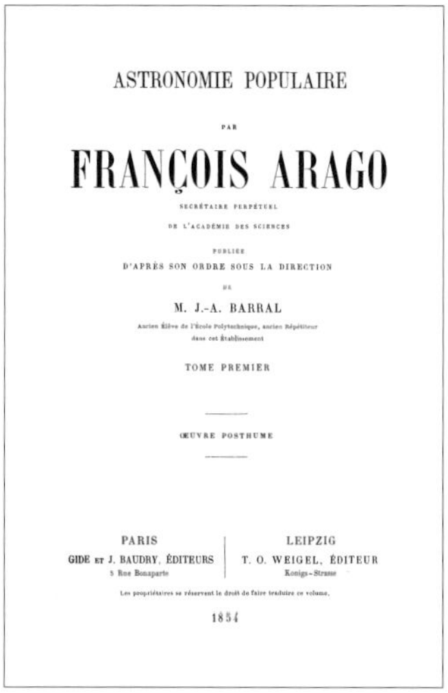
ASTRONOMIE POPULAIRE

PAR

FRANÇOIS ARAGO

SECRÉTAIRE PERPÉTUEL

DE L'ACADÉMIE DES SCIENCES

PUBLIÉE

D'APRÈS SON ORDRE SOUS LA DIRECTION

DE

M. J.-A. BARRAL

Ancien Élève de l'École Polytechnique, ancien Répétiteur dans cet Établissement

TOME PREMIER

ŒUVRE POSTHUME

PARIS
GIDE ET J. BAUDRY, ÉDITEURS
5 Rue Bonaparte

LEIPZIG
T. O. WEIGEL, ÉDITEUR
Konigs-Strasse

Les propriétaires se réservent le droit de faire traduire ce volume.

1854

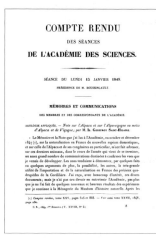

Fig. 1.7. In his final years, Arago dictated his astronomy lectures to his niece, Lucie Laugier, foreseeing that they would provide substantial royalties for his heirs.[18] They were published posthumously in four volumes.

COMPTE RENDU

DES SÉANCES

DE L'ACADÉMIE DES SCIENCES.

SÉANCE DU LUNDI 15 JANVIER 1849.

PRÉSIDENCE DE M. BOUSSINGAULT.

MÉMOIRES ET COMMUNICATIONS

DES MEMBRES ET DES CORRESPONDANTS DE L'ACADÉMIE.

ZOOLOGIE APPLIQUÉE. — *Note sur l'Alpaca et sur l'Alpa-vigogne ou métis d'Alpaca et de Vigogne; par* M. Is. Geoffroy Saint-Hilaire.

« Le Mémoire et la Note que j'ai lus à l'Académie, en octobre et décembre 1847 (1), sur la naturalisation en France de nouvelles espèces domestiques, et sur celle de l'Alpaca et de ses congénères en particulier, m'ont fait adresser, sur ces derniers animaux, dans le cours de l'année qui vient de se terminer, un assez grand nombre de communications destinées à confirmer les vues que je venais de développer. Les unes tendaient à démontrer, par quelques faits ou quelques arguments de plus, la possibilité, les autres, la très-grande utilité de l'importation et de la naturalisation en France des précieux quadrupèdes de la Cordilière. J'ai reçu, avec beaucoup d'intérêt, ces divers documents; mais je n'ai pas cru devoir en entretenir l'Académie, pas plus que je ne l'ai fait de quelques nouveaux et heureux résultats des expériences que je continue à la Ménagerie du Muséum d'histoire naturelle. Après les

(1) *Comptes rendus*, tome XXV, pages 525 et 865. — *Voir* aussi tome XXVII, 1848, page 280.

C. R., 1849, 1er *Semestre*. (T. XXVIII, N° 3.) 8

Fig. 1.8. Founded by Arago, the *Comptes rendus* reported the Academy's weekly meetings.

chapters we shall discover some of the rôle he played in the development of the transverse wave theory of light.

In the 1820s Arago's interest turned to electromagnetism. In 1830 he was made one of the two Permanent Secretaries of the Academy, where he was able to wield considerable influence. As one commentator noted, 'The President [of the Academy] is only a man of straw, whilst the secretaries direct everything and are omnipotent: they are the true masters of the house.'[14]

Foucault later wrote:

> Arago was one of the first to realize that science would not prosper in the depths of laboratories, that it would wither even in the solitude of Academies, and that after having given the world steam power, railways and the electric telegraph, science still needed to be talked about, even by the uneducated.[15]

Beginning in 1812, Arago delivered public lectures on astronomy, as required by the Act that had established the Bureau des Longitudes. They were exceptionally successful, and from 1841 until the last series in 1846 they were delivered in a 'spacious, elegant and practical' lecture theatre which had been specially constructed for the purpose at the Paris Observatory (Fig. 1.6).[16] The lectures were published posthumously, entitled *Astronomie Populaire* (Popular Astronomy, Fig. 1.7).

While approved members of the public had been allowed to attend the Monday meetings of the Academy from 1809, Arago admitted everyone who cared to turn up, and also the press. In 1835 he founded the *Comptes rendus*, or weekly printed reports of these meetings, which over time were a great influence in standardizing forms of scientific presentation (Fig. 1.8).

The rise of astrophysics – or physical astronomy, as it was also called – is often associated with the introduction of spectral analysis into astronomy following the clear formulation of empirical laws for the absorption and emission of radiation in 1859 by Gustav Kirchhoff in Heidelberg. But 'Arago introduced physics into astronomy', as a later director of the Paris Observatory noted.[17] Previously astronomers had been chiefly concerned with the movements of the stars and planets, seeking to explain them in their minutest details by Newton's laws of gravitation and motion. Arago used photometry and polarimetry to try to fathom the physical nature of celestial bodies, and showed, for example, that the Sun's surface was not solid but gaseous.

Drawn into politics by his younger brothers, Arago was elected *deputé* (representative) for the Pyrénées-Orientales and then for the 12th *arrondissement* (or district) of Paris after the 1830 revolution.[19] His politics were left-wing and republican, but he believed in public order and moderate change. He was a member of the provisional government in the newly declared Second Republic in 1848. In this capacity he promoted and signed the decrees abolishing slavery in the French colonies and flogging in the navy. We shall see in

Napoléon III

The Second Republic was short lived. Attempts to disperse the unemployed from Paris into the provinces resulted in a workers' revolt in 1848 June and a brief civil war conducted with relentless cruelty. The upshot of these 'June Days' was a general desire for strong government, which a few months later translated into the election to the French Presidency of Prince Louis-Napoléon Bonaparte. This unusual man was the nephew of Napoléon I and had been raised abroad. He entertained no doubt but that his Napoleonic destiny was to rule France. In 1836 he had crossed the border into Strasbourg and in 1840 he had landed in Boulogne, in both cases expecting to foment rebellion among the troops and be swept to power. His reward in 1840 was imprisonment in the Château of Ham, in northern France, until his escape six years later disguised as a workman. Louis-Napoléon was no common prisoner, however. While in Ham he found occasion to sire two illegitimate children and to study science (Fig. 1.9). Napoléon I had said that had he his life to live again, he would have thrown himself into the study of the exact sciences.[20] As we shall see, his nephew maintained this interest in science and later supported Foucault, Pasteur and others in their scientific endeavours.

Three years after his election as Prince-President, Louis-Napoléon overthrew democratic government in a *coup d'état* during the night preceding 1851 December 2. The *coup d'état* was legitimized in a massively favourable plebiscite, as was the restoration of the Empire a year later and Louis-Napoléon's enthronement as the Emperor Napoléon III.[22]

The Second Empire is a considerable embarrassment in France even today. Perhaps this is because it is difficult to believe that a free people should have voted for a hereditary head of state, perhaps it is because of the perceived dishonour with which the Empire ultimately disintegrated – through a hopeless war against the Prussians cleverly provoked by Bismarck, ending in military defeat at Sedan in 1870. But the two decades over which the Second Empire endured were an extraordinary time. Louis-Napoléon has variously been described as 'strange and enigmatic', an 'amiable adventurer',[23] or, most memorably, as 'Napoléon le Petit' (Napoléon the Small) by Victor Hugo, one of his regime's most vociferous exiles. Yet Louis-Napoléon's heartfelt desire was for progress and to improve the lot of the masses while maintaining social order and stability, aspirations outlined in his book *De l'extinction du paupérisme* (On the elimination of poverty). As Emperor, he built workers' villages and tried to introduce free, compulsory primary education. Despite numerous military adventures, his general aim was for a more just and less volatile commu-

Fig. 1.9. Louis-Napoléon Bonaparte (1808–73), the future Emperor Napoléon III, studying electricity in his prison laboratory at Ham in the 1840s. In a report to the Academy he complained that his readings were perturbed by the iron bars around the windows.[21]

nity of nations based on the self-determination of peoples; and he was sickened by the sufferings of war. His reign was a period of enormous technical, industrial and commercial expansion, albeit accompanied by high-level corruption. It was also a time of gaiety, crinoline and fêtes. The energetic could attend one of Paris's numerous *bals,* or public dances; one of the most famous, the Jardin Bullier, was located only 500 metres from the Observatory. Foreign visitors to Paris eyed the *grisettes* – young unmarried women of no particular fortune earning a modest living in the city and keen to enjoy themselves, which, unhindered by middle-class conventions, they did. Paris is the splendid city it is, in part because of Second-Empire replacement of dank and narrow streets by light and airy boulevards, and the installation of proper drains. The Expositions Universelles of 1855 and 1867 were showcases for France's agricultural and industrial strength and associated prosperity and modernity. Foucault died in 1868. The last twenty years of his life were essentially those of the Second Empire, and he was one of the Empire's best known and most popular savants.

The state of technology

The mid-nineteenth century was the daybreak of the technological manner in which we now live. In 1846 a Parisian newspaper, *Le National,* listed the great industrial innovations of the previous half-century: the improvement of underwater cements; the application of the steam engine to sea and land transport; the application of electricity to electroplating and telegraphy; and the discovery of the daguerreotype. The daguerreotype and electroplating feature in Chapters 3 and 4, but the other innovations contributed to the nineteenth-century's rapidly improving communications. Rivers were crossed with stronger bridges and viaducts built using better hydraulic mortars. The first railway from Paris was completed in 1837 and ran some twenty kilometres to Saint-Germain-en-Laye. It was soon followed by two competing lines to Versailles – and the first serious French railway accident – and then, in 1843, by railways to the much more distant Orléans and Rouen. By 1870, the network had extended to some 22 000 km. The Alps were pierced by tunnels at Semmering and Mont-Cenis; the sea-route to India was halved when the Suez Canal opened in 1869. From 1845 onwards, the movable paddles of optical telegraphs read from a distance through a telescope were replaced by electric wires able to transmit signals over far greater distances. Improved sanitation and nutrition reduced disease while surgery was revolutionized by anaesthesia and later by asepsis. The output of agricultural and manufactured goods grew with mechanization and the chemical industy flowered. To Foucault's obvious delight, science and industrialization were making the world a materially better place.

Chapter 2

Early years

Nantes

Foucault is a name of Germanic origin, but it is common in Brittany,[1] which in the nineteenth century extended further south than now. Foucault's high brow, broad face and prominent cheekbones (see frontispiece) are common Breton traits. Though he was born in Paris, his family had connections with Brittany's principal port, the cathedral city of Nantes, situated on the northern bank of the River Loire, some 370 km south-west of Paris. His parents bought houses there in the 1820s and it was where his grandfather died in 1829 (see Fig. 2.1). This grandfather, Jean Baptiste Marie Foucault, was a former colonial from San Domingo (modern Haiti). This colony, France's 'Pearl of the Antilles', had been one of the largest in the Caribbean, with nearly half a million slaves and an annual sugar crop of 80 000 tons.[2] A slave revolt had erupted in 1791, to which the French Revolution was of course sympathetic; but the subsequent decade was one of bloody turmoil, especially when Napoléon I tried to restore slavery on the island. At some point grandfather Foucault decided to quit the troubled colony and move to Nantes. Maybe he had family or commercial connections with the town, but it was a logical choice. Only 55 km from the sea, it was a port with links to the Caribbean; indeed, in the previous century Nantes, like Bristol, had made its fortune through slave trading. The city was a flourishing manufacturing centre, with a population of about 75 000, and although a contemporary gazette reported that 'the old parts are ugly and dirty', it added that 'the rest is elegant and ordered'.[3] There would have been plenty of opportunities for the displaced colonial – and plenty of bittersweet reminders of his former homeland.

Parents

One of grandfather Foucault's three sons had salt in his blood and became a long-haul sea captain, but another was a landlubber, and sought a career in Paris. He was Jean Léon Fortuné Foucault, who was born in about 1784,

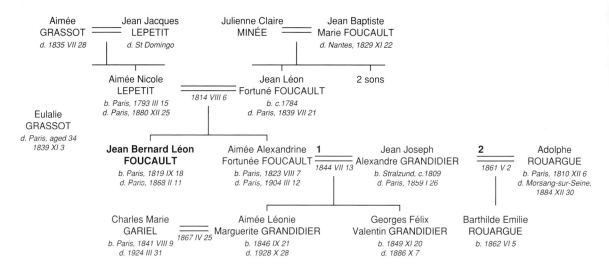

Fig. 2.1. Extracts from the Foucault family tree.[5]

Fig. 2.2. Foucault's mother Aimée in later life.

(Family collection)

perhaps in San Domingo.[4] In 1814 he married a young woman who also had San Domingo antecedents, her merchant father having died there. This 21-year-old bride was Aimée Lepetit (Fig. 2.2). Her sorrow was to long outlive not only her husband, but also their physicist son.

At his marriage Jean Léon Fortuné was already in the book trade, but in 1817 he was licensed as a *libraire*, or bookseller-cum-publisher. Between 1819 and 1829 he published two extensive collections of memoirs and reminiscences relating to the history of France, which were sufficiently well regarded that the early ones had to be reprinted (Fig. 2.3).[6] He also published volumes relating to the theatre and occasional royalist and other political tracts.[7] Part of his success may have been due to a commitment to accuracy. At a time when books were sold unbound, he advertised that any set of leaves containing even a single typographical error would be reprinted free of charge.[8]

This spirit of precision was transmitted to his son, Jean Bernard Léon (hereafter plain 'Léon', the name he used), who was born in 1819, the same year as Queen Victoria and the poet Walt Whitman. The birth took place on an autumn Saturday, and doubtless occurred at home, which at a time when people mostly got about on foot, cannot have been far from the Foucault bookshop on the Rue des Noyers in the heart of the Latin Quarter.[9] A year later his parents bought a house on the Rue de Sorbonne, and whether or not they lived there, they certainly traded from that address.

Foucault was baptized in the parish church of Saint Etienne du Mont (Fig. 2.4), only 150 metres from the Panthéon where the pendulum would immortalize his name three decades later. Possibly a brother was born who did not live. A sister did survive. She was Aimée Alexandrine Fortunée, Léon's junior by four years.[10]

Childhood

It seems Foucault spent some of his early boyhood in Nantes, where his clearly prosperous parents purchased two houses in 1823 and 1826,[11] but childhood time was also spent in Paris, where the Foucaults rented a substantial house at 5 Rue d'Assas, only a short stroll from the Luxembourg Gardens, one of the city's few public parks. The house's spacious grounds and trees will have provided amusement for the children and cool in the summer heat.[12]

Wherever spent, Léon's childhood included some common experiences, as his newspaper articles reveal. 'Who has not chased a lizard and then broken off its tail while trying to hold it captive?' he inquired in one column.[13] The horse-chestnut bears an inviting fruit: 'coloured, polished and shiny like recently varnished mahogany,' he wrote in another; 'its flesh firm and white . . . ' He bit in, only to recoil at the bitter taste. 'It is an experiment which everyone has made to their cost,' he sighed.[14]

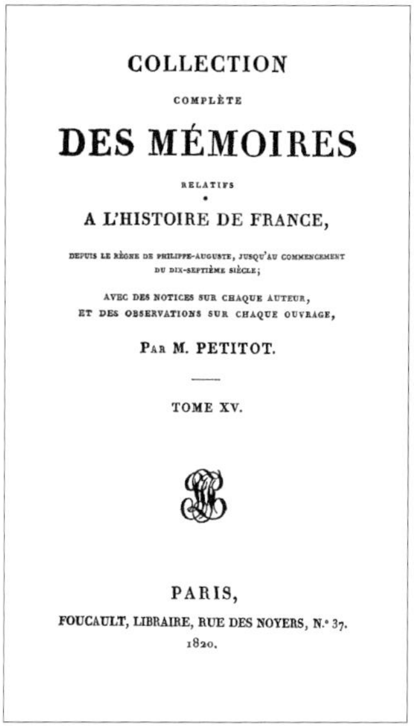
COLLECTION
COMPLÈTE
DES MÉMOIRES
RELATIFS
A L'HISTOIRE DE FRANCE,
DEPUIS LE RÈGNE DE PHILIPPE-AUGUSTE, JUSQU'AU COMMENCEMENT
DU DIX-SEPTIÈME SIÈCLE;
AVEC DES NOTICES SUR CHAQUE AUTEUR,
ET DES OBSERVATIONS SUR CHAQUE OUVRAGE,
PAR M. PETITOT.
TOME XV.
PARIS,
FOUCAULT, LIBRAIRE, RUE DES NOYERS, N.° 37.
1820.

Fig. 2.3. One of over 130 volumes relating to French history published by Foucault's father.

Father's *interdiction*

Jean Léon Fortuné Foucault resigned his bookseller's licence during the summer of 1830. Publication of the second series of volumes relating to French history had been completed the previous year, and in his mid forties he was perhaps ready for a change. Further, bookselling was undergoing a crisis. Books had been overproduced – fifty million in the first half of the decade – and the market for complete works was saturated. Booksellers were going bankrupt.[15]

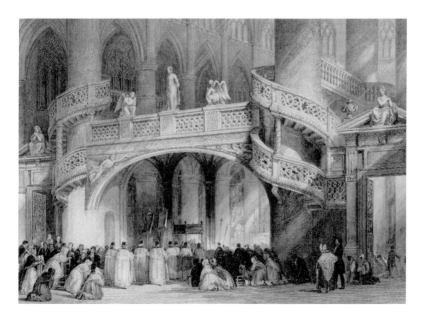

Fig. 2.4. Three days after his birth, Foucault was baptized in the local parish church of St Etienne du Mont. This church is celebrated for its magnificent rood screen between choir and nave (the last remaining one in Paris), and for the tomb of the philosopher-mathematician Blaise Pascal (1623–62).

Table 2.1. The value of money in Foucault's time: some wages and prices.

Incomes	
Workman	1–4 fr/day
Aide Astronome at Paris	
Observatory	2 000 fr/yr
Country teacher	2 000 fr/yr
Senator	30 000 fr/yr
GNP/inhabitant	1840: 300 fr
	1869: 470 fr
Prices	
Kilogramme of bread:	
Top quality	31 c
Second quality	24 c
Litre of cows' milk	50 c
Paris omnibus fare	30 c
Bachelierès sciences	
examination fee	24 fr
Typical book	6 fr
12 daguerreotype plates	
(Full size, 215 x 165 mm)	42 fr
Quarterly subscription to the	
Journal des Débats	18 fr
Carcel oil lamp	22 fr
Kilogramme of silver	200 fr
782 m^2 of land on the	
Rue d'Assas (1844)	60 120 fr

Perhaps Jean Léon Fortuné decided to close down while still solvent. But it would seem possible that his retirement was also prompted by a growing disequilibrium of mind, because only four years later he was *interdit*, or deprived of legal control over property and family because of mental incapacity. These must have been difficult years for the family as his behaviour progressed from the occasionally strange through the often peculiar to the clearly pathological. As her husband's state worsened, Aimée will have shouldered an increasing burden as *de facto* head of the family and controller of its business affairs. She opened a *cabinet de lecture* on the Rue Voltaire. *Cabinets de lecture,* or reading rooms, were a further response to the publishing crisis. Not only had books been produced in vast quantities, they were too expensive, beyond the reach of the merely comfortably off and accessible only to the rich. A volume published by Foucault's father, for example, typically cost 6 francs, a few days' wages for a workman or a day's pay for a village teacher (Table 2.1). Public libraries did not appear until the 1860s, but in a *cabinet de lecture* the literate public could enjoy books and newspapers at an affordable price. They also assured publishers of a minimum number of sales.

Aimée Foucault was a strong woman. She had not hesitated to sue an aunt and uncle over a disputed inheritance, and was blessed with the same business sense as her husband. Under her guidance the family finances remained healthy. She would later engage in property transactions, moving herself and her fatherless son into a new house a little further south on the Rue d'Assas, and then augmenting its value by acquiring adjacent land. Besides being a business-woman, she was religious. No doubt the young Léon learned his catechism and took his first communion, but the effect was not profound. In his writing there is no evidence of any serious religious belief, or that religion affected his scientific outlook. However, he was not anticlerical, and we shall see that he had no hesitation associating with people for whom faith was important.

After his *interdiction*, tutelary power over Foucault's father was accorded to his wife. A deputy guardian was also appointed. As befitted a well-to-do family, this was a man of some standing, L. J. N. Monmerqué, a judge in the Paris courts; but also a man of letters, a member of the Institute, and well-known to the family because he had edited several of the books published under the Foucault imprint. In due course, Monmerqué also became the adolescent Léon's deputy guardian.

Adolescence

His father's *interdiction* and the events leading up to it must have marked the psyche of the fourteen-year-old Léon. We will see that Foucault had a fragile side to his character, which threatened to overwhelm him twenty years later. As a teenager, he was certainly distressed when he realized that there were brown objects floating within his eyes. Unfortunately the doctor consulted was unable

Adolescence

to reassure him that this was completely normal. Here is what Foucault wrote after the Scottish physicist Sir David Brewster had published a paper on the subject of *mouches volantes*, or 'floaters' as they are known in English:

> It is likely that [*mouches volantes*] are present from the earliest age, but this age is without cares, just as it is without pity, and if it finds amusement chasing flies, they are not the sort which we are discussing at present.

(Foucault is punning on the double meaning of *mouche*, which can mean 'fly' as well as 'speck'.)

> One only notices them later, in adolescence, when many become victims to a sort of hypochondria... however little one may be infused with the spirit of observation, one soon realizes that the effect is entirely personal, and that the cause lies within oneself. These cloudy tufts... always appear the same, on the white of the ceiling, on the curtains of one's bed, on a sheet of paper; one can but think that they are due to the interposition of some opaque object; little by little the imagination works and one believes one's vision compromised at an age when one has so much need and so much desire to see everything. We speak from sharp and poignant memories, and we would entreat surgeons to be better informed than they used to be about a circumstance which is in no way worrying.[16]

Angst is common in adolescence but this episode should be read for the evidence of observation and reflection that it reveals rather than any tendency to excessive anxiety. More happily, Foucault filled teenage hours with model making – first a boat, then a telegraph. This was not an electric telegraph, whose development was to come a few years later, but an optical one, where moving paddles signalled the letters (Fig. 2.5). Foucault modelled the telegraph situated atop the nearby church of Saint Sulpice, which he could see from his window (Fig. 2.6). The paddles' silent and incessant motions clearly captivated him; perhaps he even decoded the messages. The telegraph was followed by a model requiring even greater skill and dexterity: a steam engine.[17]

It is reported that as a boy Foucault attended a small school in Nantes, rather than being educated at home; and that back in Paris he was sent as a day-boy to the prestigious, nearby and very Catholic Collège Stanislas.[11] (His sister attended a *pension,* or boarding school.) How long he remained at the Collège Stanislas is unclear. Although his adolescent character was later described as 'soft, timid and hardly exuberant',[17] it seems that at school he was 'unteachable and unstudious'. This early manifestation of the independence of mind which later was to be so characteristic would have made unimaginative, rote learning in a schoolroom intolerable. A more flexible private tutor was employed. Foucault promised to submit only perfect work, and evidently kept his word, though with no great ardour. This personalized but unusual approach

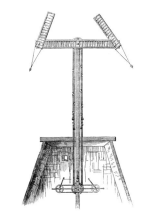

Fig. 2.5. Prior to the development of the electric telegraph in the 1840s, messages were transmitted by the moving paddles of optical telegraphs.

Fig. 2.6. The optical telegraph situated atop the southern (right) tower of the nearby Church of Saint Sulpice was a subject for the youthful Foucault's spree of model making. St Sulpice is the biggest church in Paris after Notre Dame. Foucault's parents had been married there in 1814 and it would be the site of his own funeral service in 1868.

Fig. 2.7. Léon Foucault and his sister Aimée Alexandrine Fortunée in mourning dress for their father in 1840.[22]

(Family collection)

was nevertheless sufficiently successful that his education was later described as 'very complete',[18] and in Chapter 6 we will encounter the impressive erudition that he displayed not many years later as a journalist. He duly passed his public examinations. In 1837 he was received *Bachelier ès lettres*, which required proficiency in rhetoric and philosophy.[19] His mother, it is said, had decided that he should become a doctor, and entry to the medical school required a second *baccalauréat*.[20] Eighteen months later Léon demonstrated his knowledge of mathematics, physics, chemistry, biology and zoology in front of three examiners, who, satisfied, all placed white balls in the voting urn, thereby making him a *Bachelier ès sciences physiques*.[21] One of the examiners was the newly appointed professor of physics at the Sorbonne, the Belgian-born César Despretz. Despretz would not be so satisfied when he examined Foucault's doctorate a decade and a half later.

Medical studies

The nineteenth century saw great changes in both medicine and medical education. The Revolution had closed medical schools, but as medical practice became better informed, apprenticeship to established doctors or surgeons was replaced with more rigorous tuition in medical schools and hospitals, along with qualifying examinations. Medicine was an appropriate career for an intelligent young man from the bourgeoisie like Foucault. The work was honourable, and his mother could afford the training, which was not cheap. Soon after completing his *Baccalauréat ès sciences physiques*, Foucault entered the Paris Medical School, reportedly intending to build on his dexterity to become a surgeon. Figure 2.7 shows Foucault and his sister at about this time.

Medical students enjoyed a dissolute reputation. The *carabins*, as they were popularly dubbed, were caricatured as irreligious, shabbily dressed individuals who wasted their time and health in excessive smoking and drinking, who frequented disreputable cafés and estaminets where they played dominos and cards and sang an extensive repertoire of predominantly ribald songs, and who, as a counterbalance to the ever-present death and suffering to which they were exposed, indulged in licentious pleasures with women. From his later writings in the *Journal des Débats*, and the contents of his cellar on his death, we can be confident that Foucault liked a glass of good wine,[23] but as the following letter shows, he took his studies seriously and, despite some dancing, hardly epitomized the debauched stereotype.

The letter is self-pitying and was addressed to an unidentified distant friend – perhaps in Nantes – in the spring of 1840. It is worth quoting at length not only because it indicates that Foucault had already begun to be fascinated by science, but because it is one of the few surviving documents where Foucault talks about his emotions – about the recent death of his father, and about a young man's yearning for love:

Medical studies

Fig. 2.8. The dilapidated *École pratique* in the Paris Medical School. It was probably here that the student Foucault dissected in an attempt to soothe sad thoughts.

The death of my poor father, happening unexpectedly on a day when my sister was on release from her boarding school and our heart was overflowing with joy, threw us into a dismay, into a sadness which outsiders, knowing the state of my poor father, find difficult to believe. Hardly had we recovered from this terrible blow, when death, unsatisfied, came to strike amongst us again... Our poor Eulalie, our dear cousin, succumbed in a week to a scarlet fever complicated by several other afflictions. You know, excellent friend, that she was essentially one of our household and can understand how cruel this loss was for us. Ah, this time my heart was injured full and truly...

I fell into a solitude which deeply affected my cheerfulness, and to recover from these sad thoughts, I began my dissection precisely one month later. Certainly I learned to be familiar with the sight of death, but I did not get my good spirits back at all; and so since then no more chemistry demonstrations, no more physics experiments, no more pretty model-making. Already we have had several lovely days but my interest in my plants has not been revived. In vain I spent a little time amongst the tumult of the *bals masqués* [masked balls] (I admit it); this certainly amused me, but my heart remained so empty, so hollow, so indifferent that I am hardly certain whether it is still capable of being struck by love. To sum up, I am in an ordinary state of sadness which is all the more distressing because I foresee nothing that will be able to bring me out of it. And do not think this is due to the abuse of any sort of pleasure; I have continued to work befittingly and the proof is that my first examination in medicine, which I took three months ago, was passed brilliantly...

M. Lecorbeiller was your way this winter to give concerts.[24]... Did the sight of my excellent teacher remind you of the time when you came to Paris to make his acquaintance? I certainly remember that time very clearly and I pine for it with all the force of my soul. I had all my

Fig. 2.9. Foucault's self-pitying letter to an unnamed friend in 1840 April was signed only with the flourish with which he usually underscored his full signature until the 1860s.

(Family collection)

friends around me; I was in the first flush of enthusiasm for science, to which I am now indifferent; I had just finished my schooling; I was busy with student papers under my arm. Oh! Here I go again, fallen into complaint; it is one of my sicknesses to yearn for the past. But do not think I have a gloomy and misanthropic air, all this bitterness is happening at the bottom of my heart. I keep a smiling face for those that remain of my friends.[25]

'Dear cousin' Eulalie was described on her death certificate as a *libraire* living on the Place de l'Odéon.[26] Perhaps she worked at Madame Foucault's *cabinet de lecture* a few steps away in the Rue Voltaire. Foucault's wounded-heart comment might seem to indicate that he was sweet on her; but if so, the age gap was considerable, because she was fourteen years his senior. Despite its self-pitying tone, the letter reveals a young man who fundamentally was healthy enough, with a considerable range of interests, two of which – music and plants – are not mentioned in any of the eulogies written after his death. Perhaps gardening was only an adolescent pastime; but if so, it was no shallow fad, because he later made perceptive and informed comments in his newspaper articles when discussing botany, horticulture and agriculture. As to music, instrument playing was so common that perhaps this is why no obituary mentions it. A Pleyel piano belonged to the household on his father's death, and at his own death he owned an accordion and two violins, one of which was claimed to be a Stradivarius.[27] Whether true or not (and Cremona violins have always been much-forged), these instruments are sure testimony to the importance in Foucault's life of music and, as he put it, 'the beautiful effects of harmony...'[28]

Appearance and appetites

As to Foucault's physical appearance, a commentator described him as 'small, delicate, puny'.[29] Military papers prepared in 1840 give his height as 1.65 metres (5 ft 5 in); he was exempted from conscription because he was the only son of a widow. A decade later, he was exempted from another form of military service, in the National Guard, which had its origins in the Revolution as a middle-class militia composed of those affluent enough to pay taxes. This time the grounds were poor health.[30] His hearing was acute.[31] His gaze was all-seeing ('lynx-eyed' said some[32]) but it was his right eye that did the looking because he suffered from slightly divergent strabismus, as close examination of the frontispiece, Plate I etc. reveals. Our commentator continued:

> ...he walked slowly, light-footedly, constantly putting his chestnut hair back in its place, hair that was parted in the middle and always smooth and well cared for. His head was unusual and characteristic, its lines were strong, but the face was cold. He rarely smiled; sometimes irony would twist his lips slightly.

His bodily appetites were nevertheless healthy. Besides wine, he enjoyed oysters (a 'precious mollusc'[33]) and especially coffee. He praised this 'colonial foodstuff' whose:

> tasty and aromatic qualities are able to charm our senses. Besides an exquisite flavour, coffee possesses extremely remarkable stimulant properties; it intensifies cerebral faculties in the most favourable way. What excellent words, what long nights, what sustained efforts, what assiduous works derive directly from the coffee cup.[34]

Probably he did not smoke. No pipes or cigar boxes were inventoried after his death, and when he discussed tobacco in a newspaper article, there was no coffee-like eulogy, just the bland statement that 'despite the drawbacks attributed to it, it seems rooted in our habits for a long time to come.'[35]

Friends

'I keep a smiling face for those that remain of my friends,' Foucault had written to his unidentified correspondent. Friendships were important to him, and he took time to maintain and enjoy them. His longest friendship was with Jules Regnauld (Fig. 2.10). They presumably met as medical students. Unlike Foucault, Regnauld was from a modest background, and perhaps this was reflected in the 'retiring modesty' of his character.[36] His father had been a small pharmacist in the Marais, a neighbourhood of narrow streets just north of the Seine; but like Foucault, his father had died recently, and he was interested in physics. Regnauld must have been intelligent and hard-working, because he obtained a sought-after hospital internship during the latter part of his medical studies. He was then employed by the Medical School for his knowledge of the physical sciences, and later became Professor of Pharmacy.[37]

At least two of Foucault's friendships were with men twenty years older than himself, both of whom came to take a paternal interest in their young protégé, perhaps substituting for the father who had been lost to mental illness and then death. One was the affable Dr Alfred Donné, who will be presented in Chapter 4. After Foucault's death, Donné made these comments concerning his personality and social skills:

> Foucault was not what one would call lovable: he had neither the suppleness of character nor the desire to please necessary for one to be considered agreeable by the world. He defended his opinions quietly, but without any concessions – not even to ladies – and sometimes over matters that were not scientific, but which concerned the conventions of society, in which he was not always as well versed. When I had introduced him into the intimacy of my own family, he was often felt to be disagreeable and annoying, most especially because he never lost his temper, but defended his opinions in a tone that was always cool and calm. This calm was a great strength in certain circumstances, and I have seen him – he,

Fig. 2.10. Foucault's longest-lasting friendship was with Jules Regnauld (1820–95), pictured here in later years.

(Académie de Médecine)

of weak and fragile appearance – put the most formidable opponents to flight. If he was not supple, his fidelity in friendship withstood all tests. So many times I said to women who were exasperated by his composure, 'He is not amiable, but you can count on him, and you will find that he will always hold towards you the same sentiments of affection and friendship.' And indeed, for thirty years he did not change. He became famous, but he never crushed you with his glory, he never failed to acknowledge services rendered, he never failed to express delight at seeing old friends, and he was never less-willing to explain the most abstract physical laws to laymen – and above all to laywomen – with the lucidity that he knew how to put into explanations of theories and phenomena. I see him still, during our evenings in the country, undertaking to unmask the deepest mysteries of light or electricity to a fashionable audience: a scrap of paper, a pencil, and some simple, carefully drawn figures assisting the meaning of his economical and accurate words. Because Foucault no more made concessions concerning facts than he did concerning opinions, and he remained rigorously accurate while nevertheless lowering scientific truths to a more accessible level.[38]

Another older friend, though not such a close one, was the Abbé (Father) François Moigno (Fig. 2.11).[39] Like Foucault, Moigno was a Breton. He entered the Society of Jesus as a novice aged eighteen, and during his theological studies met many Parisian scientists. He took refuge with his order in Switzerland following the anti-clerical July Revolution in 1830; later he taught mathematics at the Jesuit seminary in the Rue des Postes in Paris, and published books on mathematics, optics and mechanics which made his reputation as a savant. He then entered into conflict with his order, finally leaving it in 1844, creating, it was said, a scar that never healed. Moigno was appointed chaplain of the Lycée Louis-le-Grand in Paris, and began a career of science writing. In 1852 he was founding editor of a weekly scientific news magazine entitled *Cosmos*. It was perhaps about then that he and Foucault met as journalists and became friendly; they certainly travelled to England together in 1854 to attend the meeting of the British Association for the Advancement of Science in Liverpool, whence Foucault reported back to Regnauld that:

> The Abbé is admirable... I renounce trying to describe the wealth of his intrepid nature, drinking, singing, praying, cracking jokes, bursting into peals of laughter, reciting his breviary between two cigars, acting as an interpreter between German and English, improvising at the lectern, replying to every toast and holding high the national honour. He seems to me to be one of the consonances of this hurly-burly, this immense bear garden which is the British Association.[40]

Fig. 2.11. The Abbé François Moigno (1804–84), one of Foucault's friends and ardent supporters. After leaving the Jesuits he wrote for journals such as the monthly *Revue scientifique et industrielle* and newspapers such as *La Presse* and *Le Pays* before founding the weekly *Cosmos* in 1852.

Foucault may have foreseen nothing to break his melancholy, but his enthusiasm for scientific pastimes was soon rekindled by an exciting discovery – photography.

Chapter 3

The metallic eye: photography

The year 1839 may have been a sad one for Foucault with the death of his father and cousin Eulalie, but the great news that year – news that set the scientific and artistic worlds alight – was the public announcement of the first practical photographic process.[1]

The camera

Photography requires both camera and recording plate. The camera was an established device used by artists and travellers to help compose their pictures and transfer three-dimensional scenes onto flat canvas or paper (Fig. 3.1). In the earliest version a pinhole produced images within a darkened box, and the operation of this *camera obscura* had been explained in the eleventh century by Arab philosophers who understood, following the ancients, that light rays travel in straight lines. Four centuries later, image sharpness and brightness were improved when the pinhole was replaced by a convex glass lens. However, with a simple lens (one made of a single piece of glass) the images were still slightly blurred. It is important to understand these *aberrations*, which we shall also meet in Foucault's work as a telescope builder.

One problem arises because the glass of a lens acts like a prism, dispersing the rays into the spectrum of red-to-violet colours so admired by Newton (Fig. 3.2). Objects are not brought into sharp focus but are seen ringed with coloured bands. This defect of *chromatic aberration* arises in all optical instruments, such as telescopes or microscopes, if they are made with simple lenses.

The way to correct this chromatic aberration was an important eighteenth-century discovery, made by Chester Moor Hall, a London barrister. However the name of the optician John Dolland is often associated with the discovery because Dolland simultaneously managed to correct another image defect, called *spherical aberration*, which we will discuss later in this chapter.

Hall's invention involves a composite lens of two elements made from glasses with different dispersions. One lens is convex and converges rays while

Fig. 3.1. The *camera obscura* in one of its many forms. In 1839 all that was required was some automatic way of preserving the image it cast.

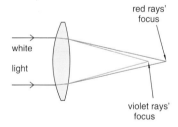

Fig. 3.2. Glass refracts violet light more strongly than red. This differing refraction is known as *dispersion*, and because of it a simple lens does not focus all colours into the same plane. The resulting image defect is called *chromatic aberration*.

Fig. 3.3. Principle of the achromatic doublet lens, first devised in the eighteenth century. A stronger convergent lens of crown glass is combined with a weaker divergent lens of flint glass to produce a combination that overall still converges rays to a focus. However the dispersion of the flint glass exactly cancels that of the crown glass and all colours come to a common focus.

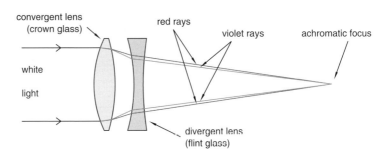

the other is concave and diverges them. The convergent lens is made stronger than the divergent one so that their combined effect remains convergent, but by using more dispersive glass for the divergent lens it is possible to arrange that the dispersions of the two lenses are equal and opposite and so cancel. All rays in this *achromatic doublet* then focus in a single plane (Fig. 3.3).

The glass used for the convergent lens was invariably crown glass, which is essentially ordinary window-glass made from sand, potash, soda and chalk.[2] For the divergent lens flint glass was used, which contains a large proportion of lead oxide and is very dense. If the lenses of the achromat were small, they might be glued together to keep them clean and aligned, and to minimize internal reflections. From about 1813, the optical cement was almost always Canada balsam, a water-soluble oleoresin collected from blisters in the bark of the Canadian balsam fir.[3]

In the early nineteenth century, the *camera obscura* was further improved with a meniscus lens that produced a flatter field and allowed larger apertures, resulting in brighter and better images right to the edges of the field of view. In 1839, the camera was an established technology readily available from opticians.

The recording plate

It had long been known that many substances change their physical or chemical nature under the influence of light; the Swedish chemist Berzelius listed over one hundred light-sensitive materials in 1808. Many people had experimented with these bodies in the hope of obtaining permanent images, and with silver chloride and silver nitrate in particular, which darken when exposed to light. The problems were, first, that the exposures required were inordinately long; and second, that some method was needed to stop the reaction so that further exposure to light did not turn the images completely black.

The first person to take a permanent picture from nature was Joseph Nicéphore Niepce (1765–1833), who lived near Châlon-sur-Saône in Burgundy. In 1822 he made a contact print of an engraving, and then in 1826, probably, he took the world's oldest surviving photograph, of a courtyard, using a bitumen-covered, polished-pewter plate.⁴ Exposed areas of bitumen became less soluble and when the plate was washed with petroleum they were left behind as grey-white highlights, or light tones, while uncovered metal produced the shadows, or dark tones, provided that the plate was angled so that the incident rays were reflected away from the viewer. The image was thus a positive one, with exposed regions bright and unexposed areas dark, but the result was only partially satisfactory because during the eight-hour exposure the shadows in the courtyard moved, confusing the illusion of depth.

In 1829 Niepce entered into partnership with L. J. M. Daguerre, who was also experimenting with light-sensitive chemicals (Fig. 3.4). Daguerre had worked as a theatrical designer and painter and had opened 'dioramas' in Paris and London. There the paying public could view huge paintings on thin linen in which, through clever changes of lighting, night gave way to day, landslides ripped down valleys, or the faithful gathered under the rood screen in the church of Saint Etienne du Mont for midnight mass.

Daguerre will have used a *camera obscura* in the production of his paintings and the link with Niepce originated through their common patronage of the best optical supplier in Paris. This was the Chevalier firm, first established in 1765.⁵ In the 1820s the company was in the hands of Vincent Chevalier, son of the founder, but it was the grandson, Charles Chevalier, who really had the gift for optics (Fig. 3.5).⁶

Fig. 3.4. Louis Jacques Mandé Daguerre (1789–1851), painter and inventor of the first practical photographic process.

Niepce and Daguerre agreed to share all discoveries and profits. Both men continued experiments with bitumen, but also with silver-coated metal plates exposed to iodine fumes, in which the resulting golden-yellow silver iodide was the light-sensitive substance.

Niepce died in 1833 and thereafter Daguerre experimented alone, concentrating on silver iodide. He still had two problems: the exposure times were long, and some way was needed to render the plate insensitive to light once the exposure was over. In the spring of 1835 he accidentally found a solution to the first problem after he left lightly exposed plates in a cupboard overnight and was surprised to find images visible on them the next day. Trial-and-error removal of the various items in the cupboard showed that a dish containing a few drops of mercury was responsible. Daguerre had made the astonishing discovery that a latent image, formed during an exposure of only half-an-hour, could be made visible, or *developed*, by mercury vapour. In the autumn, the discovery was reported in the *Journal des Artistes*, a weekly magazine devoted to artistic criticism and technique. It was claimed that 'a preparation' could be used to preserve the image,⁷ but it took two more years before Daguerre discovered that the developed image could be fixed by washing it in strong brine, which removed the unchanged silver iodide.

Daguerre tried to sell the secret by public subscription for between 200 000 and 400 000 francs, but was unable to attract enough investors. Before attempting a second subscription campaign, he thought to obtain testimonials from

Fig. 3.5. The English musical-instrument maker and physicist, Charles Wheatstone (1802–75, left) and the Parisian optician and precision-instrument maker Charles Chevalier (1804–59).

(George Eastman House)

eminent artists and scientists. At a time when the primary concern of art was still the accurate representation of nature, the painter Paul Delaroche wrote, 'The painter will find in this process a quick way of building a portfolio of sketches which he could not otherwise obtain except through the expenditure of much time and pain, and far less perfectly, whatever his talent.'[8] In the *Journal des Artistes* (described as 'well-informed in these sorts of things') the critic Jules Janin declared, 'Art has nothing left to say to this new rival...it is here a matter of the finest, most delicate and most complete reproduction to which the works of God and the products of man can aspire.'[9] Scientists whose approbation Daguerre sought included Arago, Biot, and Arago's friend, the German polymath and francophile Alexander von Humboldt.

Arago recognised the potential in Daguerre's discovery. As Permanent Secretary of the Academy and as an elected politician he had the influence to do something about it. At the Académie des Sciences on 1839 January 7 he outlined what the process could do (at that point he did not know what was involved), and he tried to dispel widespread misconceptions created by rumours and Daguerre's understandable secrecy.[10] Daguerre's procedure did not record colours, only shades of light and dark. Extraordinary detail was captured; the rods of very-distant lightning conductors were later cited as an example.[11] The exposure depended on the time and season – 'In summer at noon, eight to ten minutes suffice.' Arago noted how useful the discovery would be for recording architectural monuments, but he also foresaw uses in physics and astronomy. Daguerre had already imaged the Moon onto one of his plates at Arago's

The celestial sphere and telescope tracking

request, but an 'obvious white mark' was all that had been obtained. Arago commented that he had no doubt that an unblurred lunar image would result if the telescope was attached to 'a parallatic machine' driven by a clock.[8]

The celestial sphere and telescope tracking

To understand this remark, we need to review the diurnal or daily motion of the heavens, which we will find relevant to much of Foucault's science.

Because the stars and other celestial bodies are so distant, it is convenient to imagine that they are attached to the inside of an imaginary and stationary *celestial sphere* with the Earth at its centre (Fig. 3.6). The Earth turns from west to east about its north–south axis, but to the Earthbound observer it is the celestial sphere which appears to rotate in the opposite sense about this same axis, which intersects the celestial sphere at the *celestial poles*. The inverted bowl of the sky is the half of the celestial sphere that is above the observer's horizon at any moment; the other half is hidden. The *meridian* is the vertical north–south plane which divides the observer's east from west and passes through the overhead point, or *zenith*.

The elevation of the celestial pole above the horizon is equal to the observer's latitude (which is why latitude is easy to determine). At a given latitude, some stars are always above the horizon, some rise and set, and some

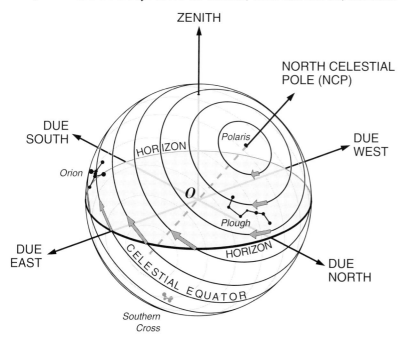

Fig. 3.6. The celestial sphere, drawn for the latitude of Paris. Constellations and other heavenly bodies can be supposed attached to this imaginary sphere of indefinitely large radius. An Earthbound observer views the sphere from its centre, O. The celestial sphere is stationary in space, but because of the Earth's west-to-east rotation it appears to the observer to rotate from east to west about the Earth's north–south axis (dashed). Arrows represent this apparent motion. The observer sees only the half of the celestial sphere that is above the local horizon. From the latitude of Paris, the Plough (or Big Dipper as it is called in North America) is always above the horizon (though hidden by blue sky during daylight hours), Orion rises and sets once per sidereal day, and the Southern Cross is permanently invisible.

are never seen. The stars take the same time to make their circuit as the Earth does to make a complete 360° rotation; this interval is the *sidereal day* of $23^h56^m04.1^s$. Stars therefore rise approximately four minutes earlier compared to successive meridian passages of the Sun whose average defines the 24^h of the *solar day*. This difference of about 1/365th of a day arises because during a sidereal day the Earth has advanced along its orbit around the Sun by 1/365th and needs to turn a further 1/365th of a rotation for the Sun to reach the meridian again.

We can now understand Arago's remark. To a first approximation, the Moon moves with the celestial sphere. To obtain an image that is stationary with respect to the daguerreotype plate, and therefore unstreaked, it is necessary for the *camera obscura* or telescope to track the Moon across the sky.

There is a cunning way to achieve this tracking. In order to access all parts of the sky, a telescope must be able to move about two axes. If one of these axes is set up parallel to the Earth's axis, and then rotated by clockwork at the sidereal rate of 360° per sidereal day, the telescope will move in step with the celestial sphere, as illustrated in Fig. 3.7. A celestial object remains stationary in the telescope's field of view and can be daguerreotyped crisply. Telescope mounts based on this general principle are now all described as 'equatorial', though in Arago's time a subtle distinction divided them into two classes, equatorial and parallatic.[12]

Arago's announcement on 1839 January 7 drew a rapid response from an Englishman, W. H. Fox Talbot. Fox Talbot had been experimenting with paper-based photography, with some success. He assumed that Daguerre's procedure must be the same, and wrote claiming priority of discovery. It rapidly became clear that the two processes were different, and that for the moment, Daguerre's produced far superior results.

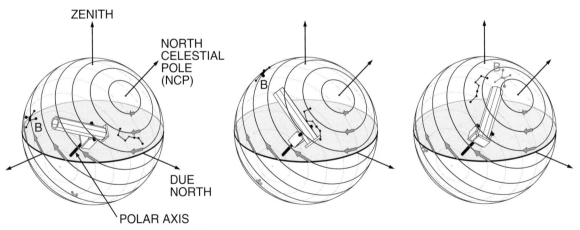

Fig. 3.7. The equatorial telescope mount. To track a star it is necessary only to rotate the mount around the polar axis in the sidereal direction at the sidereal rate. Here the telescope tracks the star Betelgeuse (marked B) in the constellation of Orion.

Endowing the world

Fig. 3.8. Daguerre's discovery announced by Arago at a joint meeting of the Académie des Sciences and the Académie des Beaux-Arts on 1839 August 19.

Endowing the world

Arago foresaw that the use and improvement of Daguerre's procedure would be hindered if it became a patented, commercial process. 'It therefore seems essential that the Government should compensate M. Daguerre directly,' he concluded, 'and that France should then nobly endow the whole world with a discovery that can contribute so much to the progress of art and science.' Eight months later Louis-Philippe signed the Act giving annual pensions to Daguerre and Niepce's heirs. Full details of the process were revealed at a joint meeting of the Académie des Sciences and the Académie des Beaux-Arts on August 19. Interest was so great that queues for the public benches formed three hours beforehand, and not everyone was admitted.[13] Daguerre was taken by a violent sore throat and a fit of nerves, and it was Arago who spoke (Fig. 3.8).[11]

The plates were thin sheets of copper covered with a layer of silver which might be only 25–50 μm thick.[14] To obtain good blacks, it was important to give a high polish to the silver face, which was then cleaned with dilute nitric

Fig. 3.9. The English Astronomer Royal, George Airy (1801–92), wrote to his wife that daguerreotypes 'surpass in beauty anything that I could have imagined'.[16]

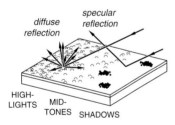

Fig. 3.10. The light and mid tones of the daguerreotype consist of varying surface densities of roughly uniform globules of mercury–silver amalgam. Occasional large clumps of irregular globules occur in the shadow regions, but there are no isolated globules. Held to the light at the correct angle, the globules diffuse light towards the viewer, but the mirror regions of pure silver do not. The image can reverse (appear negative) if the viewer sees the specular reflections.

acid and exposed to iodine fumes in subdued light. After exposure in the *camera obscura,* the plate was developed in mercury vapour at 60–80 °C. Excess silver iodide was no longer removed with brine, but with sodium thiosulphate solution, which had recently been recommended by the English astronomer Sir John Herschel. The plate was finally washed and dried.

In the following months people worked feverishly to replicate the procedure, which required great cleanliness and care for the best results. For the well-off, Daguerre had made arrangements for the manufacture and sale of the required plates, polishing pads, iodine and mercury boxes, chemicals and *camera obscura* with Alphonse Giroux, an instrument seller. (The lenses were contracted out to Chevalier.) Others made their own apparatus. Results soon flooded in to the Academy, as Arago had hoped. Even students at the École Polytechnique were taught how to take daguerreotypes.[15]

We are now so accustomed to the printed image that it is difficult to imagine the impact of the first daguerreotypes. Their beauty was admired by the English Astronomer Royal, George Airy (Fig. 3.9), while the American painter Samuel B. Morse, who later invented the Morse code, commented on the astounding precision and detail of the image.[17]

The nature of the daguerreotype image was a question that naturally arose; indeed, on August 19 Arago had announced that inspection with a microscope had shown that the tones were composed of regions of uniform little amalgam spherules of about 1/800th of a millimetre in diameter (1.2 μm). These spherules were most densely packed in the light areas, and were absent from the shadows. They gave rise to an image in just the same way as Niepce's bitumen: specular reflection from the bare silver produced shadow regions, while highlights arose from diffuse reflection from regions with a high density of spherules. Recent research involving techniques such as X-ray spectroscopy in the scanning electron microscope has revealed that there is very little else to the image structure (Fig. 3.10).[18] Daguerreotypes can record very fine detail because of the small size of the spherules, which are much smaller than the grains in modern photographic emulsions.

Hippolyte Fizeau

The amalgam spherules were fragile and it was said the image could be rubbed off with a feather. Arago had encouraged Daguerre to search for a protective varnish, while the chemist J. B. Dumas, whom we will encounter again, had proposed a dextrose wash. Improved image robustness was the first substantial advance in daguerreotype technology and came not from Foucault, but from another twenty-year-old medical student called Hippolyte Fizeau (Fig. 3.11).

Fizeau was in many ways Foucault's double. He had been born in Paris, a mere five days after Foucault. He was from a well-to-do background, with a doctor father in the town of Suresnes, just to the west of Paris,[19] and had attended the Collège Stanislas. He was a medical student who was devoting recreation hours to science, and like Foucault, he would in due course abandon medicine for physics. What he discovered was that chemical gilding hardened the daguerreotype image. Arago first made allusion to this discovery at the Academy in the spring of 1840, hardly six months after his disclosure of the daguerreotype process.[20]

The Académie des Sciences

At this point it is appropriate to say more about the structure of the Académie des Sciences and its meetings, which have hardly changed since Foucault's time. The Academy was divided into eleven sections representing different disciplines (Table 3.1). Most sections were composed of six Academicians who were required to reside in Paris. They were complemented by correspondents living outside the metropolis. The most distinguished of foreign scientists might be honoured as one of eight foreign associates. Membership was for life. Aspiring Academicians had to wait patiently for the grim reaper to create a vacancy.

The Academy met in public session on Mondays at 3 p.m. in the *Grande salle des séances* at the Institute. Figure 3.12 shows the principal features of this meeting room.

Each year the Academy elected a new Vice-President and President (the previous year's Vice-President) to preside over meetings. The real power, as already noted, lay with the Permanent Secretaries who controlled the agenda and, once elected, continued in this rôle until they died or resigned.

Communications to the Academy could take various forms. A paper might be read out in its entirety by its author, or more often in summary form. For an author who was young or little-known, the paper might be presented by an Academician. Papers that were not authored by a member of the Academy were referred to *ad hoc* commissions of typically two or three Academicians for evaluation. Sometimes these commissions might report on the paper at a later meeting, but this was a rare exception, unless the paper was of very great novelty or importance. Another way to present a discovery was to write to the Academy, and each week the Permanent Secretaries would begin the meeting with a summary of the correspondence received. The Academy might accept the deposit of *plis* or *paquets cachetés* (sealed packets). Even at the time, some considered these to be a pernicious invention. In them authors outlined discoveries or ideas that were not yet developed enough for presentation at the Academy; but if someone else later presented similar, fully developed work, the author of the *pli cacheté* could ask for it to be opened and so recover some

Fig. 3.11. Hippolyte Fizeau (1819–96) as a young man.

Table 3.1. The Academy's eleven sections. Most sections comprised six resident members and a number of correspondents. A twelfth group of Academicians were *libre*, or independent of any established discipline.

Geometry
Mechanics
Astronomy
Physics
Geography & Navigation
Chemistry
Mineralogy
Botany
Rural economy
Anatomy & Zoology
Medicine & Surgery
Académiciens libres

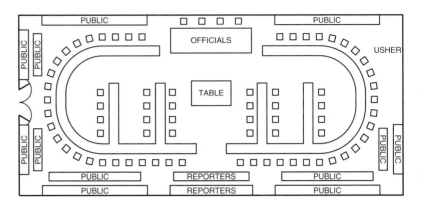

Fig. 3.12. The Academy's meeting room in the nineteenth century. Each academician had a designated place. The President, Vice-President and the two Permanent Secretaries sat on the raised officials' podium.[22] Cf. Fig. 3.8.

glory for having had the idea first. Elections of new members took place in public meetings. Finally, at the end of the public session the Academy would sometimes go into *Comité secret*, or Secret Committee, to debate the merits of candidates and confidential administrative matters. The meeting would typically finish about 5 p.m. The Permanent Secretaries edited a report, which included excerpts from many of the manuscripts presented at the meeting. The report was published within a few days as the *Comptes rendus hebdomadaires de l'Académie des Sciences* (Fig. 1.8).[21]

Gilding and sensitizing

Fizeau's gilding process was formally presented to the Academy on 1840 August 10.[23] The process was simple: after the plate had been developed, cleared of unexposed silver iodide and washed, it was heated in a mixed solution of gold chloride and sodium thiosulphite. Besides hardening the images, gilding produced brighter, more luminous pictures, and was rapidly adopted.[24] The shadow tones were darkened, too. This, said Fizeau, was because the gilding acted like a burnish on the bare silver, making it smoother and shinier.

The second substantial improvement in the daguerreotype process also came from Fizeau, at least in France. 'In general, people are fairly unwilling to admit that the [daguerreotype] will ever serve to take portraits,' Arago had told the Academy in 1839. He continued:

> In order for the image to form rapidly, that is to say within the four or five minutes of immobility which one can hope to obtain from a living person, it is necessary that the face should be in direct sunlight; but in direct sunlight the strong light would force even the most imperturbable person into continuous blinking and grimacing...

It is for this reason that the earliest daguerreotypes of humans were of corpses, or of people whose eyes were shaded by spectacles.[25] It was realized early that daguerreotype plates were most sensitive to blue and violet light, but hopes that portraiture might be possible inside pavilions glazed with blue glass had proved unfounded. Attention had turned to trying to make the daguerreotype plate more sensitive.

Silver iodide is the least light-sensitive of the silver halides. It was an obvious step to try other halides, though there were problems of supply. Iodine had only been discovered in 1812, in small quantities in seaweed ash. 'If the wracks and the kelps were not endowed with the property of condensing iodine in their tissues, the daguerreotype would probably still be waiting to be invented,' Foucault remarked.[26] The next halogen in the periodic series is bromine, a foul-smelling liquid. It too came from seaweed, in even smaller quantities, and had been discovered in 1826 by the chemist A. J. Balard (Fig. 10.4). In 1841 June it had been announced at the Academy that Daguerre's collaborator in London, Antoine Claudet, had improved sensitivity with what was called 'iodine chloride'.[27] Chlorine gas was passed over iodine crystals to produce ruby-red crystals of what is now called the interhalogen compound ICl. The iodized daguerreotype was then exposed to interhalogen vapours. Claudet had obtained some good portraits in diffuse light with exposures of 15 to 20 seconds, but it was difficult to sensitize the plates reliably and uniformly. Others were experimenting with bromine chloride and iodine bromide.

Fizeau's discovery was much simpler, and prompted by Claudet's announcement, he revealed it in a letter to the Academy. The iodized daguerreotype plate was exposed for a few instants to the vapours of a very dilute solution of bromine in water.[28] Exposures were reduced to a third of a minute. Fizeau also noted that it had been possible to obtain exposures as short as one or two minutes with an ordinary, unsensitized plate by increasing what we would now call the f-stop or focal-ratio of the *camera obscura* over the value of about f/15 adopted in the daguerreotype cameras marketed by Daguerre and Giroux.[29]

A week-end diversion

At about this time, Foucault began experimenting with the daguerreotype. The equipment and chemicals were expensive, but not unaffordable for the son of a well-off family. From the dates on Foucault's few surviving daguerreotypes and papers from this period, we find that this was mostly a week-end activity. No doubt photography provided a refreshingly modern, exciting and mind-occupying change from medicine.

Foucault inevitably tried out the various proposals for increased speed. Very likely he already knew Fizeau through the Collège Stanislas and the medical school, and he approached him for details of his bromination procedure. It involved making a red-coloured, saturated aqueous solution of bromine, and then diluting it in filtered river water until it had the same tint as *eau de vie* (distilled spirits). A quantity was poured into a covered saucer and placed in the bottom of a special box. To sensitize a plate, the saucer was uncovered

Fig. 3.13. Foucault's apparatus for the uniform application of bromine vapour. The box was small – only about 30 mm deep – and would have been used with plates a quarter of the size first employed by Daguerre. Bromine solution was introduced via the J-shaped tube puttied into the rear wall of the box. Waste solution was drained out through a second tube which is not visible. Foucault's plates fitted the box sufficiently snugly that it was only necessary to work in complete darkness when transferring the brominated plate into the camera's plate holder.

and the plate was lowered onto holding studs about 150 mm above the liquid for between 20 seconds and a minute. Fresh solution was employed for each plate. However, the brominated plates were still far from being as alike as ones that were only iodized, so trials with a first plate could not be used reliably to determine the optimum exposure time for a second one.

Foucault identified a number of deficiencies in Fizeau's procedure and set to remedy them. Although water will dissolve about 3 per cent of bromine, saturated bromine solutions can be far from identical, particularly because, said Foucault, hydrobromic acid can form which then dissolves even more bromine.[30] Instead Foucault pipetted bromine into water in order to obtain a constant, and much weaker, 0.05 per cent solution. Second, in a fraction of a minute the vapour did not fill Fizeau's large bromination box uniformly. Foucault devised a much smaller box which comprised a glass tray and lid inside a wooden case (Fig. 3.13). An iodized plate was lowered onto supporting pins above the glass lid 'where', he wrote, 'it can wait for an extended period, several hours, which can be very convenient when one is waiting for the favourable moment – a ray of sunlight, or some particular circumstance.' When it was time to brominate the plate, a fixed quantity of solution was poured into the glass tray via a tube. Foucault emphasized that it was necessary to wait a fixed interval of time in order to allow the vapour to spread before withdrawing the glass lid to start the sensitization of the plate. He advised that for maximum sensitivity, the bromination should last just less than the time required to cause fogging when the plate was subsequently developed with mercury.

Like Daguerre and Niepce, Foucault went to Charles Chevalier for his materials. He wrote up his bromination procedure in a ten-page note which was immediately published as part of a practical guide to the daguerreotype by Chevalier entitled *Nouvelles instructions sur l'usage du daguerréotype* (New instructions on the use of the daguerreotype).[31] The note finished with a compliment to Fizeau: 'But in finishing I must say it once more, the important idea, the capital idea in all this came from M. Fizeau, the idea of renewing the solution for each new plate.' Foucault added that his procedure was general and could be used with other accelerating substances such as chlorine, iodine chloride and 'the solution of iodine bromide which M. Gaudin, in his delusion that he is an inventor, considers is more active than bromine, but wrongly.'

Frustration bursts out

This unexpected sideswipe was aimed at Marc-Antoine Gaudin (1804–80), who was employed at the Bureau des Longitudes to do arithmetic. With the optician and instrument-maker N.-M. P. Lerebours, he had recently published a

48-page guide entitled *Derniers perfectionnements apportés au daguerréotype* (Latest improvements to the daguerreotype). A fortnight later Foucault used the *Journal des Artistes* to further his attack on Gaudin:

> whose name has so often echoed at the Academy that we cannot continue without saying a few words about him. We do not want to talk about his subtle procedure for splitting cobbles, nor his gigantic project of lighting the capital by means of a gas jet situated on the cut-off point of the Obelisk; we want only to discuss the worthless pronouncements that he has made concerning the daguerreotype. We still remember that about six months ago M. Gaudin announced that thanks to his new discovery, one was going to be able to give missing life to these renditions, that henceforth a quarter of a second would be sufficient to obtain a complete exposure. And this famous discovery, what is it? Why, the *verre continuateur* already announced by M. Edmond Becquerel, which M. Gaudin has only slightly modified, so that it works less well...[32]

Fig. 3.14. The physicist Edmond Becquerel (1820–91). He was the second in a dynasty of four Becquerel physicists. The third, Henri (1852–1908), discovered radioactivity.

Edmond Becquerel (Fig. 3.14) was the son of Antoine Becquerel, who was a member of the Academy and professor of physics at the Natural History Museum in Paris. Edmond had been only eighteen when he published his first paper on the heating effects of the electric arc. Using silver bromide paper, he had discovered in the autumn of 1840 that exposure to colours from dark green to red could continue the darkening commenced by previous exposure to blue-violet light. He had called these rays *rayons continuateurs*, or *continuing rays*, and the discovery had prompted an extensive report from the Academy.[33] Gaudin's suggestion was that the *rayons continuateurs* could act to develop very short daguerreotype exposures.[34]

Foucault continued:

> Ah! M. Gaudin, what time you have made us waste, not counting what we are wasting now. That is not all. M. Gaudin, tireless in his research (so much does success breed courage), has just announced to the Academy that he has been able to make a new substance which is much more sensitive than all those known previously. This new substance is iodine bromide, which M. Guérin had already had on sale for the last *three months*...Nevertheless, it must be said in M. Gaudin's favour (let us give credit where credit is due) that he prepares this substance in a certain way which has the advantage of sometimes blinding the experimenter...In the *Comptes rendus,* he declares that he has made an exposure of one nineteenth of a second. This singular fraction necessarily implies that he has a means of measuring so short a time. If it is only an approximation, one should rather say a fifteenth or a twentieth, a round number in short. We do not wish to say that it would be impossible, using some clever mechanism, to open and close the *camera obscura* for such a short and accurate interval, but if that is the case one must prove it. M. Wheatstone, who measured the duration of the electric spark – a much more surprising affair – to be one hundred-thousandth of a second, explained how he did it, and whoever has read his description

understands that it is possible and correct. But why are we tormenting ourselves so? M. Gaudin says that it is simply a cloth (an invention which he claims as his own) that he lifts and immediately drops. That is how he measures the time to one three-hundred-and-eightieth of a second, because that is the tiny difference that exists between a nineteenth and a twentieth of a second.... what astonishes us the most is that in presenting such things to the Academy, one finds no one who objects...

The attack is justified, and such criticisms would continue from others. (Two decades later Moigno wrote of Gaudin's '... strange ideas to which his imagination has long accustomed us'.[35]) Nevertheless, Foucault's article reeks of the intolerance of youth. We may also feel that his expectations were unrealistic concerning the Academy. The young need to be encouraged, and the old do not always jump in with public denunciations when foolish things have been said. It is revealing to see that Foucault was already aware of the measurement of the electric spark made by the English physicist Charles Wheatstone (Fig. 3.5), since this was a precursor to Foucault's experiments concerning the speed of light, as will be explained in Chapter 8.

Despite Foucault's comments, the *Derniers perfectionnements* by Gaudin and Lerebours proved a popular seller (it cost only 1 franc). In the third edition, published the following year, Fizeau published his bromination procedure with several improvements resulting from Foucault's criticisms.[36] Later editions expanded to include mention of Foucault's bromination box as well.[37]

Fig. 3.15. The Couvent des Carmes where over one hundred priests were massacred in 1792. This is the oldest surviving daguerreotype by Foucault, taken on 1842 May 29. The viewpoint was most probably an attic window in his house at 5 Rue d'Assas. This direct-positive image is mirror-reversed. It has been reproduced for effect against a white background, as recommended by Charles Chevalier.

(*M.-T. & A. Jammes and Sotheby's*)

Cityscapes and portraits

If the *camera obscura* contained only lenses, the image that it produced was mirror-reversed, and since this was recorded directly, the daguerreotype image was mirror-reversed too. Chevalier in his *Nouvelles instructions* reported that in 1841 September Foucault used an improved objective lens with a prism in front of it to obtain correctly oriented daguerreotypes of buildings with 15 to 20 second exposures, which were 'unquestionably superior to everything that has been done up to now'. Unfortunately none survives.[38] Only two early Foucault daguerreotypes are known, and both date from the following summer. One shows the Carmelite church and monastery on the Rue de Vaugirard (Fig. 3.15). It would have been a familiar sight to the Foucault family, located immediately south-south-east of the house that they were occupying on the Rue d'Assas, and a convenient subject to photograph. The black border around the image is specifically mentioned by Chevalier: 'M. Foucault arranges the burnish around the edges of the plate so as to produce a natural frame; the appearance of these pictures framed against a white background is a very agreeable effect.' The other surviving daguerreotype from 1842 is a portrait, so the exposure cannot have been excessive (Fig. 3.16).

Fig. 3.16. This old man was daguerreotyped by Foucault on 1842 July 2. The decoration on his left lapel, where they are always worn in France, indicates a reversing prism was used. The wide face and broad brow are Breton traits. Could he be a relative, perhaps a paternal uncle? Or is he Foucault's guardian, L. J. N. Monmerqué? There are some similarities to an engraving that shows Monmerqué twenty years younger.[39]

(George Eastman House)

Eliminating dry lines

Figure 3.16 is the only known portrait taken by Foucault, but a note found in his papers indicates he took many others. In this note he signals how to use the image defect of *spherical aberration* 'in order to produce effects imitating works of art'.[40]

To form a sharp focus, a lens or mirror must converge rays to a single point. Simple lenses and mirrors have spherical surfaces because they are easy to shape, but spherical surfaces do not in general converge rays to a single point, and the focus, such as it is, is smeared over a volume of space. This defect can be produced by non-spherical surfaces too, but is nevertheless called spherical aberration. An example is shown in Fig. 3.17 (*upper*). Spherical and chromatic aberrations occur everywhere in the image plane; in particular they affect images formed on the central or symmetry axis of the lens, as Figs. 3.2 and 3.17 show. In this they differ from other common image defects, such as field curvature, coma and astigmatism, which are absent on the symmetry axis but grow rapidly away from it. (The corners of Fig. 3.16 may have been clipped off for artistic effect, or to hide these off-axis aberrations.)

We have already noted that a doublet lens can be corrected for both chromatic and spherical aberrations, but correction for spherical aberration is an additional design step, and many daguerreotype lenses were uncorrected. Figure 3.17 (*lower*) illustrates that the outer parts of a lens cause most of the spherical aberration; the effect is additionally heightened because the outer

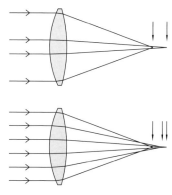

Fig. 3.17. Spherical aberration in a lens. (*Upper*) Rays passing through the lens close to its centre intersect at a different location from rays passing near the margins. Arrows indicate the intersections; there is no single, perfect focal point. (*Lower*) The intersection deviates most for rays that have passed through the outer parts of the lens.

Fig. 3.18. Foucault introduced controlled amounts of spherical aberration into his daguerreotype portraits with star-like lens stops in order to soften sharp lines and smooth rough skin.

regions of a lens also capture more of the light. Daguerreotypists were trapped between two conflicting requirements: they wanted small apertures to capture sharp images, but large apertures so that the exposures would be short.

Crisp images were not always to Foucault's taste, however.[41] He deplored 'the dryness of the lines and the roughness of the skin' in a sharp portrait. (Skin blemishes would have been additionally highlighted by the blue-sensitive emulsions.) He advocated placing a circular aperture near the camera lens in order to stop the beam down to the size where the image was sharp, and then adding in a controlled amount of spherical aberration with triangular cuts out to larger radius (Fig. 3.18). The cuts could be made deeper in order to blur each element of the image over a greater area, while they could be made broader to increase the brightness of the blurring relative to the unblurred image. It should be noted that this is much more subtle than a quite different procedure sometimes employed by photographers involving a circular or oval aperture with jagged edges. This aperture is placed much further from the lens in order to produce a circular or oval portrait with edges that fade away. There would have been no fading or reduction in size with Foucault's procedure which affected only sharpness and contrast over the entire image. Foucault in fact also experimented with this latter procedure using a rotating, jagged disc, and obtained 'really remarkable softening effects'.[42]

Foucault took up a burin and proudly and skilfully engraved his name and the date into the edge of his daguerreotypes taken in 1842. There is then a gap of two years until the next surviving one, and, more significantly, until his next publication. His advancing medical studies were doubtless monopolizing his time and energy, but by the latter half of 1843 and in 1844 he had returned to science in simultaneous collaborations with three different men. They were Fizeau, whom we have already met; Alfred Donné, whose name arose earlier; and Henry Belfield-Lefèvre, D.M.P., to whom we now turn.

Chapter 4

The 'delicious pastime' applied to science

Foucault wrote that photography rapidly became 'the delicious pastime of a multitude of passionate amateurs'.[1] However, he went beyond the amateur and mere technical improvements, and, as Arago had hoped and predicted, applied Daguerre's invention to scientific discovery.

Henry Belfield-Lefèvre, D.M.P.

Few details have survived concerning Henry Belfield-Lefèvre. An Anglo-French origin is indicated by his name and un-French use of letters to signify his 1837 doctorate in medicine (***D**octeur en **M**édecine de la **F**aculté de **P**aris*).[2] He briefly translated a French medical journal into English,[3] and wrote on ethics and Catholicism, including such thorny questions as the anatomical similarities between animals and man, but he was also interested in the less controversial but very topical issue of electrodeposition.[4]

The electrical process of *electrotyping* was a second discovery that stirred the world in 1839 – though not to the extent of the daguerreotype. The antecedents were long. Electrical reduction of metallic salt solutions to native metal was not new, but in 1836 the English industrialist and gentleman-scientist Warren de la Rue had noted that the deposited layers detached easily and carried the perfect imprint of the electrode on which they had formed. Three years later others made facsimile coins and medals by using the original to electrodeposit a mold in which a replica of the original was in turn electrodeposited. This might seem of interest only to numismatists, but a more mainstream application was in view. This was the replication of copper printing plates, which ceased to print crisply after about two thousand copies. In 1841 Belfield-Lefèvre deposited a *pli cacheté* at the Academy in which he suggested that designs or plans scratched into a wax-coated glass plate could be turned into printing plates through electrodeposition. In the following year he conceived the idea that daguerreotype plates could be made by electrodeposi-

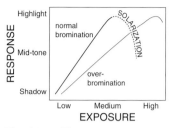

Fig. 4.1. Different exposures produce different tones on a normally brominated daguerreotype plate, but if the plate is overexposed, the response can actually be weakened in an effect called *solarization* (dashed line). If the plate is overbrominated it becomes less sensitive (a higher exposure is needed to produce a given tone) but since the plate also becomes less contrasty, a wider range of exposure levels can be recorded. (The plate's range of available tones is essentially unchanged.)

tion rather than the usual 'Sheffield-plate' method of cold-rolling a copper-silver bimetallic slab down to the visiting-card thickness recommended by Daguerre. The trick for an electrodeposited plate was to ensure it took the forming-electrode's shine so that polishing was unnecessary. Belfield-Lefèvre formed a company to make electrodeposited plates in 1842, but though the quality was good, the price was too high and the company folded.[5]

We can only speculate on how Foucault met Belfield-Lefèvre. Perhaps it was through the medical school, perhaps it was through Foucault's pious mother, perhaps it was once Belfield-Lefévre began to use his electrodeposited plates to take actual daguerreotypes and needed an experienced collaborator.[6] His religious and ethical views were unlikely to have been congenial to the sceptical and individualistic Foucault, but the daguerreotype unified them; and when Belfield-Lefèvre died prematurely a decade later, Moigno described him as Foucault's 'friend and collaborator'.[7] Their first published result together in 1843 continued Belfield-Lefèvre's concern for the daguerreotype plates themselves and discussed how best to clean and polish them.

The paper was not well received. Cleaning and polishing were critical issues: 'The beauty of the finished picture depends in great part on the perfect polish of the plate,' Daguerre advised.[8] Several applications of olive oil, pumice and nitric acid were involved, as well as heating in an oven or with a lamp. It was recognized that some fatty matter was left on the plate from the cotton swabs used for the final polish. In a letter to the Academy, Belfield-Lefèvre and Foucault claimed that, far from hindering the formation of the latent image, this organic layer was essential to it. They claimed the polish should be done with pumice and turpentine, with a final wash in absolute alcohol to eliminate some, but not all, of the turpentine.[9]

Daguerre's reaction was swift:

> Although up to now I have not judged it appropriate to respond to numerous communications which, far from announcing real improvements... have, on the contrary, only impeded the advance of the process, I find myself forced in this case to refute the theory propounded by MM. Belfield-Lefèvre and Foucault.[10]

Two other daguerreotypists felt that the theory was such 'as gravely to compromise the progress of the daguerreotype'.[11] Among the objections, it was pointed out that if a fatty layer was essential to the formation of the image, it was surprising that it had no effect on the plate's speed; however, Daguerre did begrudgingly admit that Belfield-Lefèvre and Foucault had proposed a simpler polishing procedure.

The proud Foucault will not have liked the criticism, though he will have judged its merits coolly and accurately. A few weeks later Belfield-Lefèvre wrote another letter to the Academy discussing hydrogen chloride gas as a sensitizing agent and repeating the claim that the light-sensitive layer was composed of a hydrocarbon and silver iodide.[12] Foucault signed neither this letter nor a *pli cacheté* on bromine sensitization.[13] Despite this contretemps, Foucault continued to work with Belfield-Lefèvre. Three years later he wrote a note in both their names propounding the advantages of continuing bromination of the silver iodide beyond the golden-yellow of maximum sensitivity towards a deep bluish-violet colour. Although this roughly triple dose of bromine reduced the sensitivity, it also reduced the contrast, and such plates were of practical value for scenes containing a large range of light levels (Fig. 4.1).[14] 'We have seen a little picture of this sort taken in sunshine,' commented the Abbé Moigno; 'in it we saw at the same time clouds in the sky, white houses throwing shadows with details within, and trees whose foliage was apparent in groups, rather like an artist would have drawn them.'[15] Moigno also reported a recipe by Belfield-Lefèvre and Foucault for repolishing recycled daguerreotype plates.

Alfred Donné

Even though the Paris Medical School was the biggest and best in France, the instruction offered was incomplete. Reforms had rendered the teaching more practical, and when Foucault donned the blue apron and black dissecting hat to soothe his broken heart, it will most probably have been the Faculty itself which procured the corpse rather than a private grave digger. The Faculty's underlying philosophy was the blind application of accepted therapy. Formal teaching was stilted and required encylopaedic memorization, while inquiry into underlying causes was considered superfluous.[16] There was therefore a market – and a long tradition – for a parallel system of private instructors who provided a more lively presentation of the official curriculum as well as omitted topics such as obstetrics and paediatrics. In addition, people outside the Medical School were applying science to medicine, and offering tuition in subjects like chemistry and experimental physics. Some of these *professeurs libres*, or free professors, were truly independent; others held positions in other, more science-minded faculties of the university, or occupied posts in hospitals.[17] One such was Alfred Donné (1801–78).[18]

Donné was born in Noyon, a small town in Picardy in northern France. His father was a wealthy merchant who died when Alfred was an adolescent. (It was later claimed that Donné was the only student who came to classes in a carriage.) When he was twenty the family moved to Paris. At their insistence he studied law, which he disliked, graduating as an advocate in 1826. With family duty done, he embarked on medicine. He succeeded so well that three years later he was appointed *chef de clinique*, or top junior doctor in one of the sections at the Charité, one of the Paris hospitals. In the same year, he began writing for one of the city's newspapers, the *Journal des Débats*.

To complete their doctorates, medical students needed to write a thesis of a few dozen pages. Donné decided to apply the microscope to medicine.

Fig. 4.2. It is ironic that no reliable photograph or portrait of Donné has survived, despite his important role in the early development of photography.[19]

(Family collection)

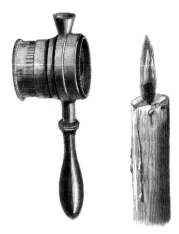

Fig. 4.3. Lactoscope invented by Donné for checking the fat content of milk through its opaqueness. This was particularly relevant at a time when milk was frequently adulterated with substances such as starch, flour, lime water, and even ground-up calf pluck or dog brain. 'And so there is progress already,' one newspaper noted, 'because although the buyer may be robbed, at least his health is not destroyed.'[24]

This was brave, because the predominant medical attitude was hostile to microscopy, or at best considered it irrelevant. There were theoretical and practical reasons for this. Disease was believed due to gross factors such as internal disequilibrium or the external environment, and medical diagnosis emphasized the use of unaided senses.[20] The study of the microscopic, wrote one anatomist, 'overlooks man to dream only of the cellular and gets lost in the abyss of the infinitely small'. The German poet and dramatist Goethe, who had an interest in science, declared, 'The microscope leads judgement astray.'

Goethe's comment was not without foundation. Since the discovery of the simple microscope by van Leeuwenhoek in the seventeenth century, the instrument had undergone many developments, but it was a complicated device that required considerable skill to use. In addition, dry-mounted specimens often lacked contrast and were difficult to interpret, and the images were blurred and surrounded by the coloured rings of chromatic aberration. However, wet mounting, fixing and staining improved the contrast between different tissues in the 1820s; and Charles Chevalier succeeded in making tiny achromatic microscope lenses.[21] (Thenceforth he signposted his shop AU MICROSCOPE ACHROMATIQUE.) Although falling in price, microscopes were expensive – they could cost several hundred francs – but this was no obstacle for the wealthy Donné. In 1831 he was awarded his doctorate with a thesis investigating 'the globules of blood, pus, mucus and the liquids of the eye'. This was a guidepost for his subsequent medical career, which was devoted to bodily fluids such as urine, saliva, sweat, semen and vaginal secretions. In the last of these he discovered *trichomonas vaginalis*, which was only the second human protozoic parasite to be found.[22] He also discovered the condition of excess white cells in blood which is now called leukaemia.[23]

The fluid to which Donné devoted most attention, however, was milk, both human and bovine (Fig. 4.3). In 1842 he published a book on child rearing, much of which concentrated on feeding, where contamination was the prime cause of infant mortality.[25] The book was empirical, realistic, and generous concerning mothers' obligations. 'I wrote it with love,' Donné stated later.[26] Following microscopic examination of the wet-nurse's milk, he won the Légion d'honneur when his advice saved the life of the sick Comte de Paris, Louis-Philippe's baby grandson.[27]

In 1837 a Medical School committee chaired by the Dean had reported favourably on Donné's research: 'Thus studies with the microscope can no longer be considered objects of pure curiosity; ...their vocation is to render great services to Medicine...' As a result, Donné was given the title of Special Professor of Microscopy, and authorized to run a *cours libre* (informal course) in medical microscopy under the aegis of the Faculty.

For his course, Donné was allocated space in the Hôpital des Cliniques, which despite a name suggesting medical wards was actually a maternity hospital.[28] Over the years attendance at Donné's course grew from a handful to more than one hundred from home and abroad. When numbers had been small, it had been easy for students to look down a few microscopes at the various healthy and sick bodily fluids that were so dear to Donné. As attendance grew, he wanted to show microscopic specimens to the whole class at once. He made some trials with the solar microscope (Fig. 4.4). This device threw a magnified image onto a screen, but sunlight was the only illumination strong enough to produce a bright image. The solar microscope could therefore not be used in the evening, which was when Donné gave his course, and its use was problematic even during the day in an often-cloudy place like Paris.

Like many lecturers since, Donné found that the students became unruly when demonstrations did not work flawlessly. The Dean was persuaded to let him set up a modified solar microscope in the lecture theatre in the Hôpital des Cliniques. The source of illumination was not the Sun, but a piece of chalk, heated to incandescence by a vigorous oxygen–hydrogen flame.[29] This light source had been devised in the 1820s by one Thomas Drummond to provide a brighter beacon-light for use in the mapping of Ireland. The Drummond light, or limelight, is bright and white; Sir John Herschel reported that 'a shout of triumph and admiration burst from all present' at its first demonstration in the Tower of London. However, the projecting gas microscope was still unsatisfactory with low-contrast specimens, such as bodily fluids, and it remained necessary to have normal microscopes available after each lecture.

It is not surprising that Foucault was one of the students who, on dark nights, slipped through the shadows of the Medical School's imposing colonnade and courtyard to attend Donné's lectures. The scientific nature of the course will have appealed to him, as well as its optics component and its application of technique to practical problems. He was also, we should remember, hardly out of that age 'when one has so much need and so much desire to see everything'.

After Foucault's death, Donné recalled their first meeting:

> Thirty years ago, a young man with dull, but deep eyes was attending the course on microscopy that I was giving at the Hôpital des Cliniques... After one lesson, I saw this young man come to the lecture bench, handle the instruments, take them apart, and examine them part by part; then, speaking to me in a cool and calm voice, he said, 'Monsieur, you said that such and such happens in a certain way; I do not think that this is correct, but that it happens in this different way.' I was tempted to find the remark impertinent, but on reflection, since it seemed reasonable, I considered it further. 'So you work with microscopes?' I asked one day – ' Oh, yes.' – 'Well then, come and see me, and we'll chat.' This young man was Foucault: he was nineteen years old at the time. 'Would you like', I asked him, 'to study closely the content of my lectures, and to use my instruments (of which I had a wide range)? Would you like to help me prepare various delicate demonstrations for my lectures?' – 'Very willingly!'

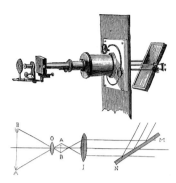

Fig. 4.4. The solar microscope was attached to a hole cut in the shutters of a darkened room. The mirror reflected sunlight into the microscope. A magnified image was projected onto a screen for all to see. The device was first presented in London in 1740 by a German physician, Dr Johann Lieberkühn.

That is how I met Foucault, how I had the ablest of assistants, how we worked together, and how we became linked by a close friendship which continued for the rest of his but short life.[30]

Time had blurred Donné's memory; more probably they met when Foucault was twenty-two or three.[31] Foucault assisted in the preparation of Donné's demonstrations for three years.[32]

Milk was not Donné's sole obsession. There was also the daguerreotype, and this fascination had begun with the public announcement of the process, which he had reported in the *Journal des Débats*.[33] Hardly more than a week later, he had succeeded in taking a daguerreotype, which, if not completely satisfactory, certainly recorded astonishing detail. 'I am henceforth convinced of the worth of M. Daguerre's invention,' he announced.[34] A fortnight later, after attending a public demonstration by Daguerre, he wrote that if he were twenty-five again, he would run off to daguerreotype the world; but since he was almost forty, he would have to be content to apply photography to his 'usual and favourite pursuits'.[35] He immediately put the daguerreotype's golden-yellow sensitive layer under the microscope and reported to the Academy that it was very uniform.[36] He tried to turn the daguerreotype into a printing plate so that multiple copies could be made, and worked on daguerreotyping microscopic objects.[37] Donné was not alone in addressing these topics. Charles Chevalier's father, Vincent (from whom he was estranged) produced photomicrographs of the scabies mite,[38] while Fizeau was working on making printing plates. Fizeau produced some reasonable results, though the range of half-tones was limited, as was the number of prints that could be drawn from the plate.[39]

The *microscope-daguerréotype*

Donné's published work on the daguerreotype petered out in mid-1840, perhaps because he was busy composing his opus on baby care. With that book published in 1842, his attention was free again, and between 1843 and 1845 he completed two new ventures in collaboration with Foucault. Both exploited recent technological advances. One, which we will discuss in the next chapter, was the construction of a projecting microscope for lecture demonstrations, which used an electric arc as its light source. The other project was to turn the material of his *cours libre* into a textbook on medical microscopy, which he published in 1843–44 as the *Cours de microscopie*...[29] Books at that time did not always include high-quality illustrations; rather, engravings might be published separately in an *Atlas*. Rather than relying on a draughtsman's skill to record the various microscopic objects, Donné wanted to use the impartial eye of the daguerreotype.

Foucault helped take these photomicrographs. Some trials were made with electric light and limelight. Bright as it seemed, limelight is actually much

Fig. 4.5. The heliostat was essential for any optics experiment requiring bright light. The mirror was oriented and then driven by clockwork in such a way as to reflect a beam of sunlight into a fixed direction. This is the heliostat devised in 1843 by Johann Theobold Silbermann (1806–65).

(Bibliothèque nationale de France)

The *microscope-daguerréotype*

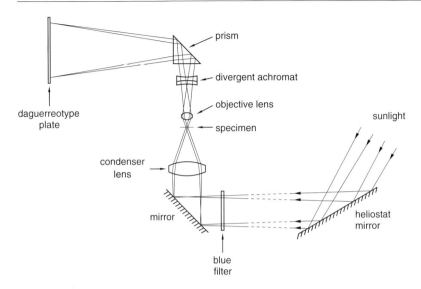

Fig. 4.6. The principal components of the *microscope-daguerréotype* used by Foucault and Donné to take the photomicrographs for their *Atlas* of medical microscopy. The exact location of the blue filter was not specified, but it was placed somewhere before the microscope stage. It absorbed heat in order to prevent specimens from burning or boiling.

fainter than sunlight (see below). Notes left by Foucault reveal that trials with the Drummond light required 'lots of gas' to keep the chalk hot for the 8–10 minutes necessary to get a daguerreotype exposure.[40] Since the source of the hydrogen and oxygen will have been chemical reactions rather than modern cylinders of compressed gas, this will have been a major inconvenience, as was the need to change the chalk frequently. A hydrogen–oxygen mixture is explosive, and in one of his numerous popular-science books, Louis Figuier claimed that Donné and Foucault caused several detonations amongst the newborns.[41] The anecdote is amusing, but it is an invention, since the gases will only have come in contact at the burner.

The electric arc was brighter: a micrograph of some animal or plant scales required only 6 seconds exposure. However, the micrographs were mostly taken using sunlight, no doubt because an exposure could be obtained even more rapidly, before fresh specimens dried or wrinkled from the heat. (Thirty years later, it was still being noted that 'only sunlight furnishes the means of obtaining useful photographs of microscopic objects'.[42]) Many of the daguerreotypes have survived, and are signed and dated by Foucault, so we know that they were taken during the summer of 1844.[29]

The apparatus was relatively straightforward. It began with a heliostat, which was a crucial device for any optics experimenter, and we will encounter it in most of Foucault's optics experiments. As its name indicates the heliostat makes the Sun, or at least a sunbeam, stationary. Its full operation will be discussed in Chapter 15, but clockwork turns a plane (flat) mirror in such a way as to reflect sunlight into a fixed direction despite the Sun's diurnal motion across the celestial sphere. Foucault and Donné probably took advantage of an improved design that had just become available (Fig. 4.5). Donné wrote:

> In the hand of the author this instrument has undergone such notable changes and such artful improvements that not only does it offer greater simplicity of usage and greater accuracy than most of those employed up to now in physics, but also its price, which used to be so high, has become completely affordable. M. Gambey's heliostats cost no less than 1200 to 1500 francs: M. Silbermann's heliostat, constructed with great care by M. Soleil, is available to physicists for 350 francs.[43]

Figure 4.6 sketches the essentials of the *microscope-daguerréotype* as Foucault and Donné called their apparatus.[44] A plane mirror reflected the heliostat beam onto a converging condenser lens which concentrated the beam from beneath onto the microscope stage and specimen. The microscope objective lenses were located above the stage and were followed first by an achromatic doublet, and then, for convenience, by a prism which reflected the beam horizontally into a *camera obscura*. Three sets of objectives were available. Their position and that of the achromat could be adjusted so as to provide any desired magnification along with a field of view that filled the daguerreotype plate. The daguerreotype plates were prepared in essentially the classical way, except that the bromination solution was even weaker at 1 in 10 000. Foucault and Donné must have waited impatiently to take the micrographs, because there was not much sunshine during the summer of 1844 (Fig. 4.7).

Rigorous fidelity

The *Atlas* contained eighty figures grouped by fours into twenty plates.[46] None of the procedures for cutting printing blocks directly from the daguerreotype plates derived by Donné, Fizeau or others could give the contrast, reliability or number of copies desired. Fizeau offered to make further trials to see if he could do better, but Donné was not keen on any procedure which destroyed the daguerreotypes, which he wished to keep intact as proof of the accuracy of the reproductions. It was therefore decided that they would be copied by engraving. The engraver employed was one of some note, Oudet, who had cut the illustrations for the zoologist Georges Cuvier's four-volume work, *The Animal Kingdom*. Engraving eighty micrographs took some time, and the *Atlas* was published in four instalments of five plates, each costing an expensive 7 fr 50. The first two sections were available in 1845 February. 'In this work everything is reproduced with a rigorous fidelity previously unknown,' an advertisement announced.[47] Figure 4.8 shows some of the actual daguerreotypes; engravings from the *Atlas* are shown in Figs. 4.9–4.11. Occasionally portions of several photomicrographs were combined into a single engraving, but more usually the micrographs were copied directly, albeit with a smaller field of view. Oudet's skill as an engraver can be judged by comparing an engraving from the *Atlas* with its original daguerreomicrograph (Fig. 4.11 vs. 4.8); Donné's opinion was

Fig. 4.7. The summer of 1844 was pretty grey, as these pictograms of the state of the midday sky over the Paris Observatory indicate, but Foucault nevertheless found enough sunshine to take almost eighty daguerreomicrographs of medical specimens.[45]

Rigorous fidelity

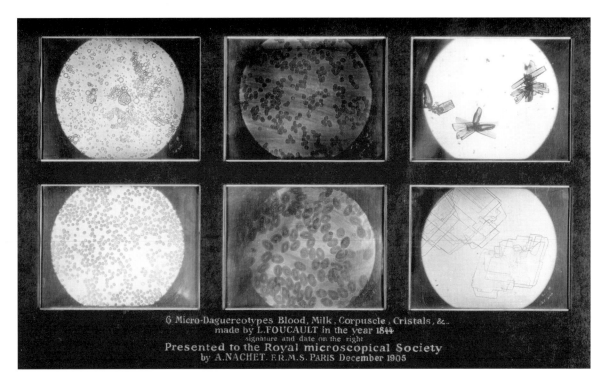

Fig. 4.8. Six micrographs, most of which are signed and dated by Foucault. The upper row shows (from left to right): ass's milk from a swollen teat; human blood; and uric acid crystals. The lower row shows: putrefying human blood; frog's blood; and cholesterol crystals.

(National Museum of Photography, Film & Television/SSPL)

that Oudet 'has almost always made a successful job of the difficult and novel task which he had to accomplish.' The magnifications of the reproductions in the *Atlas* were mostly 400 times.[48]

The specimens reflected Donné's interests – pus, blood, mucus, epitheliums (cells from organ linings), spermatozoa, ova, milk, starch and various crystals. Human blood corpuscles were compared with those from camel and frog. The change in their appearance and the expulsion of their nuclei due to the prolonged action of water and dilute acetic acid was demonstrated. The reader could examine the differing aspects of the leucocytes in pus according to the age of the suppuration, and compare them with the mucous and fatty globules in other secretions. Another figure showed pus from a syphilitic chancre, in which numerous elongated spiral objects were noted, but Donné failed to recognize these were the causative agent. One must be surprised that faeces were not imaged, because Donné certainly devoted a chapter to stools in the *Cours de microscopie*.

It had initially been planned to accompany each photomicrograph by an explanatory sketch, but the micrographs were found to be clear enough on their own. They were also, Donné found, 'the way of convincing the most unwilling spirits' that microscopic observations were not an illusion. How otherwise to explain that visual and photographic observations agreed with each other, and

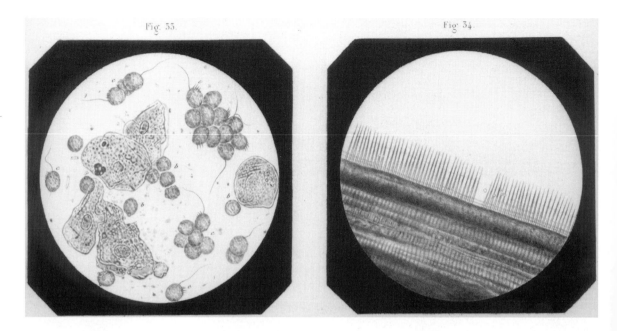

Fig. 4.9. Mucus and cilia from Donné and Foucault's *Atlas* of medical micrographs (their Plate IX). Fig. 33: Vaginal mucus containing purulent globules (*b, b, b*) and *trichomonas* protozoans (*c, c, c*). Fig. 34: Mussel cilia. The circular field of view corresponds to 0.2 mm.

(*Bibliothèque nationale de France*)

agreed between magnifications varying from 50 to 500 times?

> Hasn't the daguerreotype arrived at just the right moment to give the last proof, the most complete demonstration in favour of microscopical observations, and to destroy what might rest of thoughts of the claimed illusions of the instrument?[49]

Pretty and technically groundbreaking as they were, the photomicrographs in the *Atlas* did not captivate the medical world and were rapidly forgotten. There were several reasons for this. Staining and colouring are important adjuncts in microscopy for differentiating tissues, but these techniques were in their infancy. Although the microscope was becoming of scientific importance in medical research, it was not yet being adopted for diagnostic tests or leading to improved clinical treatments, even though Donné had devised a variety of inexpensive microscopes for doctors to carry on their rounds.[50] There was an absence of any reference to nascent ideas about the importance of cells, while the nature of the specimens was limited, with no internal tissues presented. Finally, despite his officially sanctioned course, Donné was not a member of the Faculty. His work would have had much greater influence had he been one of the great or powerful in the medical establishment.[51] The first large-scale and effective use of medical micrographs came two decades later from a surgeon in the United States Army, Lieutenant-Colonel J. J. Woodward (1833–84). Donné would have approved: Woodward distributed photomicrographs with his reports to quicken diagnoses by other doctors.[52]

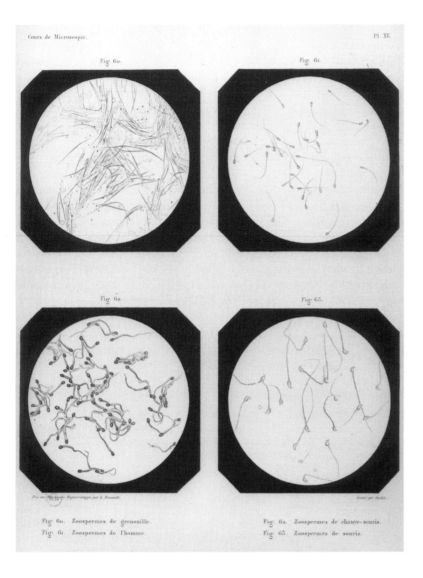

Fig. 4.10. Spermatozoa from Donné and Foucault's *Atlas* of medical micrographs (their Plate XV). Fig. 60: Frog. Fig. 61: Human (it is not recorded whether they are Foucault's). Fig. 62: Bat. Fig. 63: Mouse. The circular field of view corresponds to 0.2 mm.

(Bibliothèque nationale de France)

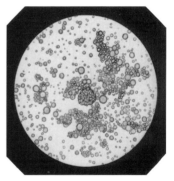

Fig. 4.11. Ass's milk drawn from a swollen teat. This engraving from Donné and Foucault's *Atlas* of medical micrographs (their Plate XIX, Fig. 82) should be compared with the original daguerreomicrograph (Fig. 4.8, upper left). Ass's milk may seem an exotic fluid, but it was reputed always disease-free and was preferred over cow's milk by the well-to-do. The circular field of view corresponds to 0.2 mm.

(Bibliothèque nationale de France)

The brightness of light

New techniques open new avenues of research. While working with Donné on the *Atlas*, Foucault was also collaborating with Fizeau to apply the daguerreotype to another topic, foreseen by Arago in 1839, which was photometry, or the measurement of the brightness of light. Although it was not explicitly stated, this collaboration in 1843–44 was motivated by the very practical issue of whether sunlight could be replaced in optical experiments and instruments

Fig. 4.12. Davy's egg. Davy initially used a 150-plate battery to power this light at the Royal Institution; a later battery comprised 2000 elements.

by the two artificial sources that might possibly rival it: the Drummond light and the electric arc.

Electric light had entered into the realm of the conceivable in 1800 when the Italian physicist Alessandro Volta had stacked together copper, zinc and acid-impregnated cloth discs to produce the first electric 'pile', or battery. Two years later Sir Humphrey Davy exhibited a carbon arc at the Royal Institution in London.[53] Davy's electrodes of sharpened charcoal burned, however, so he placed them in an evacuated glass 'egg' (Fig. 4.12). Unfortunately the electrodes gave off fumes which rapidly blackened the inside of the glass, obscuring the light.

Though Davy's egg was a popular and intriguing lecture demonstration, it was little more than a curiosity because of the large array of cells required to power it. This changed in 1842 when the German chemist Robert Wilhelm Bunsen devised an improved battery. 'Then Davy's experiment was repeated on all sides,' noted Donné and Foucault, 'and in our lecture rooms we were dazzled by the torrents of light given off by the carbon cones.'[54] The Bunsen cell will be detailed later, but it is what turned the electric arc from an amusement into a light source of potential practical utility.

Photometry abounds with terms such as intensity, brightness and flux, which though often misused, have specific technical meanings. Foucault and Fizeau measured the *intensity* (now also known as the radiance) of their three sources, since this is a quantity that enters into the design of optical instruments.

To understand intensity, consider Fig. 4.13. An extended, uniform source of light of some strength is viewed through an aperture. A patch of brightness is seen. If the source is now moved further away, the patch of brightness still looks exactly the same. If the strength of the source is altered, the patch will

Fig. 4.13. A given angular extent of a uniform, luminous source looks the same whether the source is near or far. (*Left*) Arrangement for viewing the same solid angle of two identical sources at different distances. (*Right*) Identical appearances in both cases. A small element E of the source (marked out) fills a smaller angle when the source is further away, but exactly compensating this is the fact that more such elements enter the field of view. The intensity is the physical property which quantifies this distance-independent strength of a source.

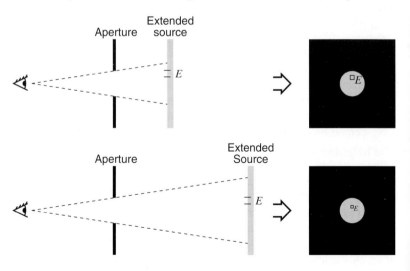

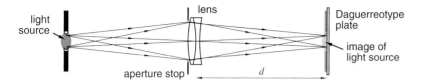

look brighter or darker as the case may be, but for a source of fixed strength, the property which quantifies this distance-invariant brightness is called intensity. Since the intensity is a property only of the surface of the source (provided there is no intervening absorption), it is quite legitimate to compare Drummond lights and electric arcs in the laboratory with the Sun, some 100 thousand million times further away.

What the daguerreotype responds to, however, is *flux* (also known as irradiance), or how much light has fallen on the plate. If a lens is used to focus the light source onto a plate, the measurement of flux can be related to the intensity of the source through knowledge of the aperture of the lens, its distance from the plate, and the exposure duration. Figure 4.14 outlines Fizeau and Foucault's experimental setup.

The relationship between flux and the resulting photographic effect was not explored in depth until the 1890s.[55] In the 1840s, the amount of flux that had fallen on the daguerreotype plate could not be read off from the different densities of light tone created. What Foucault and Fizeau instead did was to take a fixed point in the daguerreotype response and use this as a fiducial or reference level to determine the relative intensities of their three sources. They used unsensitized plates, which were more than adequately responsive for their bright sources, but which were also very similar one to another, provided they were prepared by the same person in the same way. For their fiducial point they took the flux that was just enough to start the formation of an image when the plate was exposed to mercury vapour. Of course they could not arrange the aperture stop and distance of their lens so as to produce exactly this flux in an exposure of fixed duration; rather they made a series of five or six, slightly displaced exposures on the same plate, with times between 3/5ths and 3 seconds. 'If the experiment had succeeded, the series was incomplete', they reported in their memoir presented to the Academy in 1844 April, and they were able to gauge the exposure duration that was just sufficient to form an image.[56] Tests with 'a lamp giving very stable light' showed that the effect produced was proportional to the exposure, at least for times that did not vary by more than a factor of ten; beyond these bounds the plates exhibited what is now called reciprocity failure, and the images were noticeably fainter than expected.

They began experimenting on the Sun in 1843 August and September, restricting themselves to clear days near noon, when the intensity was least reduced by the atmosphere. There were not many of these in August (Fig. 4.15);

Fig. 4.14. Intensities can be measured by using a lens of known aperture and distance to image the source onto a flux detector (the daguerreotype plate). For the Sun, Foucault and Fizeau used a lens giving $d = 1.413$ m which they stopped down to diameters between 1.3 and 3.0 mm. For limelight and the electric arc, they worked at unit magnification so as to have adequately big images of the luminous regions of the sources, which were only one or two millimetres across. Definitive measurements were made with $d = 1.125$ m and apertures between 3 and 17 mm diameter.

Fig. 4.15. The autumn of 1843 and the spring of 1844 were sunnier than the following summer (Fig. 4.7). Nevertheless, Foucault and Fizeau had to break away from their usual week-end working to measure the intensity of the Sun at 2 p.m. on September 20 and at 11:15 and 12:40 on April 2, which were weekdays (emphasized).[57]

finally, they were able to take a satisfactory measurement close to the equinox during 'a sky of a pale blue' on September 20. Thereafter they turned their attention to the artifical lights, returning soon after the following equinox to make two more measurements on the Sun 'during a time of extraordinary clarity'. For the electric arc, they first worked with what no doubt were graphite electrodes placed in a vacuum and subsequently in an inert gas, as in Davy's egg; but fumes quickly darkened the glass. Foucault found that the electrodes burned much less rapidly, and could be used in air, if they were made of *gas carbon*. This dark, dense, shiny, tightly grained material formed in abundance during the production of town gas from the distillation of coal. Gas carbon may seem esoteric to us, but it was used for tubes and crucibles in chemistry and it is not surprising that Foucault should have thought to try it.

Foucault and Fizeau experimented using Bunsen cells with smaller and larger electrodes, in series and in parallel, with freshly added acid and after a few hours use. The electric light arose from the incandescence of the carbon electrodes. The positive electrode was always brighter; while the much less luminous arc between the electrodes exhibited a purplish-blue colour. For the Drummond light they varied the pressure, and therefore flow rate of the gas mixture. It would be tedious to report all their results; a few are extracted in Table 4.1. They were surprised by the faintness of the Drummond light. Three decades later, when the relation between temperature and radiation was beginning to emerge, Fizeau used these measurements to derive a temperature of 8000 degrees for the Sun's luminous layers, or *photosphere*, as Arago had termed them.[58] Though a little high, this result was fundamentally correct.

Blandly stated, Fizeau and Foucault's experiment may not seem especially remarkable, but let us review what was involved. A dark room of some size was required, along with a heliostat. Achromatic lenses were probably bought from Chevalier or another supplier, but lens holders capable of fine adjustments were probably made by the experimenters themselves, along with screens and apertures. Measuring the diameters of the smaller apertures required a microscope and a graduated scale, possibly borrowed from Donné, who no doubt also lent the chalk, jets and other parts from his gas microscope. To avoid shifting large volumes of equipment, the experiments may well have been performed in the Hôpital des Cliniques, though this is not stated. Before using the Drummond light, Foucault and Fizeau would have had to fill the hydrogen and oxygen reservoirs, no doubt by reacting zinc with sulphuric acid, and through heating potassium chlorate.[59] For the electric arc, they undoubtedly used some of the equipment that Foucault and Donné were assembling for their projecting microscope (see next chapter). The brittle gas carbon will have had to be cut into rods and, to allow the arc to be struck, held in adjustable clips. The supplier of Bunsen cells in Paris was the instrument-maker L. J. Deleuil. The 138 required were no doubt obtained from him, along with terminals and connections.

'This excellent cell', as Foucault described it, is shown in Fig. 4.16. The Bunsen cell was similar to the earlier Grove cell, except that the expensive platinum cathode was replaced by a cheaper electrode made from densely agglomerated, pulverized coke or gas carbon. The cell gave a high voltage, and because of the conductivity of its electrolytes, which needed to be poured in just before use, it could give a strong current, which lasted for as long as one or two hours before weakening significantly. The Bunsen cell had an irritating disadvantage, however: it gave off brown nitric oxide gas. In 1843 Foucault no doubt accepted this as an occupational hazard, but some years later his tolerance had worn thin:

> volatile, irritating, caustic…which, given off into laboratories, attacks at the same time both operators and instruments…spread in the air, [it] is a real pestilence; it provokes coughing, penetrates the recesses of cupboards, where it gnaws at the most inaccessible metals, and rapidly attacks the fixtures of the cell itself…[60]

In addition to all this, there were the iodizing, mercury and wash boxes, spirit lamps, plates, polishers, chemicals and other paraphernalia needed for the daguerreotypes. It was a substantial achievement for two amateurs, and encouraged by the reception that they received, they presented some supplementary details to the Academy a fortnight later on 1844 May 6.[62] We can sense Arago's influence behind this second presentation. We cannot be sure exactly when the mighty Permanent Secretary became on regular speaking terms with the two young physicists. Perhaps it was during one of his public astronomy lectures.[63] Perhaps it was after their earlier presentations at the Academy. In any event, Arago chatted over their first memoir concerning intensities, and in so doing uncovered a vital fact.

Table 4.1. Selected intensities reported by Foucault and Fizeau to the Academy in 1844. The solar intensity on April 2 was used as a reference level, set arbitrarily to 1000. Foucault and Fizeau were aware that they had measured what they called the *chemical intensities* to which the violet-blue sensitive daguerreotype plate responded best. They made some visual comparisons and found that 'optical' ratios were about the same, which they postulated would be true for all white lights.

Noon-day Sun	
1844 April 2	1000
1843 September 20	751
Electric arc	136–385
Limelight	0.5–6.85

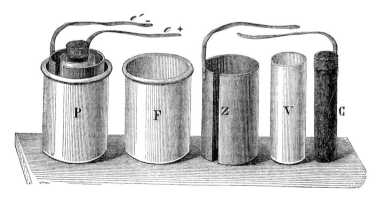

Fig. 4.16. The Bunsen cell, P, as improved by the Parisian maker L. J. Deleuil *c.*1849, and its components. It is a *constant battery*, comprising two half-cells with separate electrolytes, which reduces electrode reactions that choke off the current in a single electrolyte. Inside the porcelain or glass jar, F, the negative electrode, Z, made of zinc rubbed with mercury, was immersed in sulphuric acid. Inside this, a porous vase, V, contained nitric acid, into which dipped the positive electrode, C, made from agglomerated, pulverized coke. In 1844, however, the order of elements was the opposite, with an outer, hollow carbon anode and a central zinc cathode.[61]

Daguerreotyping the Sun

In their memoir on May 6, Fizeau and Foucault presented additional findings, such as that the electrodes turned to graphite where they had been incandescent, that light could emanate from thin wire electrodes during the electrolysis of water, and that molten silver could be used as the positive electrode in the arc. The news that excited Arago, however, was that their solar images had clearly shown a quite large sunspot towards the end of 1843 August, and that the solar disc had at all times seemed slightly fainter at the edges than at its centre. 'We do understand', the young men wrote, 'that this simple observation, made incidentally during our research, does not reflect the full importance of the matter.'

Indeed, the matter had been occupying Arago for some decades and concerned the physical constitution of the Sun. What was the nature of the incandescent material? Was it solid, liquid or gas? 'The response', wrote Arago, '... necessitates to varying degrees the examination of this critical question: are the centre and the edges of the Sun equally luminous?'[64]

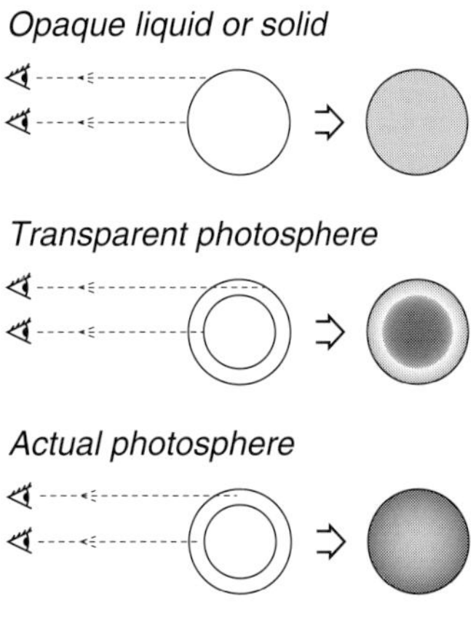

Unfortunately, the observational evidence and its interpretation were equivocal and contradictory. Our view on the photosphere varies from face-on at the centre of the solar disc to grazing at the edge. If the Sun were solid or liquid, its light should be uniformly intense across the disc because opaque, incandescent solids and liquids radiate equally intensely in all directions. Uniform intensities had been reported by scientists such as Galileo Galilei (1564–1642), Christiaan Huygens (1629–95) and J. H. Lambert (1728–77); and Huygens had believed that the Sun was a liquid ball. If the light came from a transparent, incandescent atmosphere, however, there should be a brightening away from the centre, where more luminous material lies within the line of sight (Fig. 4.17).

However, in contradiction to Galileo, Huygens and Lambert, the edge of the solar disc had been measured to be substantially *fainter* than its centre by the eighteenth-century photometrist Pierre Bouguer, Airy and John Herschel. Finally, Arago himself had measured no more than a 1 part in 40 difference.

Arago's photometry seemed to imply a solid or a liquid Sun, but the measurement was difficult because of the smallness of the solar image using a lens of any reasonable focal length. However, Arago also had evidence from the phenomenon called *polarization*.

To explain polarization, further comments about wave motion are in order. In sound waves, the molecules vibrate backwards and forwards in the same direction along which the wave is advancing. This is called a *longitudinal* wave (Fig. 4.18a). By the 1840s, however, it seemed very probable that light was a *transverse* wave: whatever it was that was oscillating was doing so in a plane perpendicular (or transverse) to the direction in which the rays were propagating (Fig. 4.18b). A water ripple is an example of a transverse wave: the water oscillates vertically while the ripple advances horizontally. For a

Fig. 4.17. The physical constitution of the Sun. If the Sun were an opaque, incandescent liquid or solid (*top*), its disc would be uniformly intense (*right*), even though rays are emitted at different angles from different parts of the surface (*left*). If the Sun were cloaked in a transparent, incandescent atmosphere (*middle*) its disc should show a brighter ring towards the outer edge where lines of sight sample more emitting material. The Sun is actually darker towards its limb (*bottom*) because the photosphere is partially absorptive. Although lines of sight near the limb sample more volume, light from the centre of the disc is more intense because it arises from hotter regions, which are radially deeper within the Sun.

light source such as a candle flame, whatever it is that oscillates does so in the infinity of possible perpendicular directions (Fig. 4.18c), but it is possible to use special crystals to select out from the beam only those oscillations in some particular direction (Fig. 4.18d). Such a beam is said to be polarized; a water ripple is a polarized wave because the fluid displacements are restricted to a single direction – the vertical. As will be discussed further in the next chapter, the discovery and interpretation of optical polarization in the first decades of the nineteenth century had provided strong evidence that light was a wave motion, and a transverse wave at that, since longitudinal waves cannot be polarized.

Polarized light abounds in nature, especially in the sky and in reflections. In about 1811 Arago had looked at the Sun with a polarization-sensitive device of his own invention, the polariscope, but he had seen no evidence for polarization anywhere across the solar disc. He had also examined gaseous flames, incandescent wrought iron and platinum, and molten cast iron and glass. He had found that the light emitted by the incandescent solids and liquids showed polarization when observed at very shallow angles with respect to their surfaces. The absence of any polarization towards the solar limb, where the Sun was also viewed at a grazing angle, indicated that the luminous regions were gaseous, in contradiction to the deduction from the roughly uniform intensity across the solar disc. However, as Laplace had realized in his *Mécanique céleste*, if there was any absorption in the solar atmosphere, the expected limb brightening could be eliminated, or even turned into a limb darkening.

There were also sunspots to be considered. They had variously been interpreted as volcanoes, flights of intervening objects, immense rocks uncovered by a liquid tide, and bodies floating in the luminous photosphere. Their east-to-west motion across the solar disc had revealed that the Sun rotated about once every $29\frac{1}{2}$ days. The picture of the Sun that had emerged was of a dark body surrounded by an inner atmosphere full of opaque, reflective clouds similar to terrestrial ones, and then above it, the luminous, cloudy photosphere. Sunspots were supposed to occur when cloud gaps in the two atmospheres aligned to reveal the dark nucleus (Fig. 4.19).

Daguerreotypes of the Sun were therefore of dual interest. They could contribute to study of sunspots, and might remove the uncertainty over whether the Sun was limb darkened or not. At Arago's behest, Foucault and Fizeau took 'a great number' of daguerreotypes of the Sun during 1844 and 1845.[65]

Sunlight was reflected by a heliostat and imaged on the daguerreotype plate by a lens.[66] The major technical difficulty was obtaining short enough exposure times.[67] Hitherto, exposures were seconds long, and could be obtained manually with a lens cap, or, like Gaudin, with a cloth. The Sun needed rapid exposures between 1/100th and 1/60th of a second. Foucault and Fizeau devised a 'quite novel' shutter whereby light was admitted through a variably

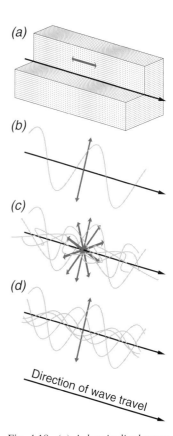

Fig. 4.18. (*a*) A *longitudinal* wave of compressions and rarefactions where the vibrations of the medium (dark-grey, double-headed arrow) are parallel to the advance of the wave (black, single-headed arrow). (*b*) In a *transverse* wave the vibrations are perpendicular to the direction of wave travel. (*c*) An *unpolarized* beam comprises oscillations in the infinity of perpendicular directions (five are illustrated). (*d*) A *plane polarized* beam made from (*c*). The beam contains a component from each vibration.

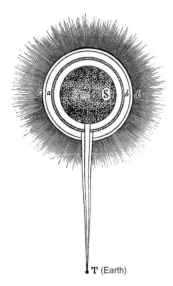

Fig. 4.19. 1840s conception of the Sun. A dark core, S, was surrounded by an atmosphere, *ab*, in which opaque, reflective clouds floated. An incandescent, cloudy photosphere, *cd*, enveloped the whole. Sunspots occurred when breaks in the two cloud layers exposed the dark body. Dimmer sunspot borders, or *penumbra*, arose when the outer cloud-gap was larger, exposing clouds in the lower atmosphere. The origin of the faint, extensive *corona* of light observed during total eclipses was still an open question (see Chapter 12).

Fig. 4.20. (*Left*) Coin-sized daguerreotypes of the Sun taken by Fizeau and Foucault in 1843–45 exhibited clear limb darkening whatever their exposure. (*Right*) The uniform images of white cardboad discs showed that the effect was not some artifact of the daguerreotype process.

sized aperture in a falling wooden slide.[68] Drops of less than a metre will have sufficed.

Two different achromatic objective lenses were used. One produced images the size of a small coin (12.8 mm in diameter).[64] Figure 4.20 represents some of these images, which, whatever their exposure, were always dimmer at the edge. The sharpness of any sunspots indicated that this dimming was not due to blurring resulting from poor focus or poor quality of the heliostat mirror.

For better detail, much bigger images were needed, and hence an objective lens with a much longer focal length. An achromatic doublet with a 9.88-m focal length was employed producing images some 90 mm in diameter.[69] Figure 4.21 presents the one daguerreotype that has survived; it shows two distinct groups of sunspots.

We may be surprised that Fizeau and Foucault did not present their conclusive result that the Sun was limb darkened to the Academy, but the research was in a sense Arago's, since he had suggested the topic, and it was up to him to publish – which he subsequently did, five years later. Further, the result was unexpected and the implications were uncertain. Though no numerical value has survived for how much darker Fizeau and Foucault found the solar limb to be compared to its edge, they must have obtained a value from their series of minimally exposed images in just the same way as they had done when comparing the Sun with artificial lights. It is unlikely to have agreed with Arago's less than 1-in-40 difference because the solar limb darkening in fact exceeds 50 per cent at the short wavelengths where the daguerreotype is most sensitive. It would not have been clear how to reconcile the different measurements, though in hindsight we can suspect the diluting effect of scattered light in Arago's instruments. Nevertheless, it seemed clear that the Sun was not limb brightened, and in due course both Arago and Foucault concluded that the photosphere had to be surrounded by an absorptive atmosphere, which would dim more at the limb where the lines of sight are longer.[68] We now understand that photospheric limb darkening is actually more subtle. Completely unsuspected by Arago was the fact that the photosphere becomes rapidly hotter inwards, so that although the light from the edge samples a larger volume of the photosphere, light from the centre arises in radially deeper and hotter regions, and is therefore more intense (Fig. 4.17).

Other Foucault daguerreotypes

Fig. 4.21. Daguerreotype of the Sun taken on 1845 April 2 at 9.45 a.m. by Fizeau and Foucault. Sunspots and the solar limb darkening are recorded clearly. The solar disc is approximately 90 mm in diameter on the plate. This daguerreotype was later reproduced as an engraving in Arago's *Astronomie populaire*, presumably because it showed the best-developed sunspots, and in this it was probably atypical, because 1845 was close to a solar minimum when sunspots are rare.

(©*Musée des arts et métiers-CNAM, Paris/Photo Studio CNAM*)

Other Foucault daguerreotypes

A few other Foucault daguerreotypes have survived. One, showing a series of solar spectra, will be analysed in the next chapter. Another is a very careful and beautifully lit daguerreotype of a bunch of grapes (Fig. 4.22). Certainly the subject is a plausible one for Foucault, with his interest in plants and later, a well-stocked wine cellar.

A third daguerreotype shows roofs (Fig. 4.23). It is unsigned, but attributed to Foucault by its original collector, Alfred Nachet, of the famous Parisian microscope-making company. The house shown may well be the one that Madame Foucault was renting in the early 1840s at 5 Rue d'Assas, with the picture taken from a viewpoint on the roof of the fore-building seen in Fig. 3.15.[70] What is most interesting is that this daguerreotype is on copper. Cuprous halides are photosensitive, and this might be a copper-based daguerreotype,[71] or a printing plate etched from the original silver plate and then strengthened for the press with a layer of electrodeposited copper according to a process which Fizeau had patented in 1843. However, Fig. 4.23 is most probably an electrotype copy on copper of a normal silver daguerreotype. Fizeau exhibited an 'admirable block' of this type at the Academy as early as the spring of 1841.[72] The copying process was relatively simple and was outlined in the handbooks published by Chevalier and Gaudin & Lerebours, but was only satisfactory with plates that had been hardened by gilding. The gilded plate was put in an electrograph cell similar to the one shown in

Fig. 4.22. Daguerreotype ascribed to Foucault, 1844. The grape variety is probably Meslier Saint-François, which used to be common in the Loire valley and elsewhere, but is now virtually unknown.

(*Société française de photographie*)

Fig. 4.23. Electrotype copy in copper of Parisian roofs, attributed to Foucault. The copy needed to be sealed immediately in a glass-fronted case to prevent oxidation.

(*Société française de photographie*)

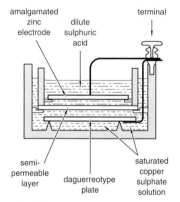

Fig. 4.24. Electrotype copying of a daguerreotype plate, as reported by Belfield-Lefèvre.[4] The semi-permeable layer could be parchment, plaster or bladder. At the daguerreotype plate, copper ions in the copper sulphate solution were reduced to the electrotype deposit of metallic copper ($Cu^{++} + 2e^- \Rightarrow Cu$). The deposition rate depended on the concentration and hence conductivity of the sulphuric acid (zinc sulphate or brine might also be used). A host of practical precautions was necessary for the best results.

Fig. 4.24. Electrolytic action slowly deposited a coherent layer of copper on the daguerreotype face. When the copper was sufficiently thick, it was possible to prise the copy off using a knife blade. The copy's hollow structure was the reverse of the original, but the scattering properties were similar, so the image produced was still a positive one. If all went well, the original daguerreotype was undamaged and could be copied again. Perhaps Foucault made this plate with his electrotyping friend Belfield-Lefèvre.

The delicious pastime

Foucault's activity as a practical photographer hardly lasted beyond the decade in which the daguerreotype reigned supreme,[73] though we will see in later chapters that he did take some photographs on glass plates in the early 1850s, and his interest in the delicious pastime remained with contributions to the Société Française de Photographie and through the invention of a heliostat for photographic enlargements.

When he first saw a daguerreotype, Delaroche is reputed to have exclaimed 'From today, painting is dead!'[74] Happily for art, it soon became clear that not everyone handled the camera with equal talent. We must regret that so few of Foucault's photographs survive, because those that do, micrographs included, reveal considerable grace and flair in their composition.

Chapter 5

The beautiful science of optics

Microscopes after his lecture were all very well, but Donné wanted to show specimens to the whole class at once. His gas microscope using limelight was unsatisfactory. Since the electric arc was so much brighter, he and Foucault set out to incorporate it in a projecting microscope.

The photo-electric microscope

We have seen that one of Foucault's innovations with the electric arc was to abandon charcoal electrodes for ones made of gas carbon, which were so hard that they needed to be fashioned by a gem-cutter. He procured rods with 3-mm-square cross sections, and found they were consumed rapidly: 20 minutes of arc light used 5 cm of the positive electrode and 2.5 cm of the negative one.[1] Nitric oxide fumes from the Bunsen cells were not the only hazard of the electric arc. On his first night of trials, the light burned his eyes. 'Intense ophthalmia until 3 a.m., which was only relieved by a foot bath,' he noted. (Welders and others now wear goggles for protection against the arc's ultraviolet radiation.)

Over the next few months Foucault developed his apparatus. He adopted a spherical condenser mirror to direct the light from the arc to the microscope stage. A mirror was cheaper than an achromatic lens, collected light from a greater solid angle, and produced a more compact design. Soon images were being cast on a screen 1.5 m and then 2 m from the microscope lenses. Heating of the microscope stage indicated that a filter was needed to absorb heat from the arc. He increased the number of Bunsen cells from 60 to 112, and controlled their current by passing it through platinum plates dipped to an appropriate depth in acid.

Though Foucault doubtless discussed progress with Donné, it appears that he experimented alone, because on 1843 December 3 he noted, 'Repeat of the previous experiments in front of M. Donné.' Donné was impressed, and a fortnight later they deposited a *pli cacheté* at the Academy.[2] In it, in Foucault's

Fig. 5.1. Cut-away view of the 'photo-electric' microscope devised by Foucault and Donné between 1843 and 1845 for lecture demonstrations. Light came from an electric arc struck between two horizontal carbon pencils located just inside the front panel of the case. The pencils were held in position by current-carrying springs seen end-on in this cross section. The concave mirror condensed the light onto the specimen, which was held in a vertical slide inserted between the front panel and the lens assembly. A tank of alum solution in front of the mirror acted as a heat filter. Louvres kept the box light-tight while allowing hot air to escape. The operator monitored the arc through dark-glass ports and manipulated numerous controls to maintain the microscope's optical adjustment.

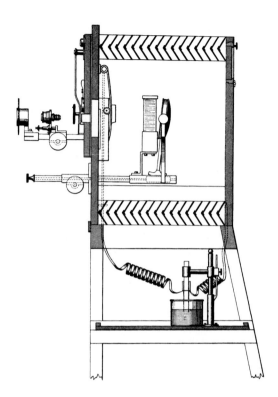

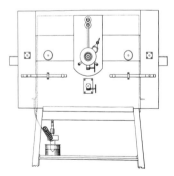

Fig. 5.2. Face view of the photo-electric microscope.

careful and legible hand, they outlined eight key developments that allowed them to project images on to a screen 3 m away. The *pli* was closed with Donné's wax seal, which appropriately pictured Romulus and Remus suckled by the wolf. In the following April they presented their work openly at the Academy,[3] but it was not until 1845 that they published a descriptive memoir after having given a practical demonstration of their device at the Société d'Encouragement pour l'Industrie Nationale (Society for the Encouragement of National Industry).[4] This society had been founded in 1801. Its aims extended beyond industry to commerce and agriculture, and meetings at its premises on the Rue Bonaparte were the ideal ground for presenting anything to do with applied science.[5]

Figures 5.1 and 5.2 show the final device, for which Foucault and Donné coined the term *photo-electric microscope*, to denote that light and electricity were involved. ('Photoelectric' has since acquired a meaning relating to electron emission.) The microscope was made by Charles Chevalier who, as Donné's instrument maker, was presumably involved in the manufacture of components for Foucault's prototype. There was a plethora of access traps and controls to manipulate: four for the carbons; two for the mirror; several for the specimen stage and microscope optics; and one for the current. 'One can see there is plenty to occupy a single experimenter,' Foucault added; 'however, it can be done, as we have proved many times.' Nevertheless, the electrode controls were duplicated on both faces of the instrument to permit two-person operation.

At the Société d'Encouragement, Donné and Foucault projected microscopic specimens as well as showing the crystallization of salts, various live microscopic animals, and the 'brilliant spectacle' of the circulation of blood, no doubt in a frog's tongue.[6] In what seems to have been a first, they also projected an image of the carbons themselves (Fig. 5.3). Most of the light was seen to come from the incandescent electrodes, and especially the fast-consumed positive one; the arc itself was much fainter.[7] Finally, they improvised some experiments involving polarized light to show how useful their device could be 'for presenting the beautiful science of optics to beginners, with all its attractiveness, which ordinarily is apparent only to a small number of adepts who have had the courage to confront the dryness of its elements.'

Used as we are to the easy projection of images during lectures, the photo-electric microscope may seem a cumbersome device, but 'the results caused quite a stir', according to a contemporary encyclopedia, and it was an important enough invention to figure on Foucault's tombstone (Fig. 16.6).

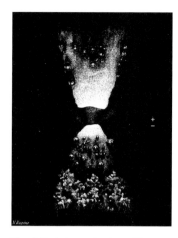

Fig. 5.3. Auto-projection of the carbon electrodes by the photo-electric microscope. The positive electrode erodes more rapidly, which is now understood to be due to electron bombardment. The 'warts' are globules of molten silica. There are less of them on the hotter electrode, where they evaporate more readily.[8]

The nature of light

Until this point, Foucault's work had been essentially technological. He had improved procedures, such as the preparation of daguerreotype plates, and built devices, such as the *microscope-daguerréotype* and the photo-electric microscope, but he had hardly broached questions concerning the underlying nature of physical reality. This changed in 1845 and 1846 when, in collaboration with Fizeau, he demonstrated the interference of light over exceptionally long path differences. But before discussing these elegant experiments, we must review mid-nineteenth century ideas about the nature of light, which also underpin Foucault's work five years later on the speed of light.

In the seventeenth century the aberrations of the newly invented telescope and microscope impelled study of the law of refraction. The mathematical relationship between the angles of incident and refracted rays was elaborated in Holland by Willebrord Snel von Royen (*c*.1580–1626)[9] and the Dutch-resident French philosopher, René Descartes (1596–1650). Their characterization of refraction included a quantity called the *refractive index*, which is higher for substances that refract more strongly. Philosophers also enquired into the fundamental character of light. On the one hand, Descartes saw interstellar space as filled by tiny, tightly packed, transparent spheres. Light was a pressure transmitted by these spheres, though Descartes occasionally used a ballistic model whereby particles were emitted and transmitted, as in his derivation of

Fig. 5.4. Sir Isaac Newton (1642–1727) proposed that light was corpuscular in nature. His ideas of space and time and his laws of motion provided the conceptual basis for dynamics for almost two centuries.

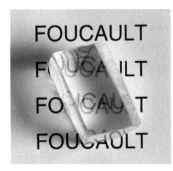

Fig. 5.5. Double refraction by a crystal of Iceland spar. The crystal splits the light from the paper into two beams, resulting in a double image. The spar, quarried near Eskifiördkr in Iceland, is a form of calcium carbonate that occurs in rhombohedral crystals (opposite faces are parallelograms, two of which have all four sides equal).

Snel's refraction law, which we will meet in Chapter 8. Newton too considered that light was composed of a stream of particles and explained additional phenomena such as the diffraction, or spreading, that occurs when light passes close to opaque objects as due to short-range forces acting on the corpuscles.

In contrast, the idea that light might be a wave phenomenon was introduced by the Englishman Robert Hooke (1635–1703). He developed this idea from his observations of colours and patterns in thin films. A similar theory was published by Christiaan Huygens in the Netherlands, who also studied double refraction, or *birefringence*, in Iceland spar crystals. In this phenomenon, a single incident beam refracts in the crystal into two beams, *ordinary* and *extraordinary*, of which only the former obeys Snel's Law (Fig. 5.5). Huygens postulated the principle now named after him that each wave disturbance is at every point a source of new disturbances, and was able to explain reflection and ordinary refraction in terms of spherically spreading secondary disturbances. Extraordinary refraction resulted when the secondary disturbances did not propagate equally fast in every direction.

The eighteenth century favoured the emissionist ideas of Descartes and Newton over the undulations of Hooke and Huygens. In France, in particular, Laplace believed that physical and chemical phenomena could be explained in terms of short-range forces between particles of light, heat, electricity and ordinary matter, claiming in 1796 that through this approach 'we shall be able to raise the physics of terrestrial bodies to the state of perfection to which celestial physics has been brought by the discovery of universal gravitation.'[10]

About 1800, however, persuasive evidence for the undulatory viewpoint was furnished by the English physician Thomas Young. Young devised an experiment whereby light emanating from two closely spaced pinholes combined to produce an *interference* pattern analogous to that produced by overlapping systems of water waves. Figure 5.6 illustrates this experiment for light passing through two adjacent slits. A subtle point is that for the experiment to work, the slits cannot be illuminated directly by the lamp because the luminous disturbances in different directions continuously change randomly and independently, and do not have a fixed alignment or *phase* of their crests and troughs. A preliminary slit therefore serves to isolate a small fraction of the light which, following Huygens' ideas, acts to produce secondary disturbances which *are* locked in step. These then propagate to the double slits, from which further disturbances propagate to the observation plane. The wiggly lines in Fig. 5.6 represent these disturbances and the relative positions of their crests and troughs at some instant. At the centre of the pattern, the waves have travelled over paths of equal length and arrive with crests and troughs aligned. They add together to produce a bright patch of *constructive interference*. Bright patches also occur whenever the two path lengths differ by a whole number of wavelengths because then too the disturbances align. At points where the waves are

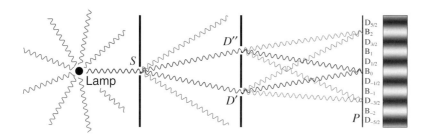

Fig. 5.6. Young's theory for the interference of light passing through two slits, D' and D''. To produce coherent (aligned) illuminating waves, the light must first be passed through a single slit, S. The resulting interference pattern in the observation plane, P, is fully bright where the disturbances arrive in step, such as at B_0, where the paths are equal, or at B_2, where the path via D' is longer than via D'' by two wavelengths. The pattern is completely dark where the waves arrive anti-aligned and cancel, such as at $D_{-3/2}$, where the path via D' is shorter by $1\frac{1}{2}$ wavelengths. The choice of observation plane is arbitrary: interference is present everywhere. Legend has it that Young's inspiration arose from noting superposition of the ripples in the wakes of swans swimming on the pond at Emmanuel College, Cambridge.

half a wavelength out of alignment crests and troughs cancel, resulting in zero disturbance and a dark patch of *destructive interference*. As Foucault commented, 'This last fact especially seemed very extraordinary, and to present it to the world in a picturesque way, one has been pleased to say that we know, today, how to produce darkness with two rays of light.'[11] For intermediate path differences, cancellation is partial and the intensity, while not zero, is reduced. Young envisaged light as a longitudinal wave like sound. He hypothesized that each colour of light has a characteristic frequency and the corresponding wave, propagating according to Huygens' principle, accounts for reflection, refraction, diffraction and interference.

For a while, corpuscular ideas seemed bolstered by the discovery of polarization by E. L. Malus.[12] In about 1807, Malus held a piece of Iceland spar up against rays reflected off the windows of the Luxembourg Palace in Paris. To his astonishment, the two images changed markedly in intensity as he turned the crystal. Previously, this had only ever been observed with light that had already passed through a crystal. Further, of the two beams that emerged from Iceland spar in such circumstances, only one was reflected from a water surface if the crystal was held at a certain angle, while the other was completely refracted into the water. If the crystal was rotated 90 degrees, the behaviour of the beams swopped. Malus described the light as *polarized*, and believed this resulted from the corpuscles having some sort of asymmetric sides or poles. In his view, reflection from a surface sorted the corpuscles according to these asymmetries. Malus might have gone on to make further discoveries had he not succumbed to tuberculosis soon after.

Malus's work on polarization was continued in Scotland by the ardent emissionist Sir David Brewster and in France by the *polytechniciens* Arago and Biot. Biot worked in the emissionist tradition, but Arago was less convinced. When Augustin Fresnel (Fig. 5.7) produced an undulatory theory of light in 1821 based on *transverse* waves, it was Arago who championed his cause. Initially few agreed with Fresnel because of the difficulty of imagining a medium rigid enough to propagate waves at light's speed, which was known to be enormous, but the ability of Fresnel's theory to explain double refraction as resulting from different propagation velocities for different polarizations convinced many of its essential correctness.

Fig. 5.7. Augustin Fresnel (1788–1827).

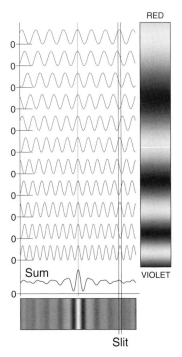

Fig. 5.8. Principle of Fizeau and Foucault's interference experiment. A real light source radiates at many different wavelengths producing many two-slit interference patterns (*graphs*) which sum to a jumbled smear except at the pattern centre (*below*). Foucault and Fizeau selected a small portion of the jumbled pattern with a slit and dispersed it into a spectrum in order to separate out the interference at each wavelength (*at right*). Note how the brightness of the spectrum matches the intensity within the slit of the adjacent individual interference patterns.

Interference at long path differences

The interference pattern sketched in Fig. 5.6 might be expected to extend indefinitely in the observation plane, but in practice the modulation petered out some number of fringes away from the centre. Foucault and Fizeau noted that a simple explanation for this came from the composite nature of real light sources. The spacing of the fringes in Fig. 5.6 has been illustrated for a particular wavelength of light. Since sources such as candles, oil lamps and the Sun radiate over a range of wavelengths, the total interference pattern is actually the sum of many differently coloured and differently spaced sets of fringes. The different patterns are aligned at the centre, but further away they rapidly slip out of step and sum to a uniform smudge, as illustrated in Fig. 5.8. If the range of illuminating wavelengths is reduced, for example by a coloured filter, the unsmeared region has a greater extent, but ultimately alignment is always lost and the fringes become smeared.

However another phenomenon might also destroy the fringes. It has been explained that interference does not occur if Young's two slits are illuminated directly by the lamp or Sun. Foucault and Fizeau also noted the non-interference 'of two rays from the same source, first orthogonally polarized, and then brought into the same plane of polarization, but without first having been polarized in a common plane'.[13] This phrase reeks of 'dry elements' and we need not understand it; the relevant point is that these facts had led to the conclusion that the emission of the 'luminous movement' was subject to very frequent perturbations. In modern terms, light comprises a series of wavetrains of finite length. Individual wavetrains can and do interfere with each other, but because the phase alignment of different wavetrains is constantly and randomly varying, the interference pattern jiggles very rapidly and the eye perceives only a uniform sensation.

How long were these wavetrains? Were they short enough to influence the extent of Young's fringes, or was the composite nature of real light sources an adequate explanation? The suspicion was that the wavetrains might last only some tens of wavelengths, in which case analogies with sound would break down and wave theory might even fail to explain diffraction patterns which required coherence over large areas.[14] These were the questions that Foucault and Fizeau set out to answer in 1845.

Their method was very simple. They used a slit to select a narrow interval of the blurred interference pattern, as shown in Fig. 5.8, which they then dispersed with a prism so that the intensity corresponding to the interference associated with each wavelength could be studied separately. The result was a spectrum that was bright at wavelengths where there was constructive interference, and dark where the interference was destructive. Figure 5.8 illustrates a slit placed only a few fringes away from the central position of zero path difference, but Foucault and Fizeau were ultimately able to examine interference patterns with path differences of over 7000 wavelengths.

Plate II shows what is perhaps part of the apparatus, while Plate III illustrates the principle of their experiment. They adopted a system that transmitted much more light than Young's slits, and this was essential, because even sunlight is very weak once it has been passed through a narrow slit and dispersed. A beam from a heliostat was admitted into their darkroom via a cylindrical lens, which produced a line focus, or one-dimensional image of the Sun, equivalent to the first slit in Young's arrangement. Two plane mirrors, which were slightly displaced and inclined with respect to each other, produced virtual images of this luminous line. (Virtual images are ones from which light appears to have come, but no actual rays have passed through them. Images in a flat mirror are always virtual.) These virtual images were analogous to Young's double slits. The resulting jumbled interference pattern was sampled by a slit in a screen about 2 metres away. A prism dispersed the light into a spectrum that was a few millimetres long and was examined through an eyepiece. Lenses near the prism played an important rôle. The first ensured that the rays falling on the prism were parallel so that all rays of a particular colour were deviated by the same angle. The second lens focused these differently inclined beams down to separate points. In this way maximum sharpness was achieved both in the spectrum and in its modulation by the interference pattern.

The actual spectrum seen by Foucault and Fizeau will have differed in a number of respects from the representation presented in Fig. 5.8. A prism does not spread wavelengths uniformly, so the blue-violet colours will have occupied a disproportionate fraction of the spectrum, which will also have been dimmer along its edges because of the solar limb darkening. However, the principal difference derives from the fact that the solar spectrum is not smooth but is crossed by instrinsic dark stripes which are simulated in Plate IV. Some of these lines and bands had been discovered in England in 1802; more were discovered in 1814 by the Bavarian glass-maker Joseph von Fraunhofer, who had lettered the strongest features from A in the red to H for a striking double line in the violet.[15]

These dark features provided valuable landmarks in the solar spectrum, and because of them Foucault and Fizeau were able to establish the path differences involved in their interference fringes. Plate IV also simulates the solar spectrum crossed by Foucault and Fizeau's interference fringes. The path difference in absolute terms, such as millimetres, depends only on the separations between the virtual slits and the interference point. For a bright fringe in the spectrum, this path difference is equal to some whole number of wavelengths. To be specific, consider two bright fringes falling close to the Fraunhofer E and F features. The corresponding wavelengths and whole numbers are different, but their products are identical because they equal this (unknown) path difference. Since the wavelengths of the Fraunhofer lines were known, this gave a first equation linking the two whole numbers. Now in Plate IV, adja-

Table 5.1. Simultaneous equations allowing solution for the unknown whole numbers n_E and n_F corresponding to bright fringes near the Fraunhofer E and F features. With n_E or n_F determined, the path difference can be calculated.

$$n_E \lambda_E = n_F \lambda_F \; (= \text{path difference})$$

$$n_F - n_E = \frac{\text{fringe count}}{\text{between E and F}}$$

where λ_E = known E wavelength
λ_F = known F wavelength

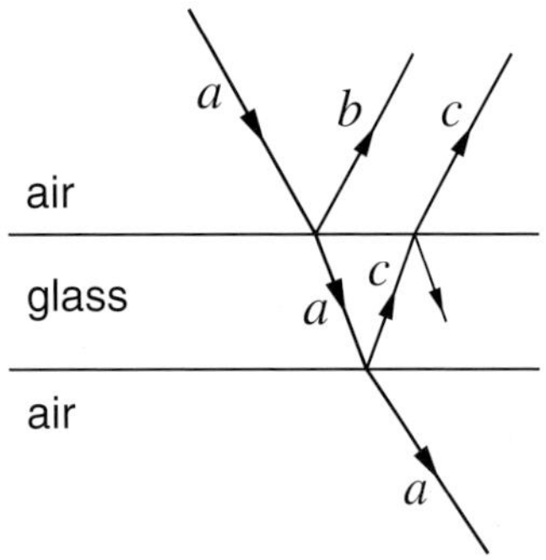

Fig. 5.9. Another way that Foucault and Fizeau obtained large path differences was to interfere the rays b and c reflected off the front and rear surfaces of a glass plate.

cent fringes correspond to a one-wavelength alignment change. A count of the number of fringes crossing the spectrum between the E and F features therefore provided a second simultaneous equation relating the two whole numbers, which could therefore be solved for, as could the path difference itself. Table 5.1 expresses the argument algebraically.

Fizeau and Foucault presented their work to the Academy in a written memoir on 1845 November 24.[16] To prevent crowding, Plate IV shows only ten fringes between the E and F absorptions, but Foucault and Fizeau reported having obtained 66 and then 141 fringes over this interval, corresponding to path differences of 813 and 1737 F wavelengths, or 0.4 and 1.4 mm. They also reported some modified experiments where the path difference was produced by passing one of the beams through a thin glass plate, in which, in the wave theory, the wavelength is reduced, or by reflecting the beams off opposite sides of the plate (Fig. 5.9). In this latter case they obtained a path difference of 3406 wavelengths. Finally, they obtained fringes using the path difference between the ordinary and extraordinary rays in crystals of gypsum, calcite and Iceland spar, all of which are birefringent. The interpretation of all their results was simple and natural in terms of the wave theory of light, and cast additional doubt on the emissionist viewpoint.

'With M. Fizeau we certainly delighted in the contemplation of the pretty phenomenon which was under our hands,' wrote Foucault, 'but our ambition goes a bit further.'[17] They realized they could use the fringe positions to measure the dispersion of the ordinary and extraordinary refraction in birefringent crystals, but despite the claim that 'we plan to set to work immediately', they never did. The work would have been tedious, and far from Foucault's liking.

Chromatic polarization

A few months later, in the spring of 1846, the two experimenters presented the Academy with what they had done instead.[18] This second part of their research did involve birefringence and its variation with wavelength, but in a different way. It too had a clear interpretation in terms of the wave theory.

The experiment demonstrated a phenomenon that in a simpler form had come to be called *chromatic polarization*. Interference was certainly evidence for the wave nature of light, but for unpolarized beams it could be explained quite satisfactorily in terms of Young's longitudinal waves. (There is nothing in Figs. 5.6 and 5.8 that requires the waves to be transverse.) Polarization had been the Rosetta stone that had enabled Fresnel to unlock the transverse nature of light, so new manifestations of polarization phenomena were eagerly sought for the confirmation – or denial – that they might bring.

Chromatic polarization had been discovered by Arago in 1811 when he held a sheet of mica up to clear sky and examined it through an Iceland spar crystal. The crystal's birefringence produced a double image of the mica disc, but to Arago's surprise, the images were tinted. Further, the tints were complementary: that is, the wavelengths that were present in one image were absent in the other, and vice versa (Fig. 5.10). This was forcefully illustrated where the two images overlapped and combined to produce white light. Arago's observation led him to realize that the blue sky was polarized (no colours were seen against clouds) and became the basis of the polariscope with which he had found no evidence of polarization in the solar photosphere (Chapter 4).

Now 'whoever has not been able to judge by experience in what way so-called polarized light differs from ordinary light, it is very difficult to imagine the position of a plane of polarization,' as Foucault himself noted a decade later.[20] 'We are even more sorry that these notions are not more widespread because they are essentially elementary,' he continued. These obstacles persist: polarization phenomena *are* fundamentally elementary; but they are usually complex in detail and involve difficult three-dimensional geometry. We will therefore limit our discussion to some summary remarks concerning polarization, and an overview that gives a flavour of what Fizeau and Foucault's polarization experiment involved. A full description is available in textbooks.[21]

Ideas concerning polarization are illustrated in the prism devised by the Scotsman William Nicol in 1828 (Fig. 5.11). The Nicol prism comprises a rhomb of Iceland spar that has been cut at a particular angle and glued back together with a layer of Canada balsam. The birefringent spar sorts the transverse vibrations of an incoming unpolarized beam into independent components along directions that are related to the crystal's internal structure and are at right angles to each other. These components constitute the ordinary and extraordinary beams, and are refracted differently. Now when light strikes an interface between a denser and a less-dense medium, in general some of the beam is reflected and some is refracted, but if the light strikes at too shallow an angle refraction does not occur and the beam is wholly reflected. (Foucault and Donné took advantage of this *total internal reflection* in their *microscope-daguerréotype*, where they used a prism as a high-efficiency mirror.) The cut in the Nicol prism is angled so that extraordinary rays are transmitted through

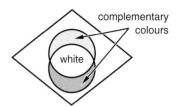

Fig. 5.10. Chromatic polarization, characterized by one mid-nineteenth century review as 'one of the most remarkable and beautiful discoveries which has ever been made in the history of optics'.[19] Two coloured images are seen when a birefringent crystal such as Iceland spar is used to examine a mica disc that has been illuminated from behind with polarized white light. The colours are *complementary*. For example, if one image of the disc is yellow, the other has a purplish hue resulting from the mix of the other colours of the spectrum: red, orange, green, blue, indigo and violet. Where the two images overlap, all wavelengths are present and white light results.

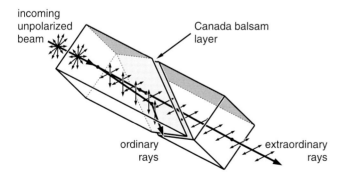

Fig. 5.11. The Nicol prism, made from Iceland spar, only transmits light that is polarized in the direction of the short diagonal of its end face and so acts to produce a polarized beam from an unpolarized one. A long crystal is needed so that the layer of Canada balsam is struck at angles that cause total internal reflection for ordinary rays but not for extraordinary ones.

Fig. 5.12. Addition of orthogonal plane waves (ones that are polarized at right angles) with a phase shift. Dots indicate the instantaneous values of the two disturbances at some point in space as well as their combined effect in the plane perpendicular to the waves' advance. A little later the disturbances have changed (*crosses*) as have their resultant, and similarly over a full cycle of variation (*heavy trace*). Depending on the relative alignment of the two orthogonal waves, the combined disturbance may constitute linearly, elliptically or circularly polarized light. The sense of rotation, or handedness, of the elliptically and circularly polarized disturbances depends on whether the phase shift is greater or less than half a wavelength. In Fizeau and Foucault's experiment, the addition of the ordinary and extraordinary waves was further complicated by variations in their amplitudes.

the less-dense Canada balsam layer, while ordinary rays are totally internally reflected there and are then absorbed by blackening painted on the long faces of the prism. The Nicol prism therefore produces a fully plane-polarized beam from an unpolarized one.

In their experiment, Fizeau and Foucault used a Nicol prism to polarize a beam of sunlight, which then passed through a plate of gypsum.[22] The birefringent gypsum decomposed the polarized beam into ordinary and extraordinary components. Unlike the decomposition in the Nicol prism, these components derived from a *polarized* beam and were *not* independent: the crests and troughs of the ordinary and extraordinary vibrations were aligned. However, the alignment slipped as the waves propagated through the gypsum where the speed and wavelength differ for ordinary and extraordinary components. On leaving the plate the components excited a combined luminous disturbance in the air, which was analogous to the figures displayed on an oscilloscope when signals of the same frequency are applied to the horizontal and vertical plates, but offset by a phase difference. The disturbance may move not only back-and-forth along a straight line but also around an ellipse or circle (Fig. 5.12).

Whether the light emanating from the gypsum slab was linearly, elliptically or circularly polarized depended on the phase difference (or alignment slip) that had built up. This depended on the difference between the gypsum's refractive indices for ordinary and extraordinary rays, which in turn depended on the wavelength of light involved, since refractive indices change with wavelength, as Foucault and Fizeau had promised to measure. The beam coming from the gypsum slab (or Arago's mica disc) appeared white to the eye, but was actually a complex assortment of polarizations. Where the beam was split into separate polarization components by a second Nicol prism, as in Arago's experiment, wavelengths that were strong in one were weak in the other, and vice versa, resulting in complementary tints.

Rather than study the polarization with wavelengths mixed, Fizeau and Foucault used the slit and prism of their interference apparatus to disperse the light into a spectrum (cf. Plate III). Then, they wrote, 'these phenomena, which are so complicated, take on a remarkably simple character which permits various consequences of the theory to be submitted to observation.' They were able to study the polarization wavelength by wavelength. Figure 5.13,

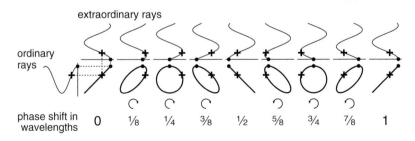

taken from their memoir, illustrates their results, in which they found that the sequence of polarizations repeated across the spectrum as wave theory predicted.

Their memoir concluded with an 'addition' in which they mentioned that 'during the clear days that attended the end of this winter' they had increased their dispersion with additional prisms and had been able to see interference corresponding to as many as 7394 Fraunhofer G wavelenths. They had also been able to detect chromatic polarization using birefringent plates one hundred times thicker than was possible in undispersed light. 'The existence of these phenomena of the mutual influence between two rays, for the case of very big path differences, is interesting for the theory of light,' they concluded, 'in that they reveal a persistent regularity on the emission of successive waves which no other phenomenon has so far indicated.' In the *Journal des Débats* Foucault added that their result 'leads us to suppose that the real limit at which interference ceases is incomparably greater yet'.[23] He was right: we now know that the wavetrain length for thermal sources such as radiating atoms may be thousands of times longer than probed by Foucault and Fizeau, and tens of millions times longer for lasers.

Because of their length, no doubt, only an extract of Foucault and Fizeau's first memoir was published in the *Comptes rendus*, and only the title of the second. As usual, the Academy appointed a team of commissioners to report on the work: Arago, of course; the bear-like Jacques Babinet (Fig. 5.14), who had worked on chromatic polarization; and the relatively young Victor Regnault (1810–78), who had been elected to the chemistry section of the Academy, but whose interests were turning to physics. As Foucault later wrote,

> We were still beginners; we had perhaps recklessly attacked a very delicate issue; and to convince our commissioners, it was absolutely essential to have them witness experiments that to us had seemed decisive and to set the novel apparatus working before them...Sunlight was indispensable. It did not fail us...[24]

Exactly where our experimenters had their laboratory can only be a matter of conjecture. Perhaps it was at 17 Rue du Cherche Midi, where at one time Fizeau had lodgings, perhaps it was at his family home in Suresnes, just to the west of Paris. If it was with Foucault, it was at about this time that he and his mother moved house.

When the *commissaire-priseur*, or registered valuer, had made an inventory of father Foucault's property at the time of his *interdiction* in 1834, his eye had no doubt passed lightly over Foucault's sister, who was only ten, but a decade later, he took her as his wife. With husband dead and daughter married, Foucault's mother decided it was time to move from the rented house on the Rue d'Assas. She did not abandon her street, however. In 1844 September she bought a plot of land only a quarter of a kilometre away, on the corner with the

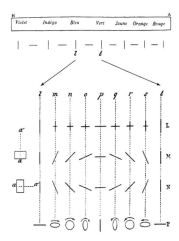

Fig. 5.13. Chromatic polarization according to Fizeau and Foucault. In their apparatus, a pattern of polarization repeated across the solar spectrum. One cycle is detailed in the lower part of the diagram. The line **P** represents the linear, elliptical or circular polarizations, which they found varied as expected.

Fig. 5.14. The physicist Jacques Babinet (1794–1872).

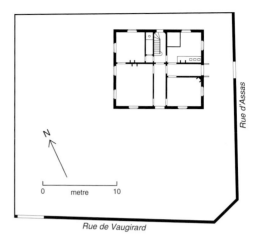

Fig. 5.15. Cadastral plan of the 'house of precision', as Foucault's friends would later call it, built in about 1845 by Mme Foucault at 26 Rue d'Assas (later renumbered 34). Foucault first saw the Earth turn in its cellar during 1851 January.

Rue de Vaugirard. The house that she and her son were to inhabit for the rest of their lives, complete with inside toilet, kitchen, and fireplaces in all rooms, was certainly finished by the end of 1846, when a survey plan was drawn (Fig. 5.15).

It was rare for the Academy's commissions to report in writing, though opinions were doubtless exchanged in corridors. In Fizeau and Foucault's case, however, a report was forthcoming, in the uneasy days following the Paris workers' uprising in 1848 June, when, as Foucault put it, 'the Academy used its sad spare time to report on the various memoirs which have long been awaiting its judgment and approval'. Jealous of a field that he had founded, Arago had planned to write the report, in which he would no doubt have emphasized his own contribution; but he was busy as a member of the provisional government, so it was Babinet who reported, and very favourably.[25] 'It is an honour and an encouragement whose worth we keenly appreciate,' Foucault wrote. The apparatus was 'an experimental invention of the first order', Babinet declared and the Academy ordered that the memoirs should be printed in the *Mémoires des savants étrangers*, which was an irregular collection of meritworthy papers by authors who were not members of the Academy. This was an order that was often made but rarely carried out. The memoirs finally appeared in the *Annales de chimie et de physique*, over four years after they had been submitted to the Academy.[26] We can see Arago's hand in this, since he was one of the *Annales'* editors, but the delays can only have been frustrating for two young scientists who were 'still beginners'. They were, unfortunately, almost inevitable for those who wanted the Academy's approval, because the Academy would not report on memoirs which had been printed. Authors were faced with the dreadful dilemma of deciding how long to wait before abandoning hope of a report and publishing elsewhere.[27]

Medicine abandoned

When his sister married in 1844 July, Foucault still described himself as a medical student, but sometime after, during his success with optics, he quit. It is reported that he was an *externe*, or non-resident junior doctor in a hospital and that medicine was quite to his taste.[28] Why then did he renounce the healing arts?

Figuier claimed that the famous surgeon Denonvilliers precipitated the change when he upbraided Foucault in front of the other *externes* for devoting too much time to physics.[29] Others ascribed Foucault's 'sudden' change of career as due to revulsion at 'the sight of blood and the distressing spectacle of suffering which could only be eased at the expense of even greater suffering'.[30] Foucault's writings in the *Journal des Débats* show that he was very matter-of-fact, and even had a certain ghoulish fascination with gore. 'A few hammer blows are enough to break the skull of a corpse...' he once wrote.[31] It therefore seems surprising that he was squeamish about blood, but in the days before anaesthesia and asepsis he would have been acutely aware of the limitations of surgery. 'It must not be hidden that every time the scalpel is taken up, the possibility of death arises,' he noted.[32] In addition, medicine had reached a dead end and was not going to progress without the application of science, which many in the profession opposed.[33] 'The way things are going,' mocked one, 'the poor human body will soon be seen as no more than a test tube of chemicals.'[34]

In constrast, Foucault revelled in the material aspects of the body. In one of his few working papers to have survived, he treats the forces on the bones of the foot (Fig. 5.16). The standard analysis took this as an example of what was known as a *lever of the second kind*, with the fulcrum (the ball of the foot) at one end of the lever, with the applied force (the tension in the Achilles tendon) at the other. Foucault objected that the corresponding mathematics placed the tension on both sides of the equation, 'which leads nowhere', and preferred to place the fulcrum at the joint between leg and foot. This certainly put the tension in the tendon on only one side of the equation and made it clear how enormous it can be, but as annotations show, Foucault subsequently realized that linear equations can be rearranged to separate the variables, and both analyses lead to the same physical reality.

Immature as they are, these jottings reflect Foucault's outlook. As his taste for science and physics developed, the anti-scientism of many of his putative medical colleagues can only have repelled him. 'For long enough the application of the positive sciences to the healing arts has been scorned,' he later wrote.[35] Great advances were soon to come, but from men outside medicine, such as Louis Pasteur (1822–95) and the physiologist Claude Bernard (1813–78). There was also the practical consideration that however honourable the profession, there were already many more doctors in Paris than patients able

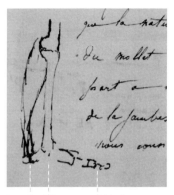

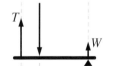

Lever of the second kind

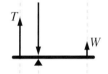

Lever of the first kind

Fig. 5.16. Anatomy and physics united: sketch of the lower leg from jottings by Foucault dated 1844 April, and analysis in terms of levers. The Achilles tendon applies a tension, T, which raises the foot slightly so that the full weight of the body, W, is taken by the ball of the foot.

(Family collection)

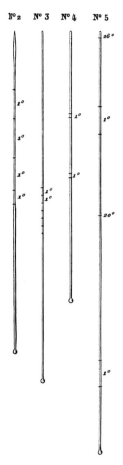

Fig. 5.17. Liquid-in-glass thermometers procured by Foucault and Fizeau for probing calorific rays, shown approximately life size. The first detection of interference fringes, recorded secretly in a *pli cacheté*, was probably made with No. 2, which contained ether. The other thermometers contained alcohol. No. 5 was used for the experiments reported openly.

to pay their fees.[36]

Whatever the cause, Foucault was now free to devote himself exclusively to science.

Calorific rays

Foucault wrote that he and Fizeau would have further developed their work on interference and chromatic polarization, but 'we were halted by the expenses that more complete experiments would have caused us'.[24] They did, however, use their techniques to examine *calorific rays*.

Calorific or heat rays had been recognized in 1800 by the astronomer William Herschel, the discoverer of the planet Uranus and father to John Herschel, whom we have already met. Herschel had been observing the Sun, but the filters in his eyepiece became hot and cracked. To investigate this, he placed small thermometers in a solar spectrum and found the hottest temperatures beyond the red at what are now called infra-red wavelengths. Later experiments showed that heat could be reflected, refracted, transmitted and scattered in the same way as light. To quote from the memoir which Foucault and Fizeau sent to the *Académie des Sciences* in 1847 August:

> The numerous analogies revealed by experiment between the properties of calorific rays and those of light rays have resulted in the idea of undulatory movements being extended to include calorific rays. This picture is generally accepted today, but for all that, it is only based on analogies, because none of the properties observed so far in calorific rays reveals any undulatory nature in them. The existence of interference phenomena would be decisive...[37]

There were two problems to circumvent: the narrowness of interference fringes and the small temperature rise that they would produce. Tiny, sensitive thermometers were essential. Figure 5.17 illustrates some of those 'procured'. They were liquid-in-glass thermometers, and probably were made by the experimenters themselves, since blowing soft glass does not require elaborate equipment. A wooden box excluded draughts (Fig. 5.18). The bulb of their best thermometer had a diameter of only 1.1 mm, while the diameter of the expansion channel was a mere 0.01 mm. The alcohol rose by about 8 mm per degree centigrade. The liquid level was read using a microscope in which one division of the eyepiece scale corresponded to about 1/400 °C. A candle half a metre away caused a seven-division change in the thermometer 'whose success is, so-to-speak, a stroke of good luck', as Foucault candidly admitted.[38]

The young men first attempted to detect the bright fringe at the centre of the interference pattern produced by two mirrors, which were set to a very shallow angle in order to increase the fringe width. Success was rapid. In 1846 May, two months after presentation of their work on optical interference, they deposited a *pli cacheté* at the Academy to record their priority. They had detected the heat of the central bright fringe and the one on either side.

Calorific rays

The maximum temperature rise was one-twelfth of a degree. 'Calorific rays are able to intefere like visible rays,' they wrote, emphasizing their conclusion with underlining.[39]

The *pli* remained unopened, since no one tried any similar experiment before our experimenters were ready to present full results sixteen months later. Their most impressive results involved chromatic polarization. Once more, the Sun was the only source strong enough to produce detectable effects. Sunlight from a heliostat was passed through a Nicol prism, a plate of gypsum 0.83 mm thick, a second Nicol prism, and finally was dispersed into a spectrum as previously. To make measurements, it was necessary to be able to sweep the spectrum across the thermometer box. An additional prism acted as a mirror, and by rotating it the spectrum could be displaced sideways.

The spectrum was first explored in the absence of any interference fringes. Figure 5.19, trace *S*, reproduces Foucault and Fizeau's sketch of this spectrum. Despite the wooden box, the ambient temperature drifted, so they resorted to an artifice that remains essential in infra-red work and is now called *chopping*. Temperatures were recorded with the solar spectrum present and blocked. The difference was the sought-for signal, and eliminated the slowly changing temperature of the surroundings. The dashed trace shows the resulting map. The maxiumum temperature rise occurred in the visible part of the spectrum, just, and was about one-tenth of a degree. With the small thermometer bulb, it became apparent that the 'invisible calorific rays' were crossed by regions of reduced intensity analogous to the Fraunhofer features in the visible part of the spectrum. As Foucault and Fizeau acknowledged, this discovery had already

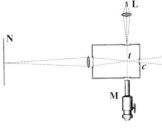

Fig. 5.18. The box used by Fizeau and Foucault to shelter their thermometers from draughts. The part of the spectrum under study was focused on the thermometer bulb at *t*. The temperature change was read with the microscope **M**. The relative position of spectrum and thermometer was determined from a screen on the front of the box near *c*, or by projection onto the distant screen **N**.

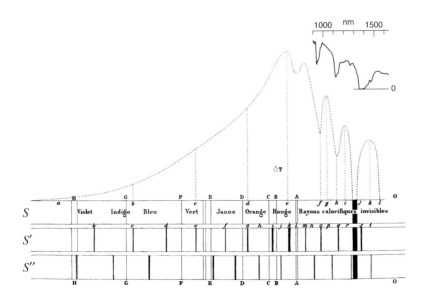

Fig. 5.19. Calorific emission of the Sun and its chromatic polarization reported by Fizeau and Foucault in 1847. The size of the 1.1-mm thermometer bulb relative to their spectrum is indicated at **T**, while the dashed line indicates the observed temperature rise. (A modern solar infra-red spectrum has been overlaid.) There was 'no detectable heat' in the strongest feature, which Fizeau and Foucault labelled **j**. The sketches S' and S'' indicate the interference fringes obtained with orthogonal polarizations. The fringes extended into the infra-red part of the spectrum, proving that calorific rays are waves.

been made by John Herschel from watching the drying of alcohol-moistened strips of black paper. It is now known that these features are not intrinsic to the Sun but arise due to absorption by water, water vapour and ice in the Earth's atmosphere.

By adjusting the polarization with the gypsum plate and Nicol prisms, it was possible to cross the spectrum with complementary polarized interference fringes. The presence of these fringes in the infra-red part of the spectrum proved the wave nature of calorific rays. It was a simple matter to extend the calculation based on the number of fringes between known Fraunhofer features in order to determine the wavelengths of Herschel's infra-red features. Yet it was Fizeau alone who presented this calculation, and this reflects a difference between the two experimenters.[40] Fizeau was fully capable of expressing physical reality in mathematical and quantitative terms. This was not Foucault's forte, as his jottings on the foot have already revealed. His grasp of nature was intuitive and qualitative, as we shall find in further cases.

The experimenters went on to observe the bright peak caused when rays diffract around an edge. It was no great surprise that calorific rays were characterized by wavelengths or that they diffracted and interfered. 'But to contribute usefully to the advance of science,' Foucault noted, 'one must sometimes not disdain from undertaking simple verifications.'[41]

The daguerreotype response

Scientists often employ equipment and procedures without fully understanding how they work, but this is always unfortunate. The spirit is left unsatisfied; and worse, misinterpretations can arise. Since they had used the daguerreotype as a photometric sensor, it is logical that Fizeau and Foucault should have decided to investigate its response to light more closely.

Others had of course had similar thoughts. That more light produces a stronger effect had been apparent from the start, as had the solarizing effect of overexposure (Fig. 4.1). Not all colours of light acted equally. The daguerreotype plate showed enhanced sensitivity towards the violet, and confirmed the presence of rays beyond the violet limit of human vision. The response ceased in the far-red, at wavelengths that were still visible to the human eye.

However, this was not the full story, and for what follows it is important to keep in mind that in normal usage, daguerreotype exposures were relatively low and produced no obvious changes to the sensitive layers, which only revealed images when developed with mercury. In contrast, vastly greater exposures to light could on their own liberate free silver without development in both daguerreotype plates and papers impregnated with silver salts or other light-sensitive substances.

Spectra are often enveloped in a diffuse halo of light of all wavelengths that has scattered within the apparatus. On its own, this diffuse light causes a partial response, but the effect is more complicated when a spectrum is superimposed. Experimenting with impregnated papers, John Herschel had found that the darkening was reinforced at violet wavelengths, as expected, but in the red it was *reversed*, with the spectrum appearing as a light band against the darker background produced by the diffuse light. Herschel called these red rays *protecting rays* because they protected the silver salts from alteration. In addition, there were Becquerel's continuing rays. Normally exposed daguerreotype plates could be *developed* with very long exposures to red light.

Starting in 1844, Fizeau and Foucault attempted to clarify some of these issues using brominated daguerreotype plates – the most sensitive available. These were first exposed for a few seconds to the white light of a Carcel lamp. This lamp epitomized luxury in the drawing room, but in the laboratory its merit was its steady light due to the constant supply of oil fed to the wick by a clockwork pump. The plate was next exposed for a few minutes to a solar spectrum, followed by development with mercury vapour. By December 9 they had amassed sufficient results to deposit a *pli cacheté* at the Academy. They had found that red rays could indeed act negatively, as they put it, on what would now be called the latent daguerreotype image, whereas green and shorter waves acted positively, as every photographer knew. The negative action of red rays possibly differed from Herschel's protecting rays in two ways. First, in Herschel's experiments the light levels were very high and there was no development. Second, his wavelengths acted simultaneously rather than sequentially, and a latent image was plausibly never produced, whereas Foucault and Fizeau had demonstrated the reversal of a previously formed latent image. They further found that orange rays would act negatively if they were faint or the exposure was short, but postively otherwise.

Where the action was negative, the Fraunhofer lines were recorded as bright instead of dark. The explanation was simple. There was little intensity at the lines' centres, and in consequence the overall light tint developed as a result of exposure to the Carcel lamp remained unaffected. Through this effect, Foucault and Fizeau were able to see that the Fraunhofer A feature is actually double (this is now understood to result from rotation of the terrestrial oxygen molecules that cause the absorption) and that other features occurred

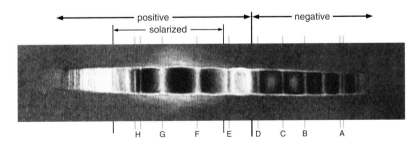

Fig. 5.20. Engraving of a typical daguerreotype of the solar spectrum obtained during Foucault and Fizeau's investigation of the daguerreotype response. Here wavelengths longward of about the Fraunhofer D line have a negative effect and the Fraunhofer features appear as light lines against a dark background instead of the reverse. Shorter wavelengths produce a positive image, such as around the E feature. However, intensities near the F, G and H lines have exceeded the dynamic range of the daguerreotype plate and the image is solarized. 'This tint, although dark to the eye has for the physicist the same meaning as if it shone most brilliantly and its presence indicates the point corresponding to the most active rays,' they wrote. A solar spectrum, like the solar disc, is fainter at the edge, and therefore the solarized region has light, unsolarized flanks. However, the bright borders in the negative part of the spectrum appear to be an invention of the engraver and are absent in the daguerreotype reproduced in Plate V. The lighter patches in the regions between the D & C and C & B lines arise because the intensities at the centre are sufficient to change these areas from negative to positive action.

beyond it, 'where our sight does not see'. Figure 5.20 represents one of their spectra where solarization is also evident. Important prerequisites for this work were that with the bromination box and careful technique, they were able to prepare plates of uniform and identical sensitivity 'with confidence', and that with lenses in front of and behind their dispersing prism, they were able to get a very pure solar spectrum so that there was no confusion due to smearing of wavelengths.

They next varied the experimental conditions. The work was time consuming and no doubt they found it repetitive. Needing sunlight, it also required clear skies. For these reasons, perhaps, it progressed slowly, while Foucault and Fizeau devoted their energies to their more exciting interference experiments. There was a further reason. Fizeau suffered from dreadful migraines. He gave up medical studies and travelled to restore his health.

It was while he was away, in the autumn of 1846, that N.-M. P. Lerebours sent a memoir to the Academy concerning achromatic camera lenses. As an aside, Lerebours noted that green and redder rays seemed to retard the photographic action of bluer light.[42]

This was the cue for Foucault to write to the Academy to claim his and Fizeau's due. He summarized their results from two years previously, requested that the *pli cacheté* be opened for corroboration, and outlined the work still to be done before a definitive memoir could be written.

This letter prompted Edmond Becquerel to admonish Foucault and Fizeau for not having summarized prior discoveries (particularly his) and for having used sensitized plates:

> they considered the daguerreotype plate as offering *a unique sensitive layer*, whereas it is only a mixture of materials which may behave differently to the different parts of the spectrum... It is therefore necessary to operate, as I have, using single substances...[43]

Foucault rarely left a rebuke unanswered:

> It would seem, indeed, that before anything else we ought to sketch the history of our subject. This, in any event, is our intention...

> M. Ed. Becquerel will not be forgotten, because we will debate the question of whether the expression *continuing rays* can still be permitted in the presence of the new facts which we have demonstrated.

> The least that can be done is to let us produce our complete Memoir; only then will I admit the right of an informed gainsayer to discuss and criticise our ideas.

> If M. Ed. Becquerel will kindly be patient...[44]

The comment about continuing rays is a rare example where Foucault was unfair, because he will have known that their experiments confirmed the effect in certain circumstances.

Fizeau and Foucault must have finished their memoir rapidly because Fizeau reviewed results publicly two months later on 1847 January 16.[45] The memoir was not printed, however. The issue became topical again in the autumn when Claudet, Daguerre's licensee in London, wrote to the Academy denying Becquerel's continuing rays.[46] Becquerel responded, as did Marc-Antoine Gaudin.[47] Magnanimously forgetting Foucault's blistering attack on him in the *Journal des Artistes* six years earlier, Gaudin graciously concluded:

> I hope, moreover, that MM. Foucault and Fizeau, whose notecase holds beautiful observations concerning the roles played by the different rays in the spectrum, will not delay publishing them any further, and that they will finally make sense of all these facts.

The hope was forlorn. The full memoir was only published thirty years later, in the *Recueil*. This was a shame, because it contained interesting results, as well as a conclusion about the photographic process which has proved correct.

Foucault and Fizeau had varied the experimental conditions in obvious ways. They had varied the exposure to the Carcel lamp, which had little effect, and to the solar spectrum, stacking five spectra on a single plate (Plate V). Here the negative effect migrated from the green to the extreme red as the exposure increased from 5 to 120 seconds. What this meant was that from the green and longwards, rays first acted negatively, but then positively as the exposure increased. Their hypothesis that all rays might act this way proved false when they experimented with shorter wavelengths.

Next they abandoned the white light from the Carcel lamp and pre-exposed the daguerreotype plates to only selected wavelengths. Foucault devised a cunning arrangement to do this which he described a few years later in an early issue of *Cosmos*, no doubt to support his friend Moigno in his new enterprise as the magazine's editor (Fig. 5.21).[49] All rays passed through an intermediate spectrum where slits could select the wavelengths transmitted to the daguerreotype plate 'in a manner as sure, as positive as the artist when he takes his paints and mixes them to splash on the canvas'. The apparatus was shown to students at the Sorbonne,[50] and used in Königsberg (modern Kaliningrad) for exper-

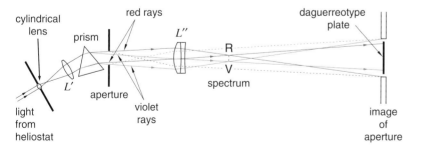

Fig. 5.21. Foucault's apparatus for pre-exposing daguerreotype plates to light of uniform tint. The lens L'' simultaneously images the aperture onto the daguerreotype plate and produces a focused spectrum, R–V, through which all the light passes. Slits across this spectrum select the wavelengths transmitted to the final image. The dashed lines do not represent real rays, but are constructions required to show the location of the final image.[48]

Table 5.2. Sensitivity of different daguerreotype plate preparations.

SENSITIVITY ↑
Iodine + bromine
Iodine + excess bromine
Iodine + chlorine
[a]Iodine + excess chlorine
[a]Iodine alone

[a]*affected by continuing rays*

iments on colour vision by the Prussian physicist Hermann von Helmholtz (1821–94).[51]

Negative effects were eliminated if the daguerreotype plates were pre-exposed to the red wavelengths between the Fraunhofer B and D features. They reappeared, in the red, if the pre-exposure was orange-yellow (between C and E) and moved as far as the green (the E line) for violet and ultraviolet pre-exposures (H and shortwards). Other things being equal, the crossover point between negative and positive action moved towards the red for longer exposures to the solar spectrum.

Fizeau and Foucault also experimented with other sensitive surfaces: simple iodized daguerreotype plates, the overbrominated ones used with Belfield-Lefèvre to record scenes with large brightness ranges, and so on. Table 5.2 lists the order of sensitivity that they found. Becquerel's continuing rays affected only the least-sensitive surface – the classical iodized-silver daguerreotype plate – and the one above.

However, tests on silver plates 'do not seem for everyone to have been completely conclusive', as Foucault and Fizeau lamented. 'People thought that we could have confused solarization with the primitive state of an unexposed plate ... and we were told to use papers while being assured that the phenomena would be completely different.' So they also experimented with silver-chloride soaked paper, using gallic acid as a developer, but being careful, as with the daguerreotype plates, to ensure that the pre-exposure was not so long as to cause any obvious darkening. The result was that they found that red rays had exactly the same negative effect.

The physics and chemistry of the daguerreotype

What conclusions did Foucault and Fizeau draw from these extensive investigations? Arago had predicted that 'perhaps thousands of beautiful pictures will have been made with the daguerreotype before its mode of action is completely explained', but in fact details were not fully deciphered until the quantum physics of solids was understood. By then, some hundreds of millions of daguerreotypes had been exposed and the technique was long obsolete.

The prevailing assumption was that photography was a chemical process; indeed, the rays to which light-sensitive substances responded were often called *chemical rays*. A chemical process meant some reconstitution of the substances present into different chemical compounds and this was patently the case for the final stage of the process, for example, when silver chloride converted to a black silver deposit under the influence of light alone, and the accompanying sharp smell of chlorine gas proved that the change was irreversible. Foucault and Fizeau's major result was that redder light was not *protecting* but *reversing*. 'The manifestation of negative properties by certain rays seems to us to be the first fact that is incompatible with any theory of the formation of photographic images which attributes everything to chemical action,' they wrote, and they went on to draw analogies with calcium sulphide, whose phosphorescence is excited by bluer colours and quenched by redder ones. They concluded that blue-violet light put the sensitive substances into some intermediate physical state which was a necessary precursor for the chemical change, which could then be effected with a developer or further irradiation. Redder illumination could reverse this physical state. Becquerel's continuing rays had little capacity for chemical effect, but could provoke chemical changes in materials that had previously been put into the intermediate state.

These conclusions are correct. What we now understand is the nature of the physical changes that produce the photographic latent image.[52]

The first stage of latent-image formation is the same in all silver-halide photographic processes. Electrons in the silver-halide crystals can only exist within certain quantum-mechanically permitted bands of energy. Impinging photons of sufficient energy raise electrons out of the so-called *valence band* of the crystals, where they normally reside, across a forbidden energy gap into the *conduction band*. These electrons roam about the crystal and may become trapped at any location where the regular structure has been damaged. Crystal defects thus acquire negative charge. The crystal also contains extra silver ions which are squeezed into the spaces between the regular latticework of atoms. These positively charged, interstitial silver ions are particularly mobile in the silver halides. They are attracted to the captured, negatively charged electrons and combine with them to form neutral silver atoms. Isolated silver atoms re-ionize spontaneously. The exposure must last sufficiently long to form more-stable agglomerations of a few silver atoms at the defect sites. These clusters are more common where the light has been brighter, and constitute the latent image. If small, they are only partially stable and can be disrupted by the lower-energy photons of light of green and longer wavelengths, whereas if the exposing light is bright and prolonged enough, the clusters grow to macroscopic size and can be seen as free silver.

Development in modern photography is an oxidation–reduction reaction, but daguerreotype development with hot mercury vapour is a chemical vapour deposition process. The mercury vapour acts as a *mineralizer*, or solvent causing the partially stable silver clusters to recrystallize as highly stable silver image particles. As for Becquerel's continuing rays, the initial image-forming exposure acts as a weak spectral sensitizer, making the plate sensitive to longer wavelengths, but the exact mechanism of the effect is still unknown and the image structure is different.

Bromination increases daguerreotype sensitivity because the bromine reacts with the silver iodide to produce the mixed salt silver bromoiodide, which is more conductive to silver ions than silver iodide. Silver clusters form more easily and so less light is needed to produce a latent image.

The Société Philomathique

The non-publication of Foucault and Fizeau's investigations into the daguerreotype response is a mystery. Foucault had certainly been very keen to publish previously, often sending papers to several journals. What is additionally intriguing is that Fizeau's review of their results on 1847 January 16 was not to the Academy (though in truth, he may have had little to add to what Foucault had already said the previous October). No, Fizeau talked to another audience, one that neither he nor Foucault had addressed before.[53]

The setting was one of the regular Saturday-night meetings of the Société Philomathique de Paris.[54] This society (*philomathique:* who loves learning) had been founded in the last year of the *Ancien régime* as a forum to review recent discoveries and talk. When the Revolution closed the Royal Academy of Sciences, Academicians made the Société Philomathique their meeting ground. The influx included men such as Laplace, the naturalist Cuvier and the soon-to-be-guillotined Lavoisier. Membership ballooned to almost sixty, which later became the membership limit.

Characterized by an attractive motto, *Étude et amitié* (Study and Friendship), the Société was a more welcoming foundation than the Academy. Intending members did not have to wait for death to create vacancies, because after ten or fifteen years members would become honorary and so liberate places. In the phrase of the time, the Société Philomathique had become 'an antichamber of the Academy'. This was not because it met on the Rue d'Anjou Dauphine, just behind the Institute, but because almost all Academicians had been a Philomathique member first. Arago, Malus and Fresnel had been members, as had the instrument-maker Henri-Prudence Gambey (1787–1847), who was one of Foucault's idols. Babinet, Donné and the astronomer Urbain Le Verrier (1811–77) were more recent members. Edmond Becquerel had been elected as early as 1841, at the age of twenty-one, no doubt in good part through his father's influence. We can see Fizeau's presentation in early 1847 as the beginning of a campaign to get himself and his friend Foucault elected to the Société Philomathique, and later to the greater glory of the Academy, but election to both companies was to prove tortuous for Foucault. In part this was because of enmities provoked by his frank comments as a journalist, to which we now turn.

Chapter 6

Order, precision and clarity: reporter for the *Journal des Débats*

> Above all, we must be accurate, and it is an obligation which we intend to fulfil scrupulously.
> LÉON FOUCAULT, *Journal des Débats*, 1848 May 30

Table 6.1. Print runs of the major daily newspapers in Paris in 1846.[1]

Le Siècle	32900
Le Constitutionnel	24800
La Presse	22200
L'Époque	11300
Le Journal des Débats	**9300**
Le National	4300
L'Univers	4200
L'Esprit public	3600
L'Estafette	3200
La Patrie	3100

The *Journal des Débats*

In the mid-nineteenth century, the *Journal des Débats* was not the daily newspaper with the widest circulation (Table 6.1), but through unswerving support of whatever government was in power, its influence had become far greater than its size would suggest. The paper had been founded in 1789 to record Revolutionary debates and decisions; a decade later it had been bought by two brothers known by the curious bynames of Bertin l'Aîné (Fig. 6.1) and Bertin de Veaux. During the First Empire, however, the paper's prudent expedient of untiringly attacking Revolutionary institutions and unashamedly flattering Bonaparte was not enough to prevent, first, censorship and a forced name change to the *Journal de l'Empire* (*Débats* sounding far too democratic), and, later, confiscation by Napoléon and his henchmen. After the fall of the Empire, the Bertins regained ownership of the paper.

The newspaper's reputation also derived from the remarkable literary talent of many of its *rédacteurs* or journalists. The poet and diplomat François René de Chateaubriand (1768–1848) is still remembered, and many of the paper's writers were or became members of the Institute. As Charles X's government became more reactionary, the *Journal des Débats* veered discretely towards liberalism, and was finally prosecuted for an article which concluded 'Unhappy France! Unhappy King!' After the July Revolution, many of the paper's writers entered parliament and the influence of the paper grew accordingly. (The paper was also rumoured to enjoy secret government subsidies.) Top people read the *Journal des Débats*.

Fig. 6.1. This famous portrait of Bertin l'Aîné (1766–1841) by the painter Ingres embodies the bourgeois solidity of both Bertin and his newspaper.

Science reporting

From its founding, the *Journal des Débats* had published pieces on science; the astronomer J. J. Lalande (1732–1807) was an occasional contributor in the last years of his life.[2] By 1829, it had become obvious that science and industry required a specialist reporter rather than men of letters. The paper plumped for youth and energy, and appointed Alfred Donné, then aged twenty-eight. Donné wrote mostly about science itself, though one commentator noted 'he is a trifle partisan'.[3] Nevertheless, Donné believed he needed to write for the typical reader, addressing topics broadly and leaving aside arid details. He widened his brief to include the Academy after its meetings were opened to the public in 1835. His report on Arago's announcement of the daguerreotype process in 1839 has already been mentioned in Chapter 3.

When Donné wrote his column on the Academy that appeared on 1845 April 2, he did not know it would be his last, but a few days later he was appointed Inspector-General of Medical Schools. Since this involved travelling all over France, he could no longer be present in Paris every Monday to report on the Academy, although he continued to provide articles on many other topics.

Donné recommended Léon Foucault as his replacement for the Academy reports. Probably he saw parallels with his own beginnings with the newspaper. Foucault too was young (twenty-five) and had studied medicine, a topic of perennial interest to readers. His brief excursion into journalism two and a half years earlier in the *Journal des Artistes,* excessive though it was, had shown that he could write with wit and colour.

Foucault's first report, on April 16, was a little stilted, but right from the start he was his own man, expressing clear personal opinions. He started with the textual reproduction of a note by the mathematician Augustin Cauchy refuting Donné's remarks in the previous column. Foucault praised Cauchy, and although he declined to comment on Cauchy's refutation, it is clear he agreed with it, saying that common sense should prevail. He next printed another refutation, this time from the Dean of the Medical School, concerning corrosive sublimate used as a means of murder. 'After having checked the facts for ourself,' wrote Foucault, 'we saw that this complaint was entirely justified.' Comments continued: 'The bizarre idea...is due to the wayward imagination of a Danish engineer,' was his view concerning a pneumatic railway. He went on to tell Arago that rather than complaining about errors in the recently published *Astronomy Lectures by M. Arago, published by one of his pupils*, he would do better to publish his lectures himself, which Arago later did (see p. 8). Paragraphs on a few other topics completed the article.

Though this first article was printed on an inside page, most were printed in the *feuilleton* or 'feature' position across the bottom of the first two pages (Fig. 6.2). These bottoms-of-page had long been used for articles of artistic and sci-

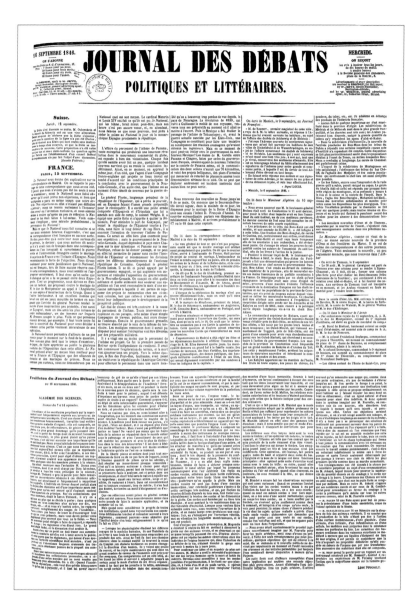

Fig. 6.2. Foucault's reports mostly appeared in the *feuilleton* position across the bottom of the front and second pages of the *Journal des Débats*. Occasionally three pages were taken when the report was very long, or only the first page when it was short. The paper was a broadsheet, sized approximately 460×600 mm, comprising typically eight or twelve pages.

(Bibliothèque nationale de France)

entific criticism and opinion, and in the late 1830s the *Journal des Débats* had expanded their scope to serialize stories by famous authors. When Foucault's articles first appeared, they were slotted between instalments of *The Count of Monte-Cristo* by Alexandre Dumas (1802–70). Other still-known serial-writers include Honoré de Balzac (1799–1850), George Sand (1804–76), and later Jules Verne (1828–1905).

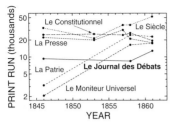

Fig. 6.3. The circulation of the *Journal des Débats* remained steady during the years that Foucault wrote for it (1845–62), while that of other newspapers changed significantly. Newspapers sold almost exclusively by subscription at the time.

Despite the *feuilleton*, the circulation of the *Journal des Débats* remained constant at about nine thousand throughout the years that Foucault wrote for the paper (Fig. 6.3). One reason was that the paper was some 50 per cent more expensive than its competitors.[4] Its price and content cemented its position as *the* paper for the bourgeoisie.

Guglielmo Libri

Foucault was not the only person aspiring to replace Donné. There was a rival, Count Guglielmo Libri (Fig. 6.4). Libri had been born in Florence in 1802 (or possibly 1803).[5] Gifted in mathematics, he published his first paper when aged nineteen and was soon appointed professor at Pisa. The July Revolution of 1830 catalysed revolutions throughout Europe. Libri was involved in a comically unsuccessful uprising in Tuscany and fled to asylum in France, where he was well received because of his double reputation as a savant and a revolutionary, especially by Arago, under whose patronage he entered the Academy in 1833 and became a professor at the Sorbonne. However, Libri soon allied himself with Louis-Philippe's austere minister, F. P. G. Guizot, and, turning to bite the hand that had helped him, attacked Arago in the *Journal des Débats*. This was probably part of a government plot to discredit the republican Arago through attacking his scientific standing. Donné too was criticizing Arago. Libri therefore already had some influence with the newspaper, augmented no doubt by a common love for books and manuscripts shared with the paper's then editor-in-chief, Armand Bertin (Fig. 6.5), son of Bertin l'Aîné. The month following Donné's departure – after Foucault had already published two articles – Libri authored a *Revue scientifique* or Review of Science.[6] In it he announced that henceforth 'a journalist' would report on the Academy while his own monthly *Revue scientifique* would enlarge on science outside the Institute. But on this occasion, Libri talked only indirectly of science, choosing to expatiate on the poor pay and career structure in higher education. Six weeks later he duly wrote another only-indirectly scientific review,[7] but there were no further *Revues* until almost a year later when he reported on the Academy's annual public meeting.[8] This account was mostly an attack on Arago.[9]

Even though Libri's articles were infrequent and elliptical, he tried to influence and control Foucault's reports as two letters that Foucault wrote to Libri in 1846 reveal.

The first letter begins, 'My mistakes and my youth make it my duty to apologize...' and ends in a similar tone, passing through a statement that he ought nevertheless to be free to describe things as he sees them, though admitting that he had written at too great a length about matters where he was not properly informed.[10]

What had young Foucault done? It is not completely clear. A week or so earlier, news had reached Paris of one of the nineteenth-century's most

Fig. 6.4. Guglielmo Bruto Icilio Timoleone Count Libri-Carucci Dalla Sommaia (*c.*1802–69).

(*Bibliothèque nationale de France*)

celebrated discoveries. Astronomers at the Berlin Observatory had found the previously unknown eighth planet, Neptune, following mathematical predictions by Le Verrier in Paris. Foucault was justifiably excited:

> First of all, let us announce some news that we did not gather at the meeting and which has just arrived. The conjectured planet whose details M. Le Verrier had so boldly predicted belongs henceforth to the realm of the real![11]

Perhaps Libri felt Foucault was trespassing on his turf by writing about discoveries outside the Academy; and he certainly wrote a *Revue* about the new planet a fortnight later.[12] However, if this was the bone of contention, it is difficult to see where Foucault was misinformed. Be that as it may, in the second letter, written the next day, Foucault was prickly almost to the point of impertinence. Libri had clearly laid down the law concerning their respective responsibilities on the newspaper, and Foucault was not happy: 'The arrangement which you have devised pretty clearly justifies the fears that I held... Apart from the little scandal that it will produce... Please be assured, Sir, of my respectful and forced submission...'[13]

However, Libri was not a man who deserved respect, forced or otherwise. These two letters have survived because they were among papers seized by the police. It may be appropriate that a man whose name means 'books' should have been a bibliophile, but Libri's mania extended to widespread theft. He had engineered appointment as Inspector-General of Libraries and was using this position to plunder books and manuscripts from repositories across France. Suspicions had inevitably arisen, and a police report was on his protector Guizot's desk when Louis-Philippe and his government were swept away by the February Revolution in 1848. The provisional government found and published the report, which must have given Arago some grim satisfaction; and Foucault too. Tipped off, Libri fled to England, but he was convicted *in absentia*, and faded from public view.

Authorship and readership

Happily, there were kinder eyes looking over Foucault's shoulder. Donné will surely have given advice to his young protégé; editor-in-chief Armand Bertin most certainly did. When Bertin died in 1854, Foucault recalled his advice: '...relate, describe, discuss, contradict, praise with abandon, rebuke with restraint' and rejoiced over 'that species of captivation which was the priceless benefit of daily contact with such a beautiful spirit'.[14] We can deduce that Foucault frequently attended the newspaper's offices in the Rue des Prêtres-Saint-Germain-l'Auxerrois, north of the River Seine. We can imagine that once the *Comité secret* began, he would walk directly there, paying the 5 centimes toll

Fig. 6.5. Armand Bertin, who was editor-in-chief of the *Journal des Débats* until his death in 1854. Oil by Delaroche, proving that painting was *not* dead.

to cross the Pont des Arts,[15] and composing his articles in the reporters' room on the second storey while the presses clattered away on the ground floor.

Who did Foucault imagine read his articles? He tells us:

> people of the world who, without pretending to be true scholars...are nevertheless looking for all occasions to acquire some positive notions concerning natural philosophy. This élite public, too despised, too unrecognized by the learned societies, is precisely the one for whom we write every fortnight, it is for these people that we perform with conviction a labour that is tougher than they realize; whose goal, when all is said and done, is to express all the results that they need to know in their own terms.[16]

Foucault made other references to the toughness of his task. He wrote of his column's 'laborious birth',[17] and envied popularizers who published in books:

> M. Figuier does not write flitting from subject to subject on throw-away sheets; he has the time to gather his thoughts, to refine his style, to win over his reader, and even if needs be, the opportunity to abbreviate.[18]

Fig. 6.6. The offices of the *Journal des Débats* in the Rue des Prêtres-Saint-Germain-l'Auxerrois.

He was aware that some of his pieces were hard going. He omitted findings on the physiological perception of dyed fabrics, which he felt were 'a bit too dry for our readers'.[19] After a couple of thousand words on electrochemistry, he explained that his next, simpler topic was chosen 'in order to refresh the mind of the reader, who has probably had difficulty following us along the meanders of this long dissertation...'[20] He was fascinated by the photographic image, and wrote approvingly of books enlivened with well-executed woodcuts, while regretting more than once that he could not illustrate his columns.[21] Nevertheless, he did sometimes present highly technical material, particularly numerical values for the parameters that describe the orbits of newly discovered minor planets and comets. After some years, he did half-attempt to justify this practice: 'As is our custom, we are going to give these elements, which have a distinct meaning for persons devoted to astronomy...'[22]

Anaesthesia

Foucault's *feuilletons* vary in style. Some are brief and dry, differing from the *Comptes rendus* only by being shorter. In others, his enthusiasm knew few bounds, as illustrated by his reaction to the discovery of anaesthesia. Thoughts that had perhaps caused him to abandon medicine resurfaced when news arrived that ether was being inhaled during dental surgery in Boston, Massachusetts:

> Could it be true that one can sacrifice an arm, a leg, some part of oneself that has become the centre of some incurable affliction, without paying the obligatory tax of cruel suffering? Will the New World, that unlimited source of riches, once more claim the honour of having provided the

unhoped-for magic whose power renders a man insensitive to the slicing of the scalpel ... ?[23]

After describing what was known about ether, and a side-swipe at the omissions of medical education, Foucault commented that he hoped to learn more at the Academy, adding pointedly:

> This will be a chance for the Medicine Section to break free from its torpor and to recall the duties of the Academician, duties which do not only consist of coming to meetings and voting to replace deceased members.

The following week there was more excitement. Debate was a rare exception at the Academy, but the Academicians were aroused by anaesthesia's potential. '... is it not so right to surrender to enthusiasm in the presence of a real conquest that is everywhere confirmed?' marvelled Foucault.[24] However:

> In the middle of this chorus a discordant voice was raised. An honourable Academician made himself the apostle of pain. He brandished the word 'immorality'; he reproached his colleagues for experimenting on men... For him, intoxication is the deepest state of degradation... 'What! In order to cure a man,' he said, 'you begin by shaming him?... and to administer the ether you use a method which I can but condemn as a surgeon and physiologist! Haven't I shown in my lectures how much more preferable it would be to administer the ether... into the carotid artery?'... M. Magendie... furnished us with the sad spectacle of a man... who, far from lauding the subsequent conquests of a science to which he had given such a strong impetus, seems incapable of allowing it to advance today without him.

Fig. 6.7. François Magendie (1783–1855), physiologist and vivisectionist.

François Magendie (Fig. 6.7) was the first truly scientific physiologist, who had made brilliant discoveries by experimenting on animals, though these vivisections had been denounced by the zoophilous English.[25] Piqued by Foucault's comments, he wrote a rejoinder, which by law the newspaper was obliged to publish. Magendie was notorious for bluntness. Responding to Foucault's report that he, Magendie, had suggested injecting ether directly into the carotid artery, he complained, 'One would need to be completely devoid of reason to imagine such a stupid thing,' and ended, 'since Monday's meeting, facts which are as serious as they are distressing have come all too soon to confirm the warning which I had sounded concerning the dangers of inhaling ether.'[26]

Foucault had fun in his reply. Surgeons at the meeting had also understood that Magendie wanted to inject ether into 'this wretched carotid artery'. 'So, academicians and journalist,' wrote Foucault, 'there we are, each justified by the other; one cannot accuse us of making deluded and invented citations.'[27]

But more importantly, the surgeons wanted to know about Magendie's serious and distressing facts. Foucault continued:

> And everyone listened; the whole world, whose closest interests we are discussing, would have wanted to be there to lend an attentive ear. Well! The whole world would have been disappointed...
>
> M. Magendie affirms it, inhalation of ether vapours is accompanied by danger. Do you know for whom? For the *surgeon*, because he runs no less a risk than finding himself in the presence of women who are having *erotic dreams* provoked by the ether.

Magendie wrote the newspaper a further letter of pompous protest, but while he may have enjoyed the last word, Foucault had come out on top.[28]

Foucault was not just a spectator with respect to his enthusiasm for ether. Undaunted by the dangers of erotic dreams, he experimented on himself and his entourage:

> The experiments which we have had the leisure to make, not only on ourself, but also on a great number of people, men and women, differing in age, character and temperament, have singularly strengthened our beliefs concerning both the efficacy of the procedure and the unimportance of the disadvantages from which people have claimed it is inseparable.[29]

The ether was doubtless administered together with his pharmacist friend Jules Regnauld, who later researched anaesthetics professionally. The mental picture of Foucault along with mother, sister and brother-in-law engaged in an ether frolic in the parlour is amusing; but ether is not to be abused. It frequently provokes feelings of suffocation, nausea and vomiting; in excess, it can paralyse the medulla causing death via respiratory and circulatory failure.

Rebuking and praising

In the early years Foucault paid little heed to Bertin's advice to 'rebuke with restraint'. He plunged in when the aged naturalist Henri Dutrochet subjected a freshwater alga to the field of an electromagnet:

> The experiment is described with great care: the dimensions of the soft iron core, the number of batteries used to develop magnetism therein, the distance at which the stonewort was placed, nothing is forgotten; and with these data anyone can, like him, confirm that nothing special happens. One may swop the poles, increase or decrease the current, and the stonewort is in no way affected and cares little whether there is a magnet strong enough to lift Mohammed's tomb beside it or not. Assuredly one must be a Member of the Institute in order to be able to present a memoir containing so little.[30]

Magendie and Dutrochet may have been old fools, but young fools were no safer from Foucault's pen. He rebuked E. H. Marié-Davy, a future colleague at the Paris Observatory, 'who for his age seems a bit too bent on changing the face of science'.[31] He rebuked Edmond Becquerel for denying the existence of diamagnetism, which had been discovered by the venerable Michael Faraday in London. ('Is this how one speaks while still young towards such an imposing authority ...?'[32]) (Since Marié-Davy and Becquerel were only a year or so younger than Foucault, these admonitions were more than a little prissy.) He rebuked a medical man for being boring.[33] Sarcasm surfaced for physiologists who claimed that the human eye does not accommodate (i.e. focus). They must have, he said, 'a major imperfection in either sight or mind'.[34]

Over time, Foucault became more tactful – and was more inclined to 'praise with abandon'. Earlier criticisms of Arago gave way to positive comments. In 1852, when it was clear that Arago had not very much longer to live, Foucault was gracious: 'After Bouguer we saw M. Arago... return to these various problems and treat them with all the elegance of a great physicist.'[35] He praised the courage of a chemist who ran the dangers of manipulating phosphorus.[36] If Arago, Edmond Becquerel and – as we shall see later – Cauchy were often at the sharp end of Foucault's criticism, Michael Faraday, the chemist Eugène Chevreul, the little-known astronomer Jean Chacornac,[37] and the physiologist Claude Bernard invariably earned his praise, as did another physiologist, the Academy's other Permanent Secretary, the polite and affable Pierre Flourens.

Flourens' great quality for Foucault was his elegance with the pen.[38] 'Style is the man,' Foucault opined,[39] and he was severe on those whose powers of expression were deficient. He upbraided Pasteur for a memoir on crystallography whose introduction 'seemed completely impenetrable to us'.[40] He rebuked the physicist and physiologist Carlo Matteucci, whose memoirs were muddled, but was mollified when the timid and sensitive Italian wrote a clear, apologetic letter, which showed that he was able to 'write in the French way, that is to say with order, precision and clarity'.[41]

Nor was Foucault happy when difficult technical words were used at the Academy. 'We admit to understanding nothing,' he wrote concerning a memoir on the reproduction of fungi which was peppered with words such as tremellales, basidium, decussate and sterigma, 'and yet we are convinced that the same ideas could be expressed quite simply and intelligibly in French.'[42] He reproached chemists for coining such 'ugly long words' as *cumylbenzoïlsulfophenylamide* and regretted that chemistry was being isolated by its burgeoning special terminology.[43]

The sharpness of Foucault's criticisms surprises us, unaccustomed as we are to unfavourable discussion of scientists' personalities in the public prints, but the prevalent journalistic trend was of the clever and cutting phrase. However, such words were hurtful and made enemies. For a journalist this mattered little, but Foucault's caustic comments singularly impeded his success with the academic establishment in Paris.[44]

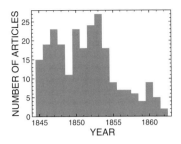

Fig. 6.8. The 246 *feuilletons* and other items published by Foucault in the *Journal des Débats*.

Humour and lyricism

Foucault decried 'the anecdotes, the quodlibets, the picturesque tales, the exaggerations with which the bad taste of popularizers believes it is essential to colour, to doll up and to disfigure the austere face of science.'[45] But despite his own austere face, he frequently enjoyed a joke. Reporting over-exploitation of oyster beds, he quipped that the solution lay 'not in putting gourmets on a diet', but in shellfish breeding.[46] He talked about 'this prejudice of respectable origin whereby man should eat bread only through the sweat of his brow, or, at the least, through that of his baker.'[47] Discussing reforestation, he wrote, '... whether they like it or not, trees must learn to grow faster.'[48] As to techniques for falsifying documents, 'There is no absolute necessity to turn our column into a forger's handbook ...'[49]

Foucault could also be lyrical. He had the nineteenth century's like of grand, regal words: 'the empire of necessity',[50] 'the sanctuary of the unknown,'[51] 'iron, this king of metals',[52] 'at the banquet of life'.[53] He liked metaphor and simile: 'the isthmus of the throat',[54] a comet's approach to the Sun 'to regild its hair',[55] and the steamship *Great Britain* 'whose propeller ploughed its tortuous way across the Atlantic'.[56] He wrote of the planet Neptune 'which gravitates slowly in its cold and pale solitude'.[57] As for nature:

> To reflective spirits, dew appears as one of the phenomena which in a soft and persuasive way best excites a feeling for the harmony of nature. With what abundance does it fall on plants when a calm and transparent night follows a beautiful summer day![58]

and 'What is more gracious and more perfect than the swan swimming with precision on the surface of water?'[56]

The reporter's lot

Foucault's reports must have provided some welcome income. No doubt it was considerably less than the 246 francs per *feuilleton* that Dumas commanded for *The Count of Monte-Cristo*.[59] More probably, Foucault's payment was similar to that received by Hector Berlioz, who was yet another former student who had abandoned medicine, in this case for music. Berlioz received 100 francs for eight or twelve columns of musical critique, and 50 francs for four columns. He noted 'Amongst the rags of its kind, the *Journal des Débats* is ... the one that pays best.'[60]

Over time, Foucault's reports varied in number, becoming rarer after his appointment to the Paris Observatory in 1855 (Fig. 6.8). In the next chapter we will discover why there was a six-month gap in the latter half of 1849.

The number of distinct topics treated within each article also varied (Fig. 6.9). For the first couple of years he attempted to mention everything that happened at the Academy, producing a sort of mini *Comptes rendus*. Inevitably,

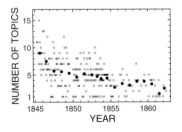

Fig. 6.9. The open circles indicate the number of topics treated in each of Foucault's reports in the *Journal des Débats*. Filled circles are yearly means.

this led to many items that contained little more than memoir titles. Over time, he became more selective, treating fewer topics in greater depth.

Outside events also brought changes to Foucault's reports. Journalists had been allowed to consult the manuscript papers and letters presented at each meeting in order to check details, but in the spring of 1853 the Academy rescinded this privilege, purportedly because it hindered preparation of the *Comptes rendus* and because memoirs were addressed to the Academy, not to the world at large. Foucault published a furious denunciation. 'The Académie des Sciences', he wrote, 'has just adopted the severest of measures affecting newspaper reporters, and hence the public,' contrasting this decision with the one some twenty years earlier which had opened the Academy's meetings to the public. The only way to check details now was by waiting the week or so until the *Comptes rendus* appeared; and even then, the desired information might not appear because the *Comptes rendus* were a selection and abbreviation composed by the Permanent Secretaries. 'We will not try to discern the motives', he added, 'which could have led the Academy to treat as enemies those who for a quarter of a century have served her to the extent that their humble capacities allowed.'[61] Moigno and others published similar protests.[62]

Foucault must have smiled wryly when he saw the next issue of the *Comptes rendus*. 'The Academy's *Comptes rendus*,' he wrote, 'the unique source which we are henceforth invited to consult, offer us, to inaugurate the new regime, two model communications...'[63] The first described how a walnut sidetable had defied gravity under pressure from an 'animal chain' of hands; in the second, an author ignorant of dynamics claimed that speeds could be measured via purported asymmetries in a pendulum's swing. 'At the instant of putting a stop to the abuse of advance publication, the Academy has certainly proceded with maturity concerning the contents of the latest *Comptes rendus*,' Foucault wrote sarcastically concerning these crackpot articles.

Throughout the following year he kept hammering away at this theme whenever the opportunity arose. Mentioning that the aging Eugène Chevreul had made some remarks concerning 'occult sciences', Foucault regretted that:

> This discussion has not been printed in the *Comptes rendus*; and...we are unable to report the thoughts of the savant academician, at least, not without entering into commerce with tables and spirits.[64]

Finally, and most substantively, Foucault deplored that he could not do justice to new, detailed and important work by the physicist Victor Regnault on the calorific properties of gases – work that though presented orally to the Academy, was represented only by its title in the *Comptes rendus*.[65]

These protests were all in vain. The Academy did not relent. Foucault continued the trend towards treating fewer topics at greater length, but whereas previously the *feuilleton* had often carried a line at the top indicating which

Table 6.2. Topics treated by Foucault in the *Journal des Débats*. The proportions did not vary over time despite the evolving nature of his reports.

Physics	20%
Medical sciences	15%
of which physiology	*7%*
Chemistry	11%
Biological sciences	8%
Earth sciences	8%
Astronomy	8%
Farming and forestry	7%
Industry	6%
Transport and communication	4%

meetings were being reported, thenceforth they listed topics treated, and, drawing from Libri, were even subheaded *Revue scientifique*.

Subjects treated

Table 6.2 indicates the distribution of topics treated by Foucault's reports. Physics of course was a favoured topic. The large amount of medicine also reflects his own inclinations, as well as that of his readership 'to indulge one of its dearest fantasies, that of talking medicine'.[66] His dog-fight with Magendie notwithstanding, Foucault was very impressed by the fast-developing field of physiology, occasionally devoting the whole of his *feuilleton* to the subject,[67] and particularly the scientific approach of Claude Bernard:

> M. Claude Bernard has performed on living animals one of these experiments which astonish by the clarity of their result and which please the spirit as if it had isolated one of the wellsprings of life.[68]

Over seventeen years Foucault wrote only twenty-one pieces on mathematics – too few to show in Table 6.2 – and of these, nine were in the first year when he was attempting to mention everything that happened at the Academy. The reason is clear. His mathematician friend Joseph Bertrand (Fig. 6.10) explained that Foucault's education in mathematics was 'very solid, but very limited'.[69] Understanding little of mathematics made it difficult to report. Concerning higher algebra by his friend, Foucault was forced to admit, 'It is impossible for us to give an idea of this work, even by an extract.'[70] It is difficult to believe that Foucault could not have understood mathematics had he tried. The problem was that he was unwilling to because he had little taste for abstraction; worse, he disdained it. Out came a sigh of relief when some mathematics *was* made clear:

> For a long time now we have despaired of understanding anything of M. Cauchy's frightful analytical lucubrations. M. Jamin has made it his duty to seize the physical meaning, when there is one...Thanks to the aid of this intelligent interpreter, it looks as though M. Cauchy will end up by showing us some small scraps of concrete truth every now and then.[71]

Fig. 6.10. Joseph Bertrand (1822–1900), mathematician and friend of Foucault. Bertrand's face was fractured in the first serious railway accident in France, in 1842, on one of the Versailles lines.

Outlook and attitude

Louis-Napoléon favoured the application of science to industry both to improve its products and for the betterment of all levels of society. Foucault shared this outlook. He approvingly reported advances in mirror-, paint- and match-making that protected the workers from the noxious effects of mercury, lead and yellow phosphorus.[72] After it became clear that goitre and cretinism were not caused by some harmful agent, but by the absence of a necessary one (iodine), he reported what countermeasures were needed, adding, 'It is about time we put these different precepts into practice.'[73]

Despite his desire for improved general wellbeing, Foucault was not enamoured with humanity in the mass. A spectacular comet had appeared in 1842, and as for many comets, its arrival had not been predicted – and could not have been – by professional astronomers. He complained acidly about the unrealism of people when this raised 'a cry of indignation from the Parisian populace who, with no instrument other than its two million eyes, shared with the Observatory the discovery of [this] new comet ... '[55] He complained when he was 'rudely elbowed' by the crowd at Arago's funeral.[74]

Others for whom Foucault had little patience were cranks and charlatans, though he was resigned to their inevitability:

> In our somewhat less credulous century, supernatural ideas have to surround themselves with the prestige of science in order to be accepted. It is because of this that we have seen the ghosts of antiquity and the middle ages give way to magnetism, phrenology and homeopathy. We no longer believe in the healing powers of holy relics, but we go and consult the somnambulist; we no longer dare draw cards and ask to be told our fortunes, but we happily let a learned doctor feel the cranium of a young son in order to predict whether he has under his fingers the making of a barrister or a general, a mathematician or a famous artist. ...It would be difficult today to foretell how the unbelievable will be presented to the masses in future centuries; but what is certain is that the scholars and the positive minds of the future will, like now, have many errors to combat and many wars to wage against prejudice ... [75]

Foucault's articles reflect some attitudes that would be unacceptable today. Although he enjoyed female company, he made occasional semi-disparaging remarks about women. After regretting some inadequate observations during a solar eclipse, he concluded, 'For the meanwhile, let us be content to know, in compensation, that a lady fainted at the instant of total eclipse, and that others shed tears.'[76] Commenting on an anthropological study of negroes, he compared 'a Caucasian individual with one of these degraded types who treads the inhospitable soil of New Holland [Australia] in a state of complete nudity', and was 'much more struck by the differences than touched by the similarities which tend to place two beings, who are so unequally endowed, within the same family'.[77] But Foucault was no blind bigot. He continued, 'But perhaps these differences only seem extreme to we other men who observe them from too nearby.' Viewed from another species, they 'would lose their importance and would give rise only to the distinction of simple variations'. He then established his abolitionist attitude towards slavery quite unambiguously:

> In fact, the conclusions of anthropological research lead this way; it is in this direction that action is being taken by governments who have allied together to abolish the unworthy trade of which the negro race has so long been the object and victim.

We must remember that with his San Domingo antecedents, Foucault will not have been speaking idly; and as a lad perhaps heard first-hand reports of the degradations of the plantations from his sea-captain uncle.

Donné wrote that Foucault had 'little concern for moral or philosophical truths'.[78] 'Today everyone philosophizes about everything,' bewailed Foucault; 'the lowliest scholar claims to apply philosophy right down to the way he knots his tie.'[79] He frequently used the word 'positive' in his articles, but only in the sense of 'certain' or 'assured', and we should not imagine he had studied Auguste Comte's *Course of Positive Philosophy*, completed in 1842, which would have been additionally unattractive to language-obsessed Foucault because of its notorious unreadability and overwhelming length.

Many of Foucault's concerns seem remarkably modern. He railed against the tastelessness of white bread and white sugar.[80] He spoke of the need for reforestation in order to protect watercourses and soil fertility.[81] When potato production was being ravaged by disease in the late 1840s, he warned against the dangers of monocultures.[82]

Politics

Foucault avoided politics in his articles, but the events of 1848 burst in. In his first column after the abdication of Louis-Philippe and the proclamation of the Second Republic, he wrote:

> The Academy did not interrupt the sequence of its meetings...In less than a week a revolution has happened, and the irresistible movement which has swept over all of us has not disunited us...Why should the members of the company have abandoned their chairs?...Do the commotions which shatter empires in any way modify the great laws of nature, whose discovery and verification form, after love of country, the first concern of true savants?[83]

One can sense Foucault's sympathy for the 'irresistible movement'. However, the Academy was not unchanged. In May, he noted that attendance had declined and that a 'gloomy indifference' permeated the meetings. 'Provisional arrangements', he wrote, 'perturb advanced study just as much as they disturb great industrial developments.'[84] A fortnight later he noted an additional phenomenon:

> A certain modesty is holding back those savants who have devoted their existence to theoretical speculation, and the debate is given over to those who have dedicated themselves more particularly to applications. It would seem that science, neglected and feeling guilty concerning the leisure of contemplation, has arisen in order to mix with those who labour, saying: 'I too aspire to be useful to the public weal;' and here are the extraction of saltpetre from the ruins of old walls, gun cotton, electroplating and a thousand other examples that show that science does

not play a useless rôle in the world, and that she does not run the risk of being accused of working-out vain theories for no purpose.[85]

An exhortation on the indivisibility of pure and applied science followed:

> May those men who are animated by the true scientific spirit cease to believe that they are obliged to consider only subjects of public utility; ...because, one must not hide it, the field of applications, so rich and fertile today, would not take long to be blighted by sterility if it ceased to be fertilized and revivified by the beneficial light which theoretical research radiates and pours incessantly upon it.

Favourable though Foucault was to Louis-Philippe's departure, he was aghast at the brutal uprising of the dispossessed the following June. The insurgents were nearby, with barricades only a few hundred metres from the Institute, and adjacent to the Luxembourg Gardens and Observatory. However, the June Days were short-lived and only one of the Academy's meetings was cancelled. The following week the Academicians were back in session. Foucault's report was brief, and preoccupied:

> Still greatly agitated as we are by the bloody battle which has wreaked destruction in our great city, we do not intend to distract public attention for long with a mere scientific bulletin...As soon as we begin to feel sure of the morrow; as soon as we no longer need fear the invasion at any moment of barbarians who seem to come from nowhere at the blast of a whistle; as soon as we can put down our arms, happy to trust our salvation to the guardianship of laws that are obeyed, believe then that scientific advance, so recently so rapid, will not take long to regain its former speed.[86]

Foucault passed over the *coup d'état* on 1852 December 2 in complete silence. Since Louis-Napoléon had recently given him 10 000 francs (see Chapter 9) this might seem surprising, but no doubt he was following newspaper policy, which was limited to publishing official communiqués. The *Journal des Débats* had suffered sufficient depredations at the hands of the first Bonaparte not to welcome the arrival of a second member of the clan. It was only after the plebiscite had legitimized the *coup d'état* that Armand Bertin revealed the paper's stance in an editorial: silence, but not submission.

Laying down the pen

Foucault's journalistic career petered out (Fig. 6.8). As his *feuilletons* became less frequent, the *Journal des Débats* began publishing columns by Babinet, as well as reports on the Academy by one of Foucault's friends, the chemist Aimé Girard (1830–98), who in due course became his successor at the paper. But there is nothing in Foucault's final *feuilleton*, published on 1862 May 11, to suggest that it was intended to be his last.

The greater part of the article was devoted to astronomy. Foucault reported the discovery of the predicted faint companion to the bright star Sirius with the new 15-inch refracting telescope at Harvard College Observatory in the United States, and its rapid confirmation at the Paris Observatory using the largest of the telescopes made by him, as will be discussed in Chapter 12. In the remainder of the *feuilleton* he wrote kindly about the third volume of a treatise on artillery just published by Colonel Ildephonse Favé, the Emperor's aide-de-camp, and professor at the Ecole Polytechnique. Favé was known to Foucault, because on at least one occasion it was Favé who had written to tell him of Louis-Napoléon's gifts of money. In addition, Favé's volume was the continuation of a series begun by Louis-Napoléon himself while imprisoned in Ham, and was advertised as 'continued with the aide of notes from the Emperor'. In such circumstances, one would not expect harsh criticism even from Foucault, but it was as well the words were favourable, because they will have eased matters three years later when Foucault and Favé found themselves contending the same vacant seat at the Academy.

How were Foucault's articles received? His friends, laughingly, called him 'the young hack-writer', but it seems that the erudition and intelligence of his words passed over the heads of many of the paper's readers. In 1853 Donné felt that Foucault's own explanations of the pendulum and gyroscope had been too complex, so he published a *feuilleton* about them for 'men and women of the world who will read me lightly...[the pendulum's] author having himself already described his works here for serious people...'[87] The sentiment was echoed during the paper's centenary in 1889: 'His style was cold, a bit dry; he wrote clearly and easily, but his descriptions lacked movement and sparkle...Above all, he wrote for a restricted circle, for his world and for the Academy...'[2]

This 'restricted circle' judged Foucault's articles by other criteria than lightness. Joseph Bertrand, who by then had risen to be one of the Academy's Permanent Secretaries, later recalled that:

> Right from the start he showed a lot of sense, a lot of delicacy and a freedom of opinion tempered by more prudence than one expects from a severe and biting personality. His first articles were remarked upon... People praised the precision of his summaries, they noted his ingenious advice, more than once they admired his new and direct analysis.

There is obviously truth in both points of view. Appendix B extracts some of Foucault's columns to help the reader form a personal opinion.[88]

Chapter 7

Mixed luck

Collaboration

Foucault must have been well-pleased with his interference and polarization work with Fizeau, and with its reception by the academic world. It was physicists' physics: their experiments explored questions relevant to the underlying nature of physical causes, and the novel results had found explanation within the influential new paradigm of the transverse wave theory of light.

Here is what Foucault had to say about working with others in 1845:

> You are working on a problem in physics; but perhaps mathematics on the one hand, and chemistry on the other, would be able to offer you very useful help. In this case, you join with a mathematician or a chemist, you double your forces, and you cannot fail to produce much better work.[1]

This is a very modern and professional point of view, and entirely reflects the reasons why Donné recruited Foucault to his research. The collaboration with Fizeau, however, was less based on complementary skills, with their subsequent careers showing that each was a skilled experimenter and quite capable of working independently. Unlike Foucault, Fizeau did not falter when faced with mathematics, but apart from this, their collaboration appears primarily to have been one of mutual joy in tasks undertaken together, though practical considerations such as space to work and sharing costs no doubt played a rôle. There was, however, some difference in outlook. Fizeau's goal was always the understanding of nature, and although Foucault shared this aim, he liked practical applications. It is thus no surprise that the two young men also worked apart. Fizeau went his own way in 1849 to measure the speed of light. Foucault's boyhood models of telegraphs and steam engines were an early indication that he was fascinated by mechanisms, and this fascination came to the fore in the late 1840s in ideas such as a telescope clock drive and a regulated arc lamp. Unfortunately for Foucault, at a point where his career seemed to be taking off, these inventions did not enjoy the same success as his experiments with Fizeau.

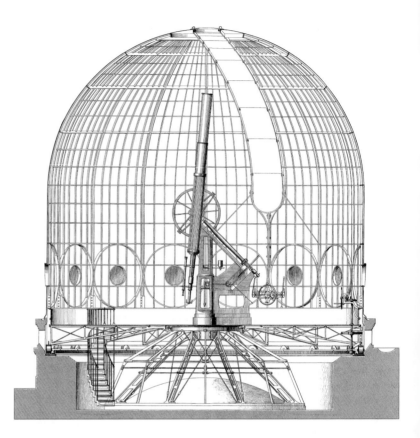

Fig. 7.1. The 12-m-diameter dome erected by Arago on Paris Observatory's east tower during 1845–47. The unusual spider-leg girder-work was designed to spread the load on the stone vaulting below as well as to provide a rigid base. The telescope was installed in 1855, but the 38-cm objective lens soon deteriorated.

The conical pendulum

In the 1830s Arago had had the Observatory's instrument maker, Gambey, construct an equatorial mount for a 10-cm diameter lens that had been made by the optician N. J. Lerebours (1761–1840). A decade later Arago decided that a much more powerful telescope was needed to examine Le Verrier's new planet for rings and satellites, and for other studies. What lens diameter might be feasible was uncertain; Lerebours had started work on a 38-cm lens before his death, but a 19-cm diameter was perhaps more achievable. Undaunted, Arago decided to push ahead with construction of a mammoth dome to house the future telescope, which to be above the rising Paris skyline had to be situated on top of the Observatory's eastern tower (Fig. 7.1).

Foucault was impressed as he watched construction advance. 'Who, casting his eyes towards the Observatory, has not noticed the immense hemispherical bubble that is arising on the left part of this severe and sober monument?' he asked in the spring of 1847.[2] The exterior copper panels will have glowed a splendid golden red.

The conical pendulum

Foucault was inspired with 'the hope of cooperating in some way in the construction of the colossal instrument which astronomy demands'.[3] What he proposed was not only his first pendulum, but also his first foray into the problem of *isochronism*, which was to dominate the last years of his life.

The heart of any clock is an isochronous physical process, which is one that repeats uniformly. In about 1583, Galileo reputedly used his pulse to time a swinging lamp in Pisa Cathedral and realized that pendulum beats are very regular, and unaffected by changes in the extent of the swing. He saw that this regularity could provide a new basis for timekeeping, but a century passed before the first successful pendulum clock was made by Huygens.

Let us consider the physics of the pendulum more closely. The core principles are illustrated by the case of a massive bob attached to a light, inextensible wire. The bob may swing to-and-fro along a line. This is called a *simple pendulum* (Fig. 7.2). In a *conical pendulum*, the bob orbits in a loop.

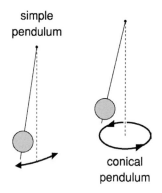

Fig. 7.2. A simple pendulum swings back and forth through the vertical (dashed). A conical pendulum orbits in a loop.

If the bob is motionless and hanging vertically, the tension in the wire exactly matches the weight of the bob. There is no net force on the bob, which remains stationary (Fig. 7.3). Now let us push the bob to one side and then let it go so that it will swing as a simple pendulum. The tension and weight no longer act in opposite directions and cannot cancel. Part of the weight acts to accelerate the bob back towards the vertical. The speed gained as the bob returns towards the vertical is enough to carry it through to the other side of the vertical, where the increasing, but now reversed, restoring force slows and then reverses the swing. The cycle repeats.

A pendulum takes longer to swing if it is longer (or if gravity is weaker) because, for a given linear displacement, the inward component of the weight is less and so the bob takes longer to cover a given distance. The mathematical relations are not simple proportions. For example, the period is doubled if the length is quadrupled. Oscillations of a pendulum of known length can be used to map the local value of gravity across the Earth.

The physics is slightly different for the conical pendulum. The bob has to be released with a certain sideways velocity for it to orbit in a circle. The inward force is then actually a component of the tension rather than of the weight, but this is a detail; what is important is that the inward, centripetal (centre-seeking) force acts to keep the bob in its orbit, in just the same way that a stone being whirled on a string is kept in its circular track by the tension in the string, or a planet is kept in its orbit by the gravitational pull of the Sun.

Let us return to clocks. Once started, a pendulum's oscillation would continue indefinitely were it not damped by air resistance and friction in the suspension. Here the *escapement* performs an essential rôle, applying compensating forces to the pendulum through the *pallets* to overcome the losses (Fig. 7.4).[4] The losses are larger if the swing is bigger, so the amplitude of the swing builds up until the losses match the energy fed in by the escapement.

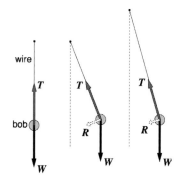

Fig. 7.3. Restoring force on the bob of a simple pendulum. (*Left*) When a pendulum hangs vertically at rest, the weight of the bob, *W*, is supported by the tension, *T*, in the wire. The net force is zero. (*Centre*) When the bob is displaced, a component *R* of the weight acts to accelerates the bob back towards the vertical, causing the pendulum to swing. (*Right*) A longer pendulum takes longer to swing because for the same displacement the restoring force, *R*, is less.

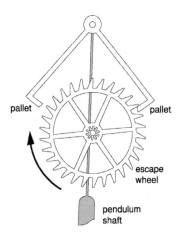

Fig. 7.4. The escapement is a key component in a clock. Through it the spring or weight of the clock motor provides gentle impulses to overcome the damping of the pendulum due to air resistance, and the rate at which the clockwork unwinds becomes linked to the regular swing of the pendulum. In this example, the train of gears from the motor (not shown) gives a clockwise torque to the escape wheel, as indicated by the arrow. The pallets are rigidly attached to the pendulum shaft and rock with it, alternately engaging and disengaging the teeth of the escape wheel, which therefore turns. Energy is fed into the pendulum when the escape wheel teeth slip across the pallets' inclined edges.

For an ordinary clock it is no disadvantage that the hands jump as the escape wheel slips from one pallet to the other. But if the clock is driving a telescope or a heliostat, the motion needs to be continuous, because that is how the stars and Sun move across the sky. For Gambey's equatorial telescope from the 1830s, the jumps from the drive clock had been smoothed by a gear train containing springs and fans. 'Although this beautiful machine...has worked in a satisfactory manner,' wrote Foucault, 'today one no longer considers that it represents the definitive solution to the problem.'[5]

Foucault probably began to think about telescope drives in the spring of 1847 when he wrote an obituary praising Gambey, who had just died, but who had planned to build the drive for Arago's telescope using a powerful clock motor connected to a centrifugal governor.[2] Centrifugal governors with their whirling weights will be discussed in Chapter 15, but Foucault had another idea. This was to use traditional clockwork, but with an escapement following the continuous sweep of a conical pendulum rather than the episodic jerks of a simple pendulum. The idea was not new; Huygens had had it, 'but, as far as I know,' Foucault wrote, 'this idea has never been the object of any serious and lasting application.'[5]

For the escapement to feed energy into a conical pendulum, the pendulum cannot be a weight flopping on a wire but must be a solid piece of metal hung from a suspension. Foucault's first thought was of Cardan's suspension, invented in the sixteenth century, which involved two downward-facing knife edges, at least one of which had to be split into two perfectly aligned halves. Instead Foucault devised a suspension that was simpler both to make and align in which two unbroken blades were clamped edge-to-edge in the form of a cross (Fig. 7.5). An important but far from obvious point is that although the suspension has the two-fold mirror symmetry of the knife edges, the pendulum is equally free to move in *any* direction and can still behave like a ball on a wire. The escapement comprised a moving finger which pressed against a peg set into the shaft of the pendulum.

Foucault presented his clock to the Academy at its Monday meeting on 1847 July 26 and to the Société Philomathique on the following Saturday. It seems likely that he had made it with his own hands, perhaps in part assembling bought components, because his memoir makes no mention of any clockmaker and he was punctilious in later papers about acknowledging his instrument-makers. The pendulum was 'a weighty piece of metal' and swung at an angle of about 20° to the vertical.

The conical pendulum can be considered as composed of two simple pendulums that swing at right angles but with lag of a quarter of a cycle between them. This is completely analogous to the circularly polarized light beam, which as we saw can be considered as composed of two orthogonal linear polarizations separated by a quarter of a wavelength phase shift (Fig. 5.12). However, the analogy breaks down if the pendulum traces out an ellipse, with different amplitudes along the ellipse's two perpendicular axes. This is because the swing of a pendulum is *not* a strictly isochronous process: the period actually increases slightly for bigger swings. So, unlike elliptically polarized light, the two swing periods are very slightly different. As we shall discuss further in the chapter concerning the Foucault pendulum, the orientation of the ellipse slowly veers around, as sketched in Fig. 7.6, but the period of a full loop is also altered. If one amplitude is 20° and the orthogonal one is 19° (or 21°), the no-longer-conical pendulum clock will gain (or lose) over a minute per day.

It was therefore very important that Foucault's conical pendulum should trace out as accurate a circle as possible. To launch the pendulum into a circular orbit he employed a circular disc placed beneath the bob. A protruding peg rubbed against the disc. By spinning the disc manually, the pendulum could be sped up to the point where it entered into a circular orbit. However, what was more essential was that if the pendulum got into a slightly elliptical orbit, then the ellipticity should die away rather than grow; or to use technical language, the circular trajectory needed to be stable. Maybe this was a thought that came to Foucault only during construction of the clock, because he wrote in his memoir that his first thought was that an elliptical oscillation would collapse into a linear one rather than revert to circularity. Foucault's argument for this was phrased in the langauge of mechanics of the time, and involved terms such as *force vive* ('live force', or *vis viva* in the Latin expression adopted in English). It would be tedious to discuss this argument, and his simplistic counterexplanation involving friction at the escapement, which was wrong; suffice it to note that air resistance stabilized the orbit. If the pendulum was given a gentle tap, the resulting elliptical path returned to a circular one 'after several minutes'.

The reaction to his clock was not what Foucault expected. We can read of his dismay in the *Journal des Débats*:

> Before starting any construction, we consulted competent persons and went through different horology books to be sure that our idea was new ... Yet since the day when we made our presentation, analogous machines have emerged from all sides, and amongst them are to be found some which have an incontestable resemblance to ours. In the presence of these facts, should we withdraw our proposal and abandon an idea for which no one is fairly able to designate the first author? Shall we suffer without complaint that a horological project applicable to astronomy is wantonly confused with the puerile conception of an *alcove clock*, or a clock *for the sick which makes no noise*? Thus, ... we have the hope that our efforts will not be counted for naught, and that one day the conical pendulum will be applied to driving parallactic telescopes.[3]

Foucault will have been further dismayed to discover that a clockmaker named Pecqueur had even deposited a *pli cacheté* at the Academy discussing how to

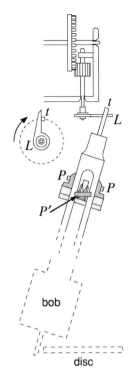

Fig. 7.5. The suspension of Foucault's conical-pendulum clock.[6] Foucault used a suspension that involved crossed upward- and downward-pointing knife edges (shown in light grey) held together by a clip. The upward-pointing blade registered against two plates, P, attached to the shaft of the pendulum. The downward-pointing knife registered against two plates, P', fixed to the clock body (which is not shown). The escapement was composed of a rotating finger, L, which pressed against a peg, t, set along the axis of the pendulum shaft. A spinning disc beneath the bob was used to launch the pendulum into its circular trajectory. The disc also prevented abrupt movements towards the vertical which could damage the delicate escapement.

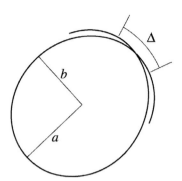

Fig. 7.6. For good isochronism a conical pendulum must be set going in a circular path. If the amplitudes in two orthogonal directions, *a* and *b*, are unequal, as illustrated here, the path becomes an ellipse. Because the periods of these perpendicular components differ, the ellipse veers around by an amount Δ each orbit. The illustrated amplitude difference and veering are very much greater than they would have been in Foucault's conical pendulum. The pendulum may move clockwise or anticlockwise; the ellipse veers in the same sense.

improve the conical pendulum's isochronism. Pecqueur went on to show the Academy a working model a fortnight after it had seen Foucault's device.[7]

The commissioners appointed to examine the pendulum were Arago, which is no surprise, and two other astronomers, Victor Mauvais and Ernest Laugier, who besides being the husband of the niece to whom Arago would dictate his *Astronomie Populaire*, had a special interest in chronometry. However, no report was forthcoming, and Foucault's design was not adopted for the telescope drive. Reasons may have been the uncertain isochronism and the several minutes that it took for any perturbation to die away. A more standard clock with a simple pendulum was chosen (visible in Fig. 7.1).[8] This was hardly a glorious outcome for Foucault, and he made no mention of his conical pendulum in the summaries of his achievements which he compiled in later years when up for election to the *Académie des Sciences*.[9]

Nor was the telescope to prove a glorious success. It was decided to buy the 38-cm Lerebours objective, which had been completed by his son, N. M. P. Lerebours, whom we have encountered in connection with the daguerreotype. An advantage for national prestige was that the diameter equalled those of the world's biggest telescopes at Pulkova Observatory in Russia and Harvard College Observatory in the United States. The price of 20 000 francs seemed reasonable, but unfortunately the quality was not, for the glass soon deteriorated.[10] Foucault's hope that conical pendulum clocks might guide telescopes was not unfounded, however, because they were adopted towards the end of the century by the Warner and Swasey Company in the United States,[11] but before then Foucault had adopted other methods for driving telescopes, as will be described in later chapters. The isochronous signals for telescope tracking are now provided by electronic oscillations.

Distributing astronomical time

It was only in 1972 following the introduction of atomic clocks that time-keeping passed from astronomers to physicists. Until then a major concern of astronomy had been the determination of time from solar and stellar observations. Observatories owned the world's most accurate clocks which astronomers continually strived to improve.

The astronomer-horologers composing the Academy's commission stayed mute over Foucault's conical pendulum, but another astronomer and newly elected Academician, Hervé Faye, felt that Foucault had something to contribute to astronomical horology. Faye was considering the two biggest remaining disturbing influences on astronomical clocks, which were variations in temperature and atmospheric pressure, and his idea was that it was better to eliminate these variations rather than try to compensate for them. He proposed putting the clock in a pressure-tight box some 25 metres underground, which is the depth at which the ambient temperature becomes steady. But how then to read the clock?

The nascent electric telegraph was the answer on everyone's mind.[12] Various systems of master clocks that controlled remote slave dials had been conceived and in some cases installed, but it was not obvious that an electrical system conceived for civil use would be accurate enough for astronomy. For this reason, Faye wrote, 'I enlisted a person used to making [electricity] work in all its forms.'[13] This person was Foucault, and the implication is that he had worked more with electricity than is recorded in his published works. Foucault's solution was very simple. Electricity entered the master-clock pendulum through its suspension. A plate with two lateral platinum buttons was attached near the top of the pendulum shaft. At the extremities of the swing the buttons made contact with two very flexible springs, one on each side. Astronomical clocks normally have pendulums that take two seconds for a complete swing, so that the escapement, which jumps twice per swing, ticks once every second. The alternate bursts of current through the two springs were led by separate wires to the slave clock.

Foucault proposed two different slaves. In one, a soft-iron plate was hinged between two electromagnets, each driven by the current from one of the contacts in the master clock. The astronomer could listen to the seconds clicking by as the plate was attracted alternately to each electromagnet.

As with some Scottish slave clocks that Foucault had reviewed in the *Journal des Débats*, it was not certain that the clicks would faithfully echo the master clock because dust or efflorescence might obstruct the contacts.[14] Foucault therefore suggested a better slave in which the electromagnets attracted an iron slug set in the pendulum shaft of a well-regulated clock. The magnetic pulses would ensure that the two pendulums swung in step, and any irregularities in the current would be smoothed over. The currents required would be very small, since they only had to provide minute synchronizing forces. This led to the considerable advantage that the batteries would last for months before needing renewal. This was the first suggestion that electricity could be used to keep two clocks synchronized, rather than just to transmit pulses which would step hands, as Foucault testily reminded a clockmaker who independently thought of the idea sixteen years later.[15]

The moment was not ripe for Faye's and Foucault's ideas. Laugier's opinion was that horology advanced by small steps and that the introduction of electricity was both unwise and unnecessary, since adequate temperature stability could be obtained by encasing clocks in stone, as at Pulkova Observatory.[16]

Automating the electric arc

We are so accustomed to being bathed in bright and steady light at the flick of a switch that it is difficult to comprehend the attention paid in the past to the question of artifical light. Nineteenth-century encyclopedias devoted scores

Fig. 7.7. The purported illumination of the Place de la Concorde by Deleuil and Foucault in 1844 December is almost certainly further misinformation from Louis Figuier, who had confused it with an earlier demonstration by Deleuil and Archereau.

of pages to lighting, while popularizers like Figuier and Moigno wrote whole books on the subject.[17]

In the 1840s Parisian thoroughfares were lit by oil, though gas was becoming increasingly important. As soon as Bunsen cells became available, the electric arc was tried in the streets, first by a young inventor the same age as Foucault named Henri-Adolphe Archereau, and then by Deleuil. Forty years later Figuier claimed that Foucault and Deleuil used electric light to illuminate the Place de la Concorde a year later, in 1844 (Fig. 7.7), but this uncorroborated claim is almost certainly false.[18]

In any case, public electric lighting was far from Foucault's preoccupations. 'It is not necessary to have handled this light for very long to form an opinion about it,' he wrote in 1845, 'and to resign oneself to the conclusion that for the moment it is only a magnificent physics experiment.'[19] The interference experiments with Fizeau focused his thoughts onto 'the poor optics experimenter, whose life is taken up with waiting for a ray of sunshine'. The arc in the photo-electric microscope could certainly provide a substitute, but its use was cumbersome because of the continual readjustment the electrodes required. 'So many times have we undertaken this constraining task,' Foucault wrote, 'and so many times we have wished that some intelligent mechanism would come and deputize for us!'[20] An automaton needed to fulfil two requirements. The carbon tips needed to be kept at the desired arcing separation and maintained at a fixed location in order to preserve alignment with the rest of the optical components in any experiment.

With the conical pendulum nearly completed, Foucault began construction of an 'intelligent mechanism'. He first saw the electrodes adjust themselves in the spring of 1847. 'My mother was a witness to this first trial,' he wrote, 'she understood its significance and believed in its future.'[21]

Foucault finished his machine at about the time of the February Revolution the following year. 'To have published then would have been the most prudent,' he later wrote; 'but we should have blushed, in the middle of the upheavals which were shaking the country, at the affrontery of trying to interest the public in secrets from the laboratory...'[22] In addition, 'I was loath to present a simple means of study, a pure tool; in my view, an instrument should but pass in second line and should only appear after having served to elucidate some new problem.'[21] Instead, he wrote, 'I amused myself with adapting the most striking and most delicate optics experiments to my new device.'

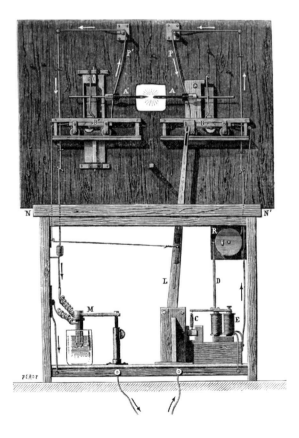

Fig. 7.8. Foucault's automatic electric-arc light completed in early 1848. The carbon crayons, **A**, **A**′, were mounted on trolleys, **B**, **B**′. Arrows indicate the path of the current. When the current fell, electromagnet **E** released spring-loaded armature **D**, which unblocked the escapement of clockwork motor **R**, so letting out a cord that allowed trolley **B** to move leftwards under the action of spring **P**. To keep the position of the arc fixed, a second cord restraining trolley **B**′ against spring **P**′ was let out in the ratio of the electrodes' erosion via lever **L**. The overall operating current and arc intensity were set with adjustments on the armature and via the acid-dipped platinum plates, **M**. The hole behind the arc is about 10 cm square and indicates the instrument's enormous bulk.

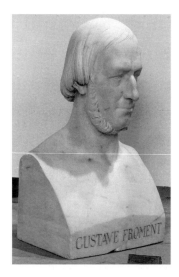

Fig. 7.9. P. G. Froment (1815–65), 'a young and very able instrument maker',[24] who provided parts for the automatic arc lamp and later built all of Foucault's most precise instruments.

(©Musée des arts et métiers-CNAM, Paris/Photo Studio CNAM)

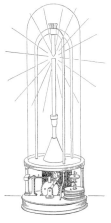

Fig. 7.10. Staite's patent electric light apparatus pictured in *The Illustrated London News*.

Foucault's automatic arc lamp is shown in Fig. 7.8. The separation of the carbon tips was controlled by the current in the arc. As carbon was consumed, the arc lengthened and the current fell. An electromagnet then released a chock allowing the spring-loaded tips to approach. To keep the position of the arc stationary, the trolleys on which the electrodes were mounted were constrained by a lever and cord to move in the same ratio as the erosion of the carbons. Electrical units were yet to be devised, but Foucault will have been controlling currents of around ten amps.[23]

Foucault may have blushed at presenting his new apparatus to a learned society, but he did show it off, rather cautiously, to his friends and colleagues: to Regnauld; to Donné; to a M. Deloge who worked at the Bibliothèque nationale; to a M. Pavé who worked at the Ministry of the Interior; to Xavier Raymond, who was a foreign-affairs journalist at the *Journal des Débats*; to a M. Delorme, a barrister; to Louis Dominique Girard, a prize-winning hydraulic engineer; and to Gustave Froment, an instrument maker, who had supplied some of the parts.

Froment (Fig. 7.9) was a new breed of instrument maker with a formal training in science. 'Born mechanician like others are born poets', Froment had graduated from the École Polytechnique in 1837, and after some time spent in Manchester studying heavy machinery, had entered Gambey's workshop in 1840.[25] Three years later he established his own workshop. His dexterity was such that reputedly he was able to split hairs in four and drill holes down the length of darning needles. Although Froment's reputation was mostly to be made with electrical machines, he made purely mechanical apparatus as well. Over the next few years, he would make all of Foucault's most important instruments.

Foucault used his arc light to show Xavier Raymond some experiments involving chromatic polarization, 'and they were much talked about in our circle at the *Journal*'.[21] It is easy to imagine what happened next. The *Journal des Débats* will have subscribed to a wide variety of foreign newspapers (it reprinted many items verbatim), and although Foucault probably paid scant attention to them, Xavier Raymond and others in the reporters' room noticed an interesting item. In its edition dated 1848 November 18, *The Illustrated London News* had published the description and engraving of an 'electric light apparatus', which like Foucault's used an electromagnet (Fig. 7.10).[26] Three weeks later, the light was even used to illuminate Trafalgar Square.[27] The English arc had been worked upon for over a decade by a certain W. Edward Staite, and the refined mechanism had been produced in association with William Petrie, an engineer.[28]

Poor Foucault! It was the story of the conical pendulum all over again. What further arc lights might suddenly outshine his own? How 'to reserve here our title of inventor, and prevent, through prompt publicity, the pain of seeing the fruit of our own work plucked from us', as he later wrote?

Acting in a rush, he submitted a brief memoir to the Academy on 1849 January 15 in which he asked that commissioners should examine his light and certify that he could not have built it in the short time since the English invention had become public.[29] He ended his memoir:

> Thus the Commission will be convinced, I hope, that I was following a well-determined line, and that in these circumstances which are so troublesome for me, I have been victim of the desire to submit to the Academy only work that is complete.

The Academy obliged and appointed two commissioners, Victor Regnault and Jean-Baptiste Dumas (Fig. 7.11).

The calm and grave Dumas was a chemist; he was professor at the Sorbonne, the École Polytechnique and the Medical School. He was one of the editors, with Arago, of the *Annales de chimie et de physique*, and was embarking on a career in politics. Dumas certainly knew Foucault, because through his influence Foucault had enthusiastically undertaken some task for the Minister of Finance sixteen months earlier.[30] Dumas was an influential man, and his opinion would carry weight.

He and Regnault performed their inspection immediately after the meeting.[31] They saw Foucault's light and an abandoned prototype. They examined receipts and talked to people 'honourably known' to the Academy. 'The procedures conceived by M. Foucault', they concluded, 'were done so in an original and independent manner from those which M. Staite on his side has devised for the same purpose.'

However, a further and more unpleasant surprise awaited Foucault. Not only had Staite and Petrie patented their light in England, Petrie had taken out a patent in France, as their representative, a M. Gaigneau, hastened to inform the Academy.[32] Foucault rushed to the Ministry of Commerce to examine the documents. He poured out his heart in the *Journal des Débats*:

> What a strange position is that of savants who naively pursue their thankless studies! From their liberal hands successively fall the battery, the voltaic flame and the electromagnet; then, when they think of bundling together the instruments which they have wrought with the sweat of their brow, a patent is enough to prohibit them from making use of it ...
>
> We will cede commercial applications and profits to you without regret, but respect teaching in lecture theatres, respect especially the humble sanctuary where discoveries unfold, building blocks of your riches.

The column continued with thanks to the instrument makers who had supplied parts. Lenses and crystals 'of superior beauty' had been provided by Charles Chevalier and the heliostat-maker Soleil, who had made most of Fresnel's optical apparatus. Essential mechanical parts had been manufactured by a clockmaker called Chaudé, and by Froment, of course.

Fig. 7.11. The chemist and statesman J.-B. A. Dumas (1800–84) confirmed the independence of Foucault's work on the regulated arc light. Foucault was grateful, and later dedicated his doctoral thesis to Dumas.

Foucault ended his column with these words:

> In thanking [these makers] here, we did not intend to be free of our debt to them; our intention was to give them our plans. Will they be permitted to use them? That is a point of law which needs to be cleared up; we will consult competent people, and we will learn from them the extent of the rigours of our laws towards science.

Foucault perhaps consulted his barrister friend Delorme. An important similarity between the English and French arc lights was their use of clockwork to move the electrodes. The legal opinion clearly was that Foucault's lamp might infringe this aspect of Petrie's patent, because although Foucault exhibited a better-made version of his lamp to the Academy in June, he abandoned work on the arc lamp for fifteen years, the lifetime of the patent. Petrie's patent also claimed 'the application of this variation... in the electric current to make this regulator work'. Here it is likely that the patent was invalid, because the idea was not novel.[33] For example, Auguste de la Rive, who was Professor of Experimental Physics in Geneva, wrote to say that he had been using an electromagnetically controlled arc light 'for a long time'.[34] Petrie's patent certainly did not stop Paris instrument makers and others from producing regulators during the 1850s which acted on the basis of the electric current but avoided clockwork. Deleuil made one; so did Jules Duboscq, who was Soleil's son-in-law, and who took over the business when Soleil retired in 1849. As presented to the Academy in 1850, the lower electrode of Duboscq's lamp was spring loaded and the upper one moved under its own weight, but subsequently Duboscq either purchased rights or decided to ignore Petrie's patent.[35] He modified his regulator to incorporate clockwork to advance the carbons (Fig. 7.12). We will return to this arc light in Chapter 15.

Fig. 7.12. Duboscq's arc light from the 1850s incorporating electromagnetic control and simple clockwork.

Sunshine and skating

Despite Petrie's patent, Foucault's arc light did reach the public – on stage.

The rage across Europe was for grand operas composed in a pre-Wagnerian style by the Berlin-born Giacomo Meyerbeer (Fig. 7.13). His opera *Le prophète* (The Prophet) opened on 1849 April 19 in Paris. Its setting was the sixteenth-century Anabaptist revolt of the poor in Westphalia and it was an immediate hit. Part of the entertainment was lavish staging: the 680 costumes, along with sets and props, cost almost 140 000 francs.[36] The chief designer, Edmond Duponchel, was always on the lookout for new effects to surprise and enchant the public, and embraced Foucault's arc lamp to simulate dawn in the third act.

The setting was the Anabaptists' overnight camp near a frozen lake. As the act ended, its musical climax was accentuated by the silent rising of the Sun, dissipating the fog and throwing its steely light on the nearby town of Münster, which the Anabaptists were about to besiege.[37] 'The effect of the rising Sun is

Fig. 7.13. The composer Giacomo Meyerbeer (1791–1864). The use of Foucault's arc light to simulate the rising Sun in Meyerbeer's opera *Le prophète* was a turning point in Foucault's relationship with precision-instrument makers.

one of the newest and most beautiful things which has been seen in the theatre,' wrote one critic. 'Thanks to the electric light, we have seen a true Sun, which cannot be stared at without blinding, and whose light penetrated to the back of the furthest boxes from the stage.'[38]

The Sun was a copy of Foucault's arc lamp, built by Froment at a cost of 350 francs.[39] This was not quite the first time that the arc had shone at the Opera.[40] The electric light in *Le prophète*, however, obtained 'the most unanimous applause',[41] but it was Duboscq's compact regulator that was preferred to create the rainbows, lightning flashes and luminous fountains that soon became regular features in operatic staging.[42]

Figuier tells us that the most unanimous applause was hard won. A reflector was required to converge the light into a beam, and Meyerbeer was a difficult task-master:

> At that time, Léon Foucault...dined with us at a *table d'hôte* [a private house where meals were served] on the Rue des Beaux-Arts. For three months, he came each day carrying under his arm the famous reflector which he had just been trying on the stage at the Opera, and through giving breath to his complaints with us, he repaired the damage to his spirits caused by the maestro's exacting demands.[43]

As with all of Figuier's writing, we must be suspicious; but perhaps there is an underlying truth that the rising Sun caused heartache, and that though still living with his mother, Foucault, at the age of almost thirty, ate elsewhere. No doubt Foucault supervised operations initially, but as we shall see, he broke down some months later and by the end of the year a certain M. Lormier was being paid 10 francs per performance to operate the light.[44]

Successful as it was, Foucault's sunrise was overshadowed, so to speak, by another piece of novel staging in the third act. An interlude of ice skating on the frozen lake was simulated by two dancers wearing roller skates. 'Prophet skates' rapidly became a fast-selling item in Paris.[45]

Electricity at the pull of a cord

Donné tells us that 'Léon Foucault was a sybarite and liked to work in comfort.'[46] The Bunsen cell was not comfortable to operate, and not only because of the nitric oxide evolved. The acids mixed via diffusion through the porous pot if the cell was left assembled, so it had to be filled just before use. Since an arc light needed sixty or more cells, this was monotonous and time-consuming. To avoid this fatiguing and eye-stinging preliminary, Foucault rearranged the cells so that they could be readied 'in less than five minutes', as sketched in Fig. 7.14. The inner porous pots were ganged together on a wooden batten. With ropes and pulleys, Foucault could raise or lower them into the nitric acid in a matter of seconds. A few minutes further sufficed for sulphuric acid to flow through a series of siphons to cover the zinc electrodes.

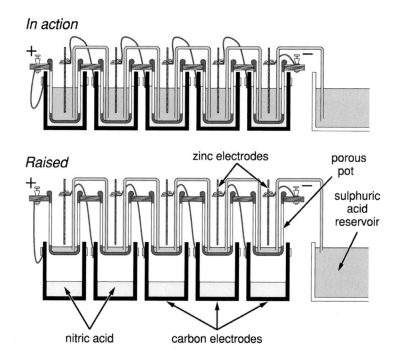

Fig. 7.14. Principle of the simple-to-use battery of Bunsen cells built by Foucault to power his arc lamp. The sulphuric acid siphoned off automatically when the porous pots were raised out of the nitric acid, and siphoned back when they were lowered. Each battery actually contained twenty cells. Exterior carbon beakers were used because Foucault devised the battery before Deleuil reversed the electrode order (cf. Fig. 4.16). The bottoms of the porous pots were impregnated with wax so that when raised they could retain sufficient sulphuric acid to keep the siphons primed.

A paper with Regnauld

The conical pendulum and the automatic arc were technological devices. Foucault completed just one scientific experiment, properly called, in the late 1840s, and it was in physiology, for which we have seen his enthusiasm. Undertaken jointly with his friend Jules Regnauld, it provides an example of a situation that often arises in science whereby a technique developed for one avenue of research finds application in another.

Foucault and Regnauld experimented on what is now called *dichoptic* (or *binocular*) *colour mixture*. It involved two differently coloured discs. The image of one disc was presented to one eye while the other disc was presented to the other eye (i.e. dichoptically). Wheatstone had found that the perception of one or other disc predominated (this is now called *colour rivalry*); the sensation was not of some composite tint.[47] However, Wheatstone noted that others had stated, 'contrary to fact', that mixed tints could arise.

Foucault and Regnauld repeated Wheatstone's experiments using coloured glasses and obtained similar results, but with the difference that after some time the colours *did* fuse to a mixed tint. This is the phenomenon of dichoptic colour mixing. The dichoptic colour was the same as would be perceived if beams of the individual colours were projected onto a white screen and looked at normally. Red and green light, for example, produced yellow. This led

Foucault and Regnauld to the thought that strictly complementary colours, dichoptically presented, should produce the sensation of white light.

To put their idea into practice, they placed two white discs in a stereoscope, one for each eye, and illuminated them with the complementary colours produced with a dichroic crystal in the same way as in Foucault and Fizeau's chromatic polarization experiment. The light source was the regulated arc lamp.[48]

They experimented 'with a great number of people' – presumably mostly the same ones that Foucault had etherized two years earlier. They found that one eye or the other often dominated initially and that only one of the discs was perceived, but that most people soon accustomed to composite perception, especially if little figures were drawn on the discs to focus attention. The colour mixing was easier when the lights were weaker.

For complementary colours, the composite perception was of whiteness, as they had predicted, and for some people the effect was sufficiently strong to persist through the entire range of complementary colours available by rotation of the dichroic crystal. 'The mixing of colours', they wrote, 'is a fact denied by M. Wheatstone, but which we confirm occurs.' The clear conclusion was that colours arose neither when light was mixed, nor simply in light's interaction with the retina, but 'in the most intimate recesses of the encephalon'.

The experiment was definitive and one supposes they had fun recruiting their experimental subjects, especially the laywomen, but since the perception of depth clearly arose in the brain, it was no great surprise that this was also at least in part the case for colour. Their work was confirmed a decade later in Leipzig by Ewald Hering, and subsequently by others. The occurrence of colour mixing implies that there must be colour mechanisms in the cortex of the brain as well as in the retinas of the eyes, but even today the full theoretical significance of binocular colour mixture remains unclear.[49]

Reversal of the D lines: a discovery unexploited

Foucault was less pressed when he presented his regulated arc light to the Société Philomathique on the Saturday following his rushed memoir at the Academy.[50] He brought the bulky lamp with him, but perhaps did not show it in action because the Bunsen cells would have been complicated to transport. His discussion revealed more details. The mechanism adjusted the carbons every four or five seconds, and could keep the light shining for an hour or more. The light was sometimes irregular because of silica and other impurities in the gas carbon (apparent in Fig. 5.3), so he had experimented with making better electrodes by heating sugar under pressure. (Once more he was following the same route as Staite.)

Foucault examined the arc that formed between metal electrodes as well as the incandescent electrodes themselves. He focused the arc onto a screen to examine its physical aspect, or onto a slit for dispersion into a spectrum with a

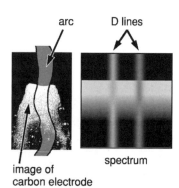

Fig. 7.15. (*Left*) Foucault reflected an image of one of the incandescent carbon electrodes onto the electric arc and then re-imaged a vertical slice of the light into a spectrograph. (*Right*) Near the D lines, the spectrum shows that the incandescent carbon radiates at all wavelengths while the arc emits only in the D lines. To Foucault's surprise, he found that where the D lines overlapped with the continuous spectrum they did not add to the brightness but appeared *dark*.

prism, just as in his experiments with Fizeau.

The spectrum of the carbon arc was 'furrowed' with bright lines, including a bright orange-yellow pair reminiscent of the dark Fraunhofer D lines in the solar spectrum. (In Plate IV the resolution is too low to reveal the dual nature of the lines.) To check whether the features were identical, Foucault focused an image of the Sun onto the arc and examined the composite spectrum. The bright lines of the arc coincided exactly with the dark lines in the solar spectrum. He should not have been surprised by this, since Fraunhofer had already noted the coincidence of the solar D lines with lines radiated by flames.[51]

Foucault was a meticulous observer, however, and noticed some unexpected effects. First, the electric arc was extremely transparent, casting only the lightest shadow over the solar spectrum. Second, he noticed that the D lines in the solar spectrum were darkened even further where they overlapped with the bright lines from the carbon arc. He confirmed this by using a concave mirror to focus an image of one of the carbon electrodes onto the arc. The incandescent electrode gave a continuous spectrum uninterrupted by any emission or absorption lines, but where it overlapped with the arc, dark D lines were seen, as simulated in Fig. 7.15. He expected the opposite, that the light from the arc would add to the light from the incandescent electrode rather than dimming it. 'Thus the arc offers us a medium which emits the D rays on its own account and at the same time absorbs them when these rays come from elsewhere,' he said.

Foucault found that the D lines were present with varying brightnesses in the light given by different metal electrodes and were considerably brightened if the electrodes were touched with potash, soda or chalk. 'Before concluding anything from the nearly universal presence of the D line, it is no doubt necessary to be sure that its appearance does not derive from some material which is present in all our conductors,' he wrote, presciently, since it is the omnipresence of sodium contamination that makes the D lines ubiquitous. His next sentence was just as prophetic:

> Nevertheless, this phenomenon seems to us today to be a pressing invitation to the study of the spectra of stars, because, if one should be fortunate enough to find the same line there, stellar astronomy would certainly draw profit from the fact.

He went on to note that the arc spectrum of silver was dominated by a single, very intense green line that could be used for optics experiments involving only a single wavelength, which previously had only been imaginable in theory.[52] He concluded:

> All these facts ... need deeper study, but in the circumstances in which I find myself, having been forestalled by the publication of an analogous device to mine in England, I wanted, by all means in my power, to show

Reversal of the D lines: a discovery unexploited

that I had a seed in my hands which can germinate and which, though it may bring harvests in the fields of industry, will at least have offered its first fruits to science.

Surprisingly, Foucault abandoned this line of enquiry, at least as far as published work was concerned. No doubt he set up some interference experiments using the monochromatic light of the silver arc and showed them to visitors, but this was material for the lecture theatre rather than the Academy. The systematic study of the sparks of a plethora of electrodes would have been no more appealing to Foucault than the abandoned systematic study of birefringent crystals. The reversal of the D lines continued to occupy his mind, however. When he went to London in 1855 to receive the Royal Society's Copley Medal (see Chapter 10), it was the subject of dinner-table conversation with the British physicist G. G. Stokes. 'That a medium was simultaneously D-emissive and D-absorptive struck me with all the freshness of originality when Foucault told me of it,' Stokes later wrote, adding that he had considered translating the report of Foucault's Société Philomathique remarks into English.[53] Had he done so, the D lines' reversal might have become more widely known and understanding achieved sooner. But in any case, understanding was not long coming. Fox Talbot, Wheatstone and others had suggested that spectral lines were characteristic of different substances and could be used in chemical analysis. The ubiquitous D lines seemed to contradict this notion, but the objection was removed in 1856 when it was shown that less than one ten-millionth of a gramme of common salt was enough to tinge a flame with bright D lines.[54] The D lines' reversal was rediscovered in 1859 by the Heidelberg physicist Gustav Kirchhoff in flames. Unlike Foucault, Kirchhoff was able to deduce why the reversal occurred. In an equilibrium situation, the atoms must emit as much D light as they absorb; this is now known as Kirchhoff's Law of Emission and Absorption, and it makes emission a necessary concomitant of absorption. In Foucault's experiment (Fig. 7.15) the radiation from the carbon electrode comes from only one side. The sodium atoms absorb the D wavelengths from this beam but re-radiate them in all directions. Because of this geometrical dilution, the strength of the D lines relative to adjacent wavelengths is reduced, though their strength is in fact increased, albeit very slightly, compared to the arc alone. In the Sun, the dark lines arise analogously. Radiation from the hotter, brighter inner layers of the photosphere is absorbed by the cooler layers above.

With the formation mechanism of the Fraunhofer lines understood, it was safe to use them for chemical analysis. In 1860 Kirchhoff published a landmark paper with his colleague Robert Bunsen (Fig. 7.16) in which they compared solar and laboratory line positions and concluded that iron, calcium, magnesium, sodium, nickel and chromium were present in the solar photosphere, while

Fig. 7.16. Gustav Kirchhoff (1824–87, standing) and Robert Bunsen (1811–99) photographed in 1862. Kirchhoff rediscovered Foucault's reversal of the D lines and realized its importance. He was then able to make the first chemical analysis of the solar atmosphere with Bunsen, twenty-five years after Comte in his *Course of Positive Philosophy* had declared the chemical composition of the stars was something forever unknowable.

common terrestrial elements such as aluminium and silicon were undetectable.

The chemical analysis of a heavenly body was a romantic exploit, and excited widespread interest. Dumas wrote an article for *Le Moniteur*,[55] and Foucault devoted one of his now-rare *feuilletons* to it:

> All these vapours vibrate like harps with a particular harmony, emitting into space luminous notes endowed with an unalterable timbre, and capable of crossing the greatest distances. Of what importance, then, the 30 million leagues that separate us from the Sun?[56]

Kirchhoff and Bunsen's work was a springboard to chemical analysis of starlight, as Foucault had almost predicted, and to the discovery of new elements. It was also the culmination of research that had involved numerous scientists.[57] Many claimed that some of the glory should reflect on them, including Brewster, the Swedish physicist A. J. Ångström, and even Moigno.[58] The importance of Fox Talbot and William Thomson (1824–1907, later Lord Kelvin) was claimed by their friends, while Stokes and others championed Foucault's cause, reprinting his memoir from 1849.[59] Foucault himself was more modest, and nowhere reminded readers in the *Journal des Débats* that he had been involved in the examination of arc spectra. He had made what proved to be a crucial observation, but he was well aware that he had been unable to recognize its importance or seize its meaning.[60]

Sensibilities affronted at the Société Philomathique

Both Fizeau and Foucault were regularly reporting their work to the Société Philomathique.[61] Fizeau's most famous paper was delivered there on 1848 December 23, and concerned the shift in frequencies when a source of sound or its observer are moving. The report ended with Fizeau's opinion concerning the possibility of detecting an analogous effect in light: 'the difficulties are not such that one cannot hope to overcome them'.[62] The prediction proved spectacularly apposite, and the Doppler–Fizeau effect is enshrined as the key tool for probing motion within the universe; but no doubt it played only a minor rôle in Fizeau's election to the Society three weeks later.

The election procedure was similar to the Academy's. Once a place fell vacant, a commission was appointed to propose and report on candidates. Two or three candidates would be ranked in Secret Committee. From the voting patterns it is clear that members knew their duty was to vote for the first-ranked candidate; other candidates would then move up the ranking and become members at a subsequent election. The dribble of votes that the losing candidates received were no doubt an arranged propriety.

In the month following Fizeau's election, a commission was formed to recommend candidates for another vacant place. Among the five commissioners were César Despretz, Foucault's *baccalauréat* examiner; and Fizeau, who sat because it was the custom to include the most recent electee. Despretz was absent when it came to reporting the commission's recommendations, which top ranked the physicist Jules Jamin (1818–86). Second equal were Foucault and an apothecary called Boutigny. The ranking was as it should have been. Jamin had come second to Fizeau, so it was now his turn to be elected, which he was on March 3.

Three months later there was another place to be filled. The nominating commission was completely different, composed of Balard, the discoverer of bromine; two further chemists, one of whom, Adolphe Wurtz (Fig. 10.4), later became friendly with Foucault; Jamin, as the newest member of the Society; and Edmond Becquerel. This commission broke convention and recommended neither Boutigny nor Foucault, but two chemists.

The membership was scandalized. The minutes for 1849 July 21 record:

> The vote for the election...was cancelled because of an insufficient number of voters, despite the fact that the attendance list showed 30 members present.

Dissatisfaction having been made clear, the membership obediently elected one of the chemists the following week.

Was the sidelining of Foucault and Boutigny due to the chemists looking after their own? Possibly, even probably, because at the end of the year, with sensitivities having been heightened, appointment of a nominating committee was postponed owing to the dearth of chemists present.[63] However, one can imagine that there had been sufficient irritations between Foucault and Edmond Becquerel for the latter to have been muted in any opposition he may have raised to the chemists' nominees.

Depression

As time passed, Donné as both man and doctor must have become aware that his young friend and collaborator had a fragile side. Some years later he had cause to convey these thoughts to the new Director of the Paris Observatory, Urbain Le Verrier:

> You know that he and I are linked by friendship... I have followed his career and on several occasions I have had cause to worry about his state of health.
>
> It is not the health of his body that I mean; without being strong, he is rarely ill, but the same is not true for his mind. The son of a father who was struck by mental alienation, he is himself, despite his remarkable faculties, of fairly weak mental state. Following periods of unremitting work, or simply after too long a stay in Paris, I have seen him fall into a state of dispiritedness which has scared me.[64]

A true friend, Donné took necessary action:

> I only got him out of this state by taking him away for some time from all concerns and from all occupations of the mind. It is thus that I took him sometimes to England, sometimes to bathe in the sea, sometimes into the countryside. After this rest of several weeks, he returned strengthened and was able without danger to give of his fullest once more.

There is a blank in Foucault's life starting after his non-election to the Société Philomathique in 1849 July. Some jottings by Fizeau survive of inconclusive 'Trials for determining the refractive index of gold' made in common with Foucault on September 10,[65] and in the same month Foucault completed work on an auxiliary regulator to isolate his arc light from variations in the supply voltage as the battery ran down.[66] But apart from this, Foucault disappeared from view. His *feuilletons* ceased, he presented nothing at the Academy, his name is absent from the minutes of the Société Philomathique. There is little doubt that he was suffering one of these periods of dispiritedness. He had done good things with Fizeau, but what had he accomplished on his own? Arago was not interested in his conical pendulum, or his ideas for electrical distribution of time. His work on automating the electric arc had come to nothing thanks to an unknown foreigner's patent. And what about his manual skills? In January, he had given just a verbal description of his arc light to the Academy. Six weeks after the opening of *Le prophète*, he exhibited an actual instrument. This was not the original instrument, but one more carefully built by Froment; it was perhaps the one being used at the Opera. Talented as Foucault was with his hands, he could see that a professional instrument maker, with a fully equipped workshop, could do much better. He had not been elected to the Société Philomathique when in the rightful order of things he should have been. And to make his self-doubt even more poignant, in 1849 July his friend Fizeau had received laurels for having moved measurement of the speed of light from the realm of astronomy into that of laboratory physics, as will be discussed in the next chapter. It was all too much. He had to get away.

Jeanneton

We cannot know how much country living or how many sea baths were necessary to revive Foucault's spirits, but it seems they were restored by December 1, helped perhaps by a love interest. On this date he received a curious little note (Fig. 7.17). It transcribes the words of a well-known bawdy song in which the young Jeanneton's plans to cut rushes are interrupted by three male passers-by. The note was given to Foucault 'for his edification'. The clear inference is that he was taken with a young woman, but was insufficiently assertive in his suit.

Extraordinary success at the Société Philomathique

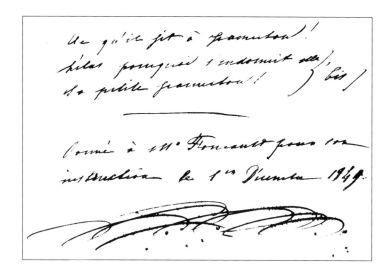

Fig. 7.17. What luck in love? The final words of the ribald song *Jeanneton* 'donné à M. Foucault pour son instruction le 1er Décembre 1849' (given to M. Foucault for his edification, 1st December 1849).

(Family collection)

Love, if that is what it was, did not end in marriage for Foucault. Was the woman 'exasperated by his composure', as Donné put it? Or was Foucault hesitant? His father had lost his reason; his own state was weak. Was it fair to risk putting a wife through the sufferings his mother had had to bear? Or visit his own fragility on children? One can imagine that ethical considerations of this sort could have held back an introspective person like Foucault.

Extraordinary success at the Société Philomathique

The Société Philomathique took a long vacation over the autumn. In November a third nominating commission was formed including Balard, Fizeau, the physicist Paul Desains (1817–85), and the aged Charles Cagniard-Latour, who fifteen months later would beat Foucault to election to the Académie des Sciences. This commission reverted to established custom, and placed Foucault top of the nominees. His work on light and electricity was outlined to the members by Fizeau. What happened then was quite extraordinary. No doubt some members were still angry about the previous election. No doubt others remembered the society's motto of *Étude et amitié* and decided a public act of support was needed for someone who had survived some difficult times. No doubt others wished to elect a promising young scientist who wrote perceptive and thought-provoking newspaper articles. No doubt some were motivated by all these concerns. In any event, three times as many members as was usual for an election turned up to cast an unprecedented ninety-three votes for Foucault. Finally, he was accepted (Fig. 7.18)! The election had been difficult, but election to the Académie des Sciences was to prove far harder.

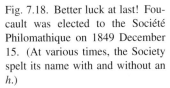

Fig. 7.18. Better luck at last! Foucault was elected to the Société Philomathique on 1849 December 15. (At various times, the Society spelt its name with and without an *h*.)

(Family collection)

Foucault participated fully in the Society. The attendance sheets for 1851 have been preserved and show that he was assiduous, missing only six of the forty-two meetings that year. He described almost all his discoveries to the Société, and he regularly commented on others' presentations when there was a link to his own work, such as concerning the luminous emissivity of heated platinum, a new battery, mucus, or how to distinguish different forms of sulphur. He helped administer the Society, chairing meetings for the last quarter in 1851, being delegated to report on advances in physics during 1860, and serving on several nominating committees.[67]

Foucault's luck had begun to turn for the better. To borrow a term indissociable from Newton's life, he was on the threshold of an *annus mirabilis*, in which he would obtain his first major result, in optics, and then go on to conceive the esteemed pendulum experiment.

Chapter 8

The speed of light. I. Demise of the corpuscular theory

It is poetically appropriate that Paris, the City of Light, should have played an early rôle in the development of modern ideas concerning the nature of light. It was in Paris that the speed of light was discovered to be finite, as will be discussed below; and Malus's discovery of polarization in the Luxembourg Palace windows underpinned Fresnel's development of the transverse wave theory of light, as did the discovery of chromatic polarization by Arago. Additionally, the physics of the speed of light became a Parisian monopoly for about forty years in the mid-nineteenth century, with Foucault making two crucial experiments: one in 1850, which is the principal subject of this chapter; and another in 1862, which is the subject of Chapter 13. Though the experimental procedures were similar, the two experiments were addressing separate and unrelated questions; but before these questions can be understood, some introductory ideas must be developed concerning the speed of light.

The speed of light

Most scholars up until the seventeenth century considered that luminous phenomena propagated instantaneously. They had arguments to support this view, of course; here are two examples. The Greeks appreciated that the stars were very distant because their pattern in the sky remained the same – there was no observable shift due to parallax – whatever the viewpoint across the then-known world. The philosopher Hero of Alexandria (fl. *c.* first century AD) noted that he saw stars immediately upon opening his eyes. Since in Greek thought vision travelled *from* the eye to the object being perceived, and back again, Hero concluded that the propagation was instantaneous. A millennium and a half later, the German astronomer Johannes Kepler (1571–1630) reasoned that since space is immaterial, it can offer no resistance to light, which must therefore travel infinitely fast.

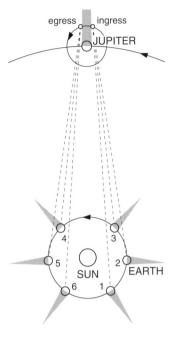

Fig. 8.1. To understand Rømer's explanation of the varying intervals between eclipses of a Jovian moon, it is sufficient to imagine Jupiter frozen in its orbit around the Sun and to consider only the relative positions of the moon and the Earth. When the Earth–Jupiter distance is shortening (1→2→3), it is only possible to see the beginning (ingress) of any eclipse because the exit (egress) is hidden behind Jupiter. Since light takes less time to cross these shortening distances, ingresses are separated by shorter intervals than when the Earth–Jupiter distance is increasing (4→5→6) and only egresses are visible.

Galileo was the first to address the issue with experiment. From his *Discorsi e dimostrazioni*, published in 1638, it seems that Galileo and a co-experimenter equipped themselves with lanterns and stationed themselves somewhat less than a mile apart. One experimenter uncovered his lantern; on seeing this, the other responded likewise, and finally the first experimenter estimated the round-trip delay. Galileo was unable to decide whether light travelled instantaneously, but concluded that if not, it was certainly exceedingly rapid.

The speed of light was shown to be finite in 1676 by the Dane Ole Rømer while working at the Paris Observatory. Rømer had been brought to Paris in 1672 by Cassini's assistant Jean Picard. In the days before accurate portable chronometers, it seemed that eclipses of Jupiter's moons might provide the timepiece needed for the determination of longitude; Galileo himself had made the suggestion in 1612 only two years after his discovery of the four brightest Jovian satellites. By 1668 Cassini had published a table of the moons' motions. Rømer joined Cassini and Picard in further observations of these satellites.

In 1676 September, Rømer presented the Academy with a prediction that the egress, or end, of the eclipse of the innermost Jovian moon expected on the following November 9 would occur ten minutes late compared to the time expected from averaging all eclipses. Observations confirmed this prediction, and soon afterwards Rømer presented memoirs in which he explained the delay as due to the light travel time across the space between Jupiter and Earth. He noted that the geometry is such that ingresses, which are when a moon disappears into Jupiter's shadow, only occur when the Earth is approaching Jupiter (Fig. 8.1). He further noted that the intervals between ingresses are shorter than the average value, whereas egresses, which only occur when the Earth is moving away from Jupiter, are separated by intervals that are longer than average. Rømer recognized that the changing eclipse intervals reflected the finite speed of light and the varying distance that it must cover between Jupiter and the Earth – which is always decreasing for ingresses and increasing for egresses. From the observed timings, he calculated that light takes 22 minutes to cross the diameter of the Earth's orbit.

Rømer's explanation was violently contested by Cassini but was readily accepted by Huygens, Newton and others. Any remaining doubt about the finiteness of the speed of light was laid to rest in 1728–29 when the Oxford astronomer James Bradley announced his discovery of the *aberration of starlight*. It is useful to understand this effect, which we shall encounter again in the next chapter, because it also provided confirmation that the Earth orbits the Sun.

The pattern of the constellations may not have altered for observation points across Hero's Hellenic world, but following Kepler's discovery of the laws of planetary motion in the early seventeenth century, there was little doubt that our viewpoint on the stars changed every six months by the much larger diameter of the Earth's orbit (Fig. 8.2). Bradley had been making measurements of the star Gamma Draconis in the hope of finding parallax changes. He failed to find any parallax shifts, which even for the nearest star amount to a maximum difference of only about $1\frac{1}{2}$ arcseconds, or about the angle subtended by a pea at a distance of a kilometre. (There are 60 arcseconds in an arcminute, and 60 arcminutes in a degree.) What Bradley *did* find is that the direction to the stars shifts, or is aberrated, because we observe them from a moving observation post (the Earth), in an exactly analogous way to which raindrops strike a windscreen at a varying angle depending on the speed and direction of the car. As the Earth orbits the Sun its direction of motion changes, and therefore so does the direction of the stellar aberration (Fig. 8.3).

The maximum extent of the change, which exceeds 40 arcseconds, depends on the ratio of the Earth's speed to that of light, and was further evidence of both light's finite speed and the Earth's orbital motion. Stellar parallax was finally detected in 1838 by Freidrich Bessel, director of the Königsberg Observatory.

Rømer's 22-minute Earth-orbit crossing time was soon improved to around 17 minutes by the English astronomer Edmond Halley (1656–1742). A numerical value for the speed of light was obtained when this time was divided into the diameter of the Earth's orbit, which was measured via difficult triangulations, as will be described in Chapter 13. An early 1850s result, quoted by Arago in his *Astronomie populaire*, was 77 076 leagues per second, or in modern terms, 308 300 km/s. The important points to retain are that the speed of light, though enormous, is finite; and that in Arago's time the numerical value derived from astronomical measurements.

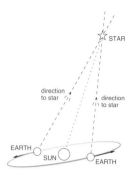

Fig. 8.2. The viewpoint on a star changes as the Earth moves in its orbit around the Sun. The resulting wobble in the direction to the star is called stellar parallax. The effect has been enormously exaggerated: the distance to even the nearest star is almost 300 000 times the separation of the Sun and Earth.

Undulation versus emission

To most scientists, the evidence for the transverse wave theory of light seemed overwhelming. Nevertheless certain diehard emissionists refused to concede defeat, such as Brewster and his dilettante compatriot Lord Brougham, though their opposition was becoming increasingly anaemic, as Foucault told his readers in the *Journal des Débats*:

> While some, as fervent disciples of Huygens, Descartes and Fresnel, confidently advance their research in a direction favourable to the wave theory, bringing proof after proof to support a system which is henceforth unshakable, others, still faithful to the errors of the immortal Newton, endeavour if not to resuscitate the old theory of emission, at least to embarrass the new school; it is particularly from the other side of the Channel that these innocent skirmishes originate. Recently it was Lord Brougham, today it is Sir David Brewster who arrives armed with very pretty, but very minor phenomena, who describes and develops them and sets them like stumbling blocks with which he hopes to trip the undulationists for an instant at least.[1]

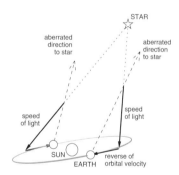

Fig. 8.3. Stellar aberration explained non-relativistically. For an Earthbound observer the rest of the universe appears imbued with a velocity that is the opposite of the Earth's own motion. Rays from the star appear to move with both their own velocity and this reflex velocity. In consequence the direction from which they come is shifted, or *aberrated*. The shift has been greatly exaggerated: the Earth's velocity is actually only one ten-thousandth of the speed of light.

However, despite the wave theory's success at explaining more and more optical phenomena, the sad and irritating fact was that there was no experimental result which was unquestionably incompatible with emissionist ideas.

Arago's *experimentum crucis*

In 1834, Wheatstone had attempted to measure what he called the 'velocity of electricity'.[2] The term causes amusement among modern physicists, who understand that electrical signals advance at different rates under different circumstances, but the issue was of crucial relevance to the practicability or otherwise of electric telegraphy, because if electrical propagation was a diffusion phenomenon, like the conduction of heat, long-distance signalling might be impractically slow. Wheatstone's apparatus involved the transmission of an electrical pulse through half a mile (0.8 km) of wire. The circuit was interrupted by small gaps between pairs of terminals inserted half-way along the wire and at each end. The terminals were mounted side by side on a board, and as a pulse of current advanced around the circuit, it jumped across each gap with a spark.

The delays between the sparks were indicative of the speed of electricity, but were very short. What Wheatstone did was to examine the sparks' reflection in a small mirror that was rotating at 800 r.p.s. (revolutions per second, Fig. 8.4). Any temporal difference was thereby converted into an angular separation, since the mirror turned slightly during the tiny interval between the sparks, resulting in slightly displaced reflections. Switchgear rotating on the same axle as the mirror synchronized the sparks with the mirror, so that the reflections appeared immobile and continuous because of the persistence of vision. The smearing in the reflected images indicated the duration of the sparks and their relative displacement yielded a value for the speed of electricity.

Fig. 8.4. Overview and detail of the one-inch (25-mm) spinning, polished-steel mirror, E, with which Wheatstone immobilized the reflection of sparks in his 1834 determination of the 'velocity of electricity'. The mirror's rotation was powered via a cord and pulley. The glass plate, R, is an eye shield to prevent injury.

Arago was in Britain for the 1834 Edinburgh meeting of the British Association for the Advancement of Science and may have learned of Wheatstone's mirror then; in any case, he was much taken with the idea, and in 1838, despite ridicule from his fellow Academicians, he suggested using a rotating mirror for a decisive test of the nature of light.[3]

Arago's idea was to compare the speed of light in air with that in water, because here the corpuscular and undulatory theories gave very different explanations of why rays refracted towards the perpendicular when they passed from air into a more refractive medium like water. According to the emission theory, no force acts parallel to the interface, so the component of the corpuscles' velocity parallel to the interface is unaffected by the change of medium. However, the perpendicular velocity is increased because of an attractive impulse given by the water. The total velocity is therefore larger, and in the emissionist view, light travels *faster* in water than in air (Fig. 8.5).

The reverse is the case in the undulatory explanation. Here the luminous disturbance must remain continuous from one point to another. The array of wavefronts deviates when light refracts, but it is not made discontinuous. The geometrical consequence is that the wavefronts are spaced closer together in water than in air (Fig. 8.5). Since the *frequency*, or number of wavefronts passing per second, is unchanged, a smaller wavefront separation means that waves travel more slowly. The velocity of light in water is therefore *less* than in air according to the wave theory. The unequivocally different corpuscular and undulatory predictions could be used in an experiment to test between the two theories.

To test between these competing ideas, Arago conceived of using the rapid flash of light from a vertical electric spark. Light from the lower portion of the spark would pass through a tube of water while light from the upper part would pass through air. The rays would then strike a mirror spinning about a vertical axis, where the slower ray, arriving later, would be reflected through a greater angle. If the air beam were deflected more, light would have passed more slowly through air and be shown to be corpuscular. A greater water-beam deviation would indicate that light was undulatory.

Arago attempted the experiment in 1843, presumably after he had become Director of the Paris Observatory and master of increased resources. To build the equipment he had recourse to Louis Breguet, who was grandson of the famous watchmaker Abraham Breguet (1747–1823).[4]

To increase the deviation, Arago's first idea was to reflect the beam off several spinning mirrors with eight or ten faces. An encircling team of observers was required, since the reflection could be in any direction. Breguet's first apparatus included three spinning mirrors which could attain 1000 or 2000 r.p.s.[5] The last cogs in the drive train were cut with special helicoidal teeth for smoother running and reduced friction. The light was faint after three reflections, so Arago and Breguet tried to triple the speed of a single mirror (Fig. 8.6). Progress was slow because Breguet was busy building equipment for the first electric telegraph in France (from Paris to Rouen), but with the mirror removed, the shaft attained an astounding 8000 r.p.s. It was presumed that air resistance slowed the mirror. A vacuum chamber was built for a new device with a 10-mm diameter mirror, but the mirror spun hardly any faster. At this point diabetes began to dim Arago's eyesight and he ceased to experiment.

Given their connections with Arago, it is natural that Foucault and Fizeau should have begun to ponder how his experiment could be made to work. According to Fizeau, an advance was made when Foucault thought of using a plane mirror to reflect the deviated beam back to the spinning mirror and thence close to the departure point, so obviating the need for an encircling team of observers.[6] Foucault and Arago, however, both claimed this idea came from Bessel, but whatever its origin, the idea was flawed.[7] The problem is that the position of the returned beam depends on the angle at which it reflects off the spinning mirror, as illustrated in Fig. 8.7. Reflection at just one azimuth

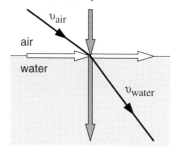

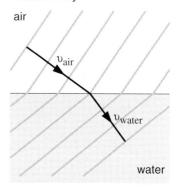

Fig. 8.5. (*Upper*) According to the emission theory, the parallel component (white arrow) of the velocity of the ray of particles (black line) is unchanged when the particles enter water, but the perpendicular component (grey arrows) is increased by the attraction of the water. The total velocity of the particles is therefore *increased* in water. (*Lower*) In the wave theory, wavefronts (grey lines) are deviated but not ruptured when they pass into water. This deviation shortens the separation between wavefronts. Since the same number of wavefronts must pass per second, their reduced separation translates to a *lower* velocity in water.

Fig. 8.6. One of Arago's spinning mirrors, built by Louis Breguet.
(©*Observatoire de Paris*)

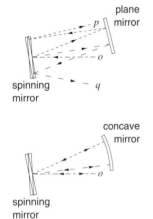

Fig. 8.7. (*Upper*) A ray originating at o is reflected by the spinning mirror. When the ray is returned with a plane mirror, its final destination, p or q, depends on the azimuth of the spinning mirror. (*Lower*) When a concave mirror is used, the ray returns to its point of origin, o, irrespective of where it reflects from the concave mirror. (In both diagrams the spinning mirror has not turned between the outward and return reflections.)

(angle around the vertical) could in principle be selected by narrowing the plane mirror down to a thin strip, but then only an unobservably small quantity of light would return to the experimenter. Foucault and Fizeau abandoned their cogitations in despair.

A light beam across Paris

At this point, Fizeau devised a grand experiment by which he measured the speed of light – in air – directly and absolutely in the physical units of leagues per second. This had nothing to do with whether light was a wave or a particle, but did provide a check on the coherence of the 308 300 km/s value obtained from astronomy.

Fizeau's experiment is represented schematically in Fig. 8.8. The heart of the apparatus was a spinning wheel cut with very fine teeth in its rim. A beam of light was brought into the apparatus by reflection off an inclined glass plate located just in front of the rim. This thin plate cannot be seen in Fig. 8.8 because it lies within the telescope tubing, as does a lens which focused the beam into a tiny spot on the eyepiece side of the rim, where the teeth and the equally sized spaces between them chopped the beam into a series of pulses.

The objective or front lens of the telescope projected the pulses out from Fizeau's home station in a roof lantern in his father's house in Suresnes, west of Paris, towards a second station 8633 metres away in a telegraph building on the Montmartre hills to the north of Paris.[8] There a second telescope objective focused the pulses onto a mirror from which they reflected back along the same path through the two telescopes to form another tiny spot on the rear side of the wheel teeth in Suresnes. Fizeau observed this reflected pinprick of light using an eyepiece focused through the inclined glass plate.

If the wheel was stationary or turning very slowly, as illustrated in the upper left view in Fig. 8.9, the pulse of light transmitted by the gap between any particular pair of teeth would return to the same point before the gap had moved, and a bright spot appeared in the eyepiece. If the wheel was turning faster, however, the adjacent tooth began to move into the position previously occupied by the gap and some of the returning light was blocked, as shown in the upper right view in Fig. 8.9. When the wheel speed was great enough, the tooth exactly filled the gap, completely eclipsing the light (bottom view). At a greater wheel speed yet, the next gap replaced the first one, and light could be seen once more through the eyepiece. At ever greater wheel speeds, there was an alternating succession of transmissions by gaps and eclipses by teeth. From the wheel speeds at which these occurred, the time taken for light to travel the known round-trip distance between Suresnes and Montmartre could be calculated, and hence the speed of light determined.

It took Fizeau only six months to complete a prototype apparatus and demonstrate the practicability of the method. The apparatus was built by

A light beam across Paris

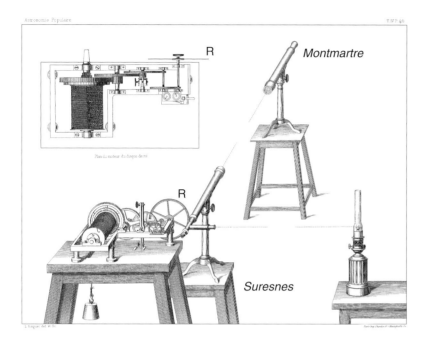

Fig. 8.8. Fizeau's method for measuring the speed of light, taken from Arago's *Astronomie populaire*. A beam of light was chopped into pulses by a rotating toothed wheel, R, in his father's house in Suresnes. The pulses were projected across Paris to Montmartre where they were collected by a telescope and reflected back to Suresnes. The time for the round-trip flight was estimated stroboscopically.

Froment with helicoidal teeth on the final gears (Fig. 8.8). Experiments were carried out in the evening 'when the atmosphere is pure and calm'.[9] A Drummond lamp was the actual luminous source. The occulting wheel carried 720 teeth and the first eclipse occurred when the wheel was turning at 12.6 r.p.s. On 1849 July 23, Fizeau reported to the Academy that based on a series of twenty-eight observations he had found the speed of light to be '70 948 leagues [per second] of 25 to the degree', or in modern terms, 315 300 km/s, close to the astronomically determined value. Sunlight and artificial light were thus found to propagate at essentially the same rate. Fizeau considered that his experiment demonstrated the possibility of measuring the velocity of light terrestrially, but set no great store by the numerical result he had obtained.

When Foucault returned to Paris the following December refreshed after his mental problems, he devoted his first *feuilleton* to his friend's measurement. 'The publication of this work is a great scientific event,' he wrote, 'and the time of its appearance will figure amongst the most memorable dates of history.'[9]

'It is wonderful to hear M. Fizeau outline the advantages of his method himself and tell of the luck he has had,' Foucault continued, concluding, 'That he has sometimes been lucky, I concede, but it is the same luck that supplies rhyme to the poet and discoveries to men of genius.'

Others too were impressed by Fizeau's experiment. For it, he and Froment were made *chevaliers* (knights, the entry grade) of the Légion d'honneur (Legion of Honour).[10] Some years later in 1856 the experiment won Fizeau

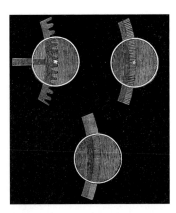

Fig. 8.9. Eyepiece views for Fizeau's 1849 speed-of-light experiment. (*Upper left*) The star-like reflection of light from Suresnes when the wheel is not moving. (*Upper right*) As the wheel speeds up, its teeth begin to move into the light path during the time taken by the beam to make the round trip to Suresnes. This blocks a portion of the returning beam and dims the reflection. (*Below*) At a greater speed, teeth have moved to block the returning beam completely.

the Triennial Prize recently instituted by Napoléon III for 'the work or discovery... which most honours or serves the country'. This prize was no trifle: it was worth 30 000 francs.[11]

The steeple chase

It would seem likely that by the time of Foucault's *feuilleton*, he and Fizeau had returned their thoughts to Arago's experiment. 'To complete the downfall of this poor theory of emission,' Foucault wrote, 'to give it the fatal blow, it was only a matter of performing [Arago's] famous experiment.'[1] Fizeau realized that the experiment would become practicable if the light was returned to the rotating mirror by a spherical mirror rather than a plane one. With the spherical mirror positioned with its centre of curvature on the spinning mirror, the light reflects back along the same path, and hence, if the spinning mirror has not turned, to its point of origin, and this irrespective of the azimuthal angle of the spinning mirror (Fig. 8.7). The returned image has thus been immobilized; in addition, because the spherical mirror can be large, the image is bright and easily measurable. According to Fizeau's protégé Alfred Cornu (1841–1902), '...the day when this solution became apparent was the last of the collaboration; the two eminent physicists separated...' The secondary literature of the time refers to an unspecified falling-out, but why Foucault and Fizeau should have argued is unclear. Perhaps the contention was over the question of how to drive the spinning mirror, because the two experimenters certainly adopted different methods as they raced in a 'steeple chase' (Cornu used the English words) to be the first to complete Arago's experiment. Whatever the cause, Foucault appears to have regretted the argument because in the posthumous inventory of his papers, documents pertaining to the spinning mirror are followed by a 'Letter from Foucault to Fizeau (conciliatory)', suggesting that he tried to repair the breach. The following year Foucault deposited a *pli cacheté*, to be described in the next chapter, which implied that he hoped to work further with Fizeau, but the rupture was irreparable.[12]

To work on Arago's experiment, Foucault 'retired to his laboratory' – no doubt the new house on the Rue d'Assas. His experience with Froment over the arc-light regulator, and the perfection of the toothed wheel and delicate gearing which Froment had cut for Fizeau's speed-of-light experiment convinced him that a professional could build far better equipment than an amateur, so he employed Froment to make the most critical part of the apparatus, the spinning mirror. Fizeau meanwhile worked with Breguet at the Observatory using one of the mirrors built for Arago (Fig. 8.6). The equipment was set up in the Meridian Room on the first floor, where Arago himself had worked (Fig. 8.10). This was an impressive location, where a north–south brass strip laid into the floor divided French east from west, and where Cassini and his successors had made long series of measurements on the meridian passages of Sun and Moon.

The optical arrangement was broadly similar in both sets of apparatus. Foucault's is sketched in Fig. 8.11, which is not to scale, and where only the central ray of each beam is drawn.

Sunlight from a heliostat illuminated a 2-mm square entrance aperture. In its initial form, the aperture was crossed by a vertical grid of eleven fine platinum wires, but later Foucault used only a single wire, and this arrangement will be described since it accords with an engraving which he later published (Fig. 8.12).

Let us consider the air path first with the spinning mirror stationary. Within a certain range of azimuth, this mirror reflects rays from the wire towards the air-path concave mirror, where an image is produced owing to a converging lens placed earlier along the optical path. The concave mirror reflects the rays back towards the platinum wire, where they would refocus, except that Foucault introduced a beam-splitting glass plate near the aperture to reflect this final image into an eyepiece. To emphasize a point already made, because a concave mirror was used, the position of the image in the eyepiece remained the same whatever the azimuth of the spinning mirror, though of course no image appeared if the azimuth of the spinning mirror was outside the range that fed rays to the concave mirror. A ruling in the eyepiece marked the undeviated position of the image (Fig. 8.12a).

When the mirror was spinning, it turned through a certain minuscule angle during the time it took light to make the trip to the concave mirror and back. The final image was therefore shifted slightly sideways in the eyepiece. The size of the deviation depended on how much the spinning mirror had rotated, which in turn depended on the mirror speed and the delay between the outward and returning beams. With such a complicated path, Foucault reported that the

Fig. 8.10. The Meridian Room in the Paris Observatory where many experiments were set up, including Fizeau and Breguet's spinning-mirror experiment in 1850, and one of Foucault's pendulums in early 1851.

(©*Observatoire de Paris*)

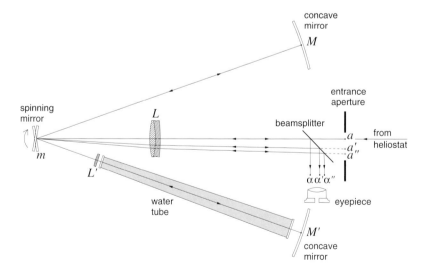

Fig. 8.11. Plan view of the optical layout of Foucault's 1850 rotating mirror experiment. Not to scale. The water tube was 3 m long while the entrance aperture was 2 mm square. The optical system re-imaged a platinum wire at a to point α, where it was examined with an eyepiece. When the mirror, m, was spinning, the time delays in the air and water legs, mM and mM', deviated the image to α' and α'', respectively. Only the central ray in each beam is shown. The lens, L', compensated for refraction by the water.

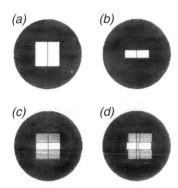

Fig. 8.12. View down the eyepiece in Foucault's air-and-water speed-of-light experiment. (*a*) Air-path image only. A vertical fiducial scratch in the eyepiece is adjusted to coincide with the image of the entrance-aperture wire. (*b*) Air-path image screened narrow. (*c*) Water-path image dimly flanking the screened air-path image. (*d*) The images deviate rightwards from the fiducial position when the mirror spins. The water-path deviation is greater.

principal difficulty was obtaining a sharp image.

The spinning mirror was held in a barrel-like fixture mounted on a spindle (Fig. 8.13). To turn the spindle, Foucault abandoned his beloved clockwork, which he felt was too self-destructive at high speeds and did not allow the mirror speed to be varied in a continuous manner or held constant for sufficiently long. Instead, he adopted the siren devised by the aged Cagniard-Latour (Fig. 8.14), though we must also suspect influence from Foucault's friendship with the hydraulic engineer L. D. Girard, who designed turbines to harness the power of France's long rivers. Foucault adapted the siren into a 24-bladed turbine driven by steam (Figs. 8.13 and 8.15).

A host of practical precautions were necessary for smooth running of the mirror and its supporting armature. A solid body can be constrained to spin about any axis, but the mass distribution about the axis will in general be lop-sided and cause sideways knocking at the bearings.[13] The body needs to be *dynamically balanced*, like a car wheel. Physically, there are special 'principal' axes within a body about which it can spin smoothly without causing any lateral forces and juddering at the bearings. Foucault needed to adjust the mirror armature so that a principal axis aligned with the spindle. 'It is the only tricky part of the machine,' Foucault noted.[14]

The dynamic balance was achieved via a triangular plate and three screws which can be seen immediately above the mirror barrel in Fig. 8.13. Principal axes are mutually perpendicular and pass through the centre of gravity, which is the point about which the stationary body balances and through which its entire weight can be considered to act in many circumstances. Foucault suspended the spindle horizontally by its ends on two inclined glass plates and filed away the triangular balance plate until the spindle showed no tendency to roll, indicating that the spindle's centre of gravity then coincided with its

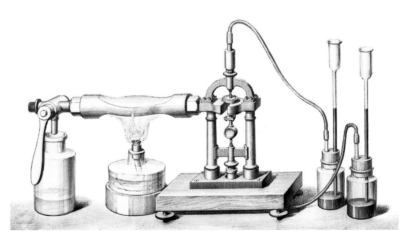

Fig. 8.13. The steam-powered, 14-mm diameter rotating mirror used by Foucault for his air-and-water speed-of-light experiment in 1850.

geometrical axis. The spindle was then *statically balanced*. To obtain dynamic balance, he rotated the principal axes by driving the three screws further into or out of the balance plate. He gauged the sense and amount of adjustment required by adding little balls of wax to and opposite each screw, and observing whether the knocking was improved or worsened. Steam came from a 25-litre boiler and was reheated by a spirit lamp to dry it just before use. Pressure bottles fed oil to the bearings. Screens were necessary to shield the optics from condensed steam, flying oil and hot air currents.

Foucault first saw the image of the wire deviate on 1850 February 17.[15] He will then have known that the experiment was going to work. However, it took a further two months to set up the water-path leg of the equipment, in which the light passed through a 3-m long tube of water. To get a satisfactory final image it was essential that the windows at the end of the tube had accurately parallel sides; luckily there was a supplier of optical plates in Paris, MM. Radiguet and Son. It was also essential that the water was limpid. The tube shed flecks of zinc oxide, so Foucault varnished the interior with copal, a resin obtained from a tropical tree. Distilled water was surprisingly murky because of microorganisms; water from the public supply provided much superior transparency. The final image of the wire was nevertheless very dim – and green – because of absorption by the long column of water. For this reason, both Foucault and Fizeau were forced to operate with sunlight, and to increase throughput, Foucault mounted two glass mirrors in the barrel, back to back. The mirrors were at first made in the conventional way for the time. Their reflecting layers were made of a mixture of tin and mercury, but this amalgam was soft, and at speeds exceeding 200 r.p.s. centrifugal forces broke the layers apart. In a precursor to his later invention of the silvered-glass telescope, Foucault deposited a tough silver layer on the mirror glass using the chemical process 'such as is beginning regularly to be applied commercially', as will be discussed further in Chapter 12.

So as to be able to see the air- and water-path images simultaneously, Foucault masked the air-path concave mirror with a screen pierced by a narrow, horizontal slit (Fig. 8.16). This reduced the height of the air-path image (Fig. 8.12b), allowing the water-path image to be seen dimly flanking it (Fig. 8.12c). The experiment finally worked on April 27, a Saturday. Foucault observed the air- and water-path deviations successively, and then simultanously, as in Fig. 8.12d, where a vertical scratch in the eyepiece marked the position of no deviation. The rightwards displacement of the image of the wire was greater for the water path, as illustrated. Further, the ratio of the two deviations was as expected given the refractive index of water. The emission theory was dead, incontestably incompatible with the experimental results! Within three hours, Foucault had had four others peer into the eyepiece and confirm his result.

Fig. 8.14. Cagniard-Latour's siren. The gears at the top drive counters for the measurement of the rotor speed. The device was named because it could sing underwater, like the sirens of mythology.

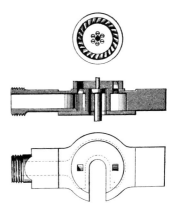

Fig. 8.15. Details of the steam feed and turbine rotor which drove Foucault's spinning mirror.

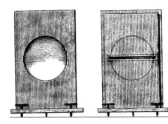

Fig. 8.16. (*Left*) Concave mirror used to return light to the spinning mirror. (*Right*) The air-path mirror screened down to narrow the corresponding image.

The result made public

Two days later Arago must have known that the 'steeple chase' was approaching the finishing line – or that it was over – for at that day's meeting of the Academy he presented an unexpected 'Note on the experiment, suggested in 1838, for deciding definitively between the wave and emission theories'.[7] Primarily the note reiterated his authorship of the method, but, reading between the lines, it also made clear that it was Fizeau, not Foucault, who enjoyed the warmth of Arago's support:

> ...This was how things stood when M. Fizeau determined the speed of light in the atmosphere by a very ingenious experiment. The experiment was not suggested in my Memoir [of 1838], and the author thus had the right to undertake it without the slightest hint of indelicacy.
>
> Concerning the experiment on the relative speeds of light in a liquid and in air, M. Fizeau wrote to me: 'I have made no trials of it, and I shall do so only upon your formal invitation.' This loyal reserve only adds to the esteem which the character and work of M. Fizeau have inspired in me, and I lost no time in authorizing M. Breguet to lend him one of my rotating mirrors.
>
> M. Foucault, whose inventive spirit is known to the Academy, came personally to tell me of the desire he had to submit a modification of my apparatus to experimental test.
>
> With the present state of my sight, I can only give my best wishes to experimentalists who wish to take up my ideas...

Foucault was present at this meeting of the Academy, on the journalists' bench, but he could not interrupt the proceedings to announce his success. After the tribulations concerning priority over the conical pendulum and arc lamp, he was, however, hardly willing to keep the news private until the following Monday, by which time Fizeau and Breguet might have succeeded with their spinning mirror. Instead, he staked his claim in a short letter which was published in the *Journal des Débats* the following day. ('Judge, Sir...of the emotion with which I had to listen to [M. Arago]...')[16] Four days later, on the Saturday, some further paragraphs appeared in the newspaper:

> The Sun having reappeared these last days, the experiment has been repeated on several occasions in the presence of a number of French and foreign savants, and already the procedures which resulted in success are beginning to be known and spread among the public.[17]

The 'foreign savants' were probably those present at the Academy the previous Monday, namely, the die-hard Scottish emissionists Sir David Brewster and Lord Brougham, and Auguste de la Rive from Geneva.

On Monday, May 6, Foucault reported to the Academy.[14] The mirror speed was estimated from the pitch of the knocking of the bearings, but was not accurately determined, which prevented an absolute determination of the velocity of light. With 600–800 r.p.s., the deviations were 0.2 to 0.3 mm. Foucault went on to suggest how to make an absolute measurement and adapt the method to calorific rays using the tiny thermometers devised with Fizeau.

Foucault's memoir was followed by a shorter one from Fizeau and Breguet.[18] They too were operating with a 3-m tube of water but with a shorter, but optically equivalent, 2.25-m air path:

> If the sky had been clear yesterday or today, we would have been able to make the observation and present the result today to the Academy; if our experiments are not yet completed, it is because we waited before beginning them for M. Arago to authorize us to embark on a topic of research which belonged to him.

This last remark is a clear insinuation that Foucault made an ungentlemanly early start by not seeking Arago's approval; indeed Arago's statement the previous week had been that Foucault had come in person to *tell* him what he was doing, not to solicit permission. In the modern scientific ethos, an idea is fair game for everyone once it has been made public, and although this view was developing in the mid-nineteenth century, it was not universally held. Arago was liberal in suggesting his ideas to others, but he was always jealous that their paternity should be acknowledged. Was this the cause of the falling-out between Fizeau and Foucault? Arago had been the inspiration of most of their work together, and quite apart from the ethics of the matter, Fizeau could reasonably feel that his future relations with Arago would be compromised if he continued to collaborate with Foucault after this perceived caddish behaviour. Unfortunately, we are unlikely ever to know for certain, because the papers that could have resolved the matter are lost. In any case, as Cornu wrote, 'Science has nothing to gain from uncovering the reason for this separation, all the more so because the discretion and courtesy of the two rivals has kept it from the public.' Nevertheless, in an attempt to placate the ageing Permanent Secretary, and perhaps Fizeau as well, Foucault concluded his presentation to the Academy with a florid acknowledgment:

> Should physicists favourably receive the fruits of my first labours, all the honour belongs to M. Arago who, in an idea of admirable boldness, showed that questions relating to the speed of light ought to pass from the realm of astronomy into that of physics, and who, by a generous act of abnegation, allowed young scientists to embark on the route which he had pointed out to them.

The spring and early summer of 1850 were indeed dreadful for scientists requiring sunlight (Fig. 8.17), but in fact Fizeau and Breguet had to replace their zinc water tube with a shorter, 2-m one made of crystal before they were able to report success on June 17, seven weeks after Foucault.[19]

Foucault showed off his apparatus to French as well as foreign savants. Mindful no doubt of Dumas' exonerating report concerning his arc light, he

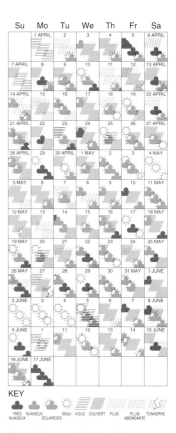

Fig. 8.17. Paris Observatory records showing the awful weather during the spring of 1850. Glimmers of sunlight on April 27 enabled Foucault to show that light travels faster in air than in water. Fizeau and Breguet announced confirmation on June 17.

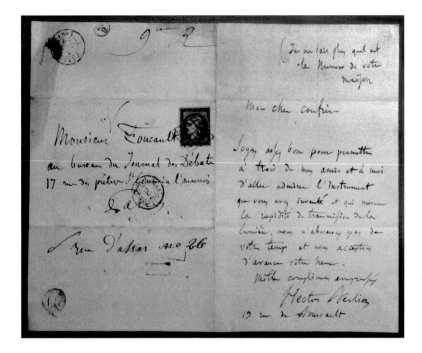

Fig. 8.18. The composer Hector Berlioz wrote asking to see Foucault's air–water speed-of-light experiment. Berlioz sent the note via the *Journal des Débats* and added, 'I don't know what your house number is any more,' because Paris streets were being renumbered.

(Family collection)

invited the busy Minister of Agriculture and Commerce (as Dumas had become) to see the experiment. But on the day when Dumas should have called by, something broke down. 'Since then I have been working non-stop to eliminate all possibility of failure,' Foucault wrote apologetically.[20] Non-scientists wanted to see the image deviate too. Hector Berlioz asked to bring along three friends. 'We shall not waste your time and accept in advance whatever hour you propose,' he wrote (Fig. 8.18).

Docteur ès Sciences

In 1850 December, Foucault obtained the first official recognition of his work. For having delivered the fatal blow to the emission theory, he was appointed *chevalier* in the Légion d'honneur and authorized to sport the order's red ribbon on his lapel. Two years later, in the spring of 1853, he submitted his air-and-water experiment to the Sorbonne for a doctoral degree, the *Doctorat ès Sciences Physiques*. By then he had also authored the pendulum and gyroscope, for which he is far better remembered, so it may seem odd that he did not write his thesis around them, but as we shall see in the next chapter, they were demonstrations of the already-known rotation of the Earth, and in that sense not novel. The spinning-mirror, however, was an experiment with decisive consequences that produced a new result, however expected that result may have been, and as such was much more suitable for a thesis.

Nevertheless, Foucault nearly failed. There were three examiners. Their president was Dumas, to whom Foucault dedicated his thesis in 'testimony of admiration for the scientist and the teacher'.[21] There was Balard, the discoverer of bromine, whose 'youth-like vivacity' will have contrasted with Dumas' gravitas.[22] Finally, there was the physicist Despretz, who had examined Foucault for his *Baccalauréat ès sciences*, though quite possibly neither he nor Foucault remembered this. For the *Doctorat* Foucault could either present theses on two topics, or present one thesis and reply to questions set by the Faculty.[23] He chose the latter option.

Despretz was not especially gifted, but had succeeded through hard work and perseverance. 'His life was regular to excess,' wrote Moigno, and although he was a kind examiner, he was solitary and jealous of the academic success which he felt others enjoyed in his place.[24] Notes have survived of a speech given by Despretz opposing Foucault's fourth candidature for election to the *Académie des Sciences* in 1861.[25] In his speech Despretz recalled Foucault's doctoral examination. He had what might be called examinerly objections, such as Foucault's not having cited prior work, and the use of absolutist words such as 'best' and 'irreconcilable'.

Fig. 8.19. Foucault's doctorate awarded in 1853.

(Family collection)

> I also asked him several other questions while he was describing his apparatus. All politely. He replied to everything; that is, he talked about my remarks, but in reality he said nothing ...[25]

Despretz tried to get Foucault to correct errors in his printed formulae by re-deriving them: he could not;[26] nor were satisfactory replies forthcoming to the Faculty's questions in chemistry. It was then time to vote:

> My two colleagues gave the white ball for the experiment. I gave my red ball for the thesis which I found neither well done nor well defended ... As for the questions, we all three were very indulgent. I should obviously have given him a black ball because he was unable to reply to any of my questions. I did not want to fail him; and of course my two colleagues could not give white balls either ... M. Balard asked M. Chasles [a mathematician present] what his impression was. He replied, 'Most unfavourable.'

Despretz's opinion that Foucault's experiment was poorly defended may be right, but his view that it was not well done is extraordinary.[27] Perhaps this was jealousy at the success of others, but Foucault is unlikely to have endeared himself to his examiner with earlier comments in the *Journal des Débats* that some of Despretz's work was either not new, or was 'patient and insipid drudgery'.[28]

To the reader who has never examined a student, the examiners' red balls may seem equally extraordinary, but pass rates would be abysmal if examiners were never indulgent. Perhaps Foucault made better responses to Dumas' and

Table 8.1. The quantum-mechanical relation between momentum and wavelength for a photon.

$$\text{momentum} = \frac{h}{\lambda}$$

where h = Planck's constant
λ = wavelength associated with photon

Balard's chemistry questions, but since the questions had been in Foucault's hands for over a fortnight, polished answers might reasonably have been expected. We can sense that Foucault will not have liked being examined, that he will have seen it as an affront to his pride and dignity; and that perhaps he will have judged, accurately, that the examiners could hardly fail the author of the celebrated pendulum experiment. Yet he could just as easily have considered his answers to the chemistry questions, which were very standard, as just another *feuilleton* in the *Journal des Débats* requiring the utmost precision and clarity; and a less-egotistical person would certainly have done so.

The wave–particle duality

An eminent Harvard historian of science, the late George Sarton, emphasized the technical audacity and decisive consequences of the air-and-water experiment when he declared, 'The story of the Arago–Foucault experiments will always be one of the most beautiful in the history of mankind.'[29] However, the experiment was not quite the *experimentum crucis* that Arago imagined. It proved that the corpuscular theory as formulated was incompatible with the experimental facts, but it did not follow that the wave theory was compatible with every other fact. Study of the radiation emitted by hot bodies showed that wave theory was inadequate too and led in the first decades of the twentieth century to the development of quantum theory with its undulatory and corpuscular duality. These developments are beyond the scope of this book, but it is of interest to discuss how quantum theory explains the *reduced* speed of light in water when light is considered in terms of photons, the quanta of light, which it must be appreciated have little in common with the emissionists' corpuscles.

In corpuscular terms it was argued that the velocity of the luminous particle parallel to the air–water interface remains constant because no force acts in this direction to change it. This was a slightly loose use of words, because forces actually alter the *momentum* of a particle. For speeds much less than the speed of light, momentum is just the product of mass and velocity, and discussion in terms of velocity or momentum are equivalent. However, momentum must be used when velocities are larger, as for the photon. Quantum mechanics indicates that the momentum of a photon is equal to a constant of nature, Planck's constant, divided by the wavelength associated with the photon (Table 8.1). Recast this way, the corpuscular argument remains valid (Fig. 8.5). The component of the photon momentum perpendicular to the interface does increase as the photon passes into water, as does the total momentum; but the wavelength is thereby *reduced*. Since the frequency is unchanged, the velocity, which equals the product of frequency and wavelength, is lessened too, in accord with the purely wave analysis.

Chapter 9

The rotation of the Earth: pendulum and gyroscope

The world was an advanced place in 1851 when the pendulum experiment made Foucault famous. In the same year, the British held their Great Exhibition in an innovative Crystal Palace of iron and glass, electric telegraph service began between London and Paris, and Herman Melville published *Moby Dick*. Scholars entertained no doubt but that the Earth turned daily on its axis. Why then did Foucault's demonstration of this commonplace fact cause such a sensation in both the learned and everyday worlds?

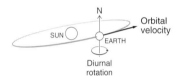

The Earth in motion

First suggestions that the Earth is in motion date back to the Greeks in the fifth century BC.[1] Later, Aristarchos of Samos (*c.* 310–230 BC) provoked accusations of impiety when he conjectured that the Earth might both turn on itself daily and orbit around the Sun annually (Fig. 9.1). However, the Greeks had reasons to favour a stationary Earth. Stars showed no parallax. Other objections were raised by Aristotle (384–322 BC). In Aristotelian physics, objects made of the elements earth and water return to Earth because that is their natural position, just as air and fire rise. Because of this inward tendency of the weighty elements, the Earth must be at the centre of the universe and therefore motionless. This *dynamics*, or physics of motion, was in accord with the observation that an object projected vertically upwards falls back to its point of launch, whereas Aristotle reasoned that if the Earth were rotating towards the east, the projectile would be left behind, falling to Earth far west of its launch point. Clouds and birds would also shear westwards. Aristotle further deduced that the Earth should be spherical because the agglomerated mass must be similar on all sides. Again, this agreed with observation: sailors approaching land saw mountain peaks before the shore because of the Earth's curvature; the shadow of the Earth, seen during eclipses of the Moon, was circular; the

Fig. 9.1. The Earth's most striking motions: its daily (diurnal) rotation and its annual orbit around the Sun. As discussed in Chapter 3, the Earth completes a full 360° rotation once every $23^h56^m04.1^s$. No Foucault pendulum yet built has been accurate enough to be affected by any of the Earth's minor motions: the precession of the equinoxes, nutation, polar wobble, variations in the length of the day, or the orbit about the Galactic centre. In addition, the solar system moves at 370 km/s with respect to the cosmic microwave background. (The microwave background defines a standard of rest with respect to the phenomena of the universe, but in current understanding it is not fundamental with respect to the laws of physics.)

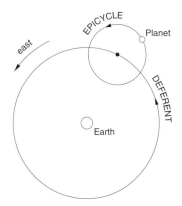

Fig. 9.2. The Ptolemaic system of the world, viewed from the north. Simple circular motion about the Earth was insufficient to explain the complex motion of the planets. Rather, each planet was attached to a circular epicycle which rotated about a point on the circumference of another circle, the deferent. The deferent in turn revolved around the Earth. The whole apparatus slipped eastwards with respect to the westward diurnal motion of the stars. Retrograde or westwards motion with respect to the stars occurred when a planet's epicycle brought it closest to Earth.

elevation of the stars and Sun varied with latitude; and elephants were to be found to both east and west. Such improbable creatures could only occur in a single location, showing that the Earth closed round on itself. (Only this last argument is invalid. African and Indian elephants are different beasts.)

Aristotle's astronomy and physics were immensely influential for almost two millennia because of their coherence and scope, and were the foundation underlying the geocentric world-view of Ptolemy (AD c. 90–c. 168), which represents the summit of classical cosmology. An immobile Earth was surrounded by the spheres of the Moon, Sun and planets. Enveloping the whole was the sphere of the stars, which Aristotle had imagined was made of some weightless, crystalline substance that powered the motion of the inner spheres through rubbing and slipping. To Ptolemy, the reality of the spheres was less clear, and he invoked a system of circular *deferents* and *epicycles* to explain the planets' sometimes backwards path, or retrograde motion, with respect to the stars (Fig. 9.2).

The rejection of geocentric cosmology, and the consequent dethronement of man as the centrepiece of creation, began in the sixteenth century with the Polish cleric and doctor, Nicolas Copernicus (Fig. 9.3). Copernicus outlined a cosmology where the Earth was neither central nor stationary, but had, along with the planets, both an orbital revolution around the Sun and a rotation about its own axis. The Copernican system explained naturally why Mercury and Venus never stray far from the Sun, and why the other planets can exhibit retrograde motion. The planets' orbital periods fell into a pattern, increasing with increased distance from the Sun. But the system was not perfect. Copernicus was still attached to classical ideas of circular motion around deferents and epicycles, and spheres. In particular, Copernicus attached the Earth's axis rigidly to a Sun-centred sphere. The unfortunate consequence was that the terrestrial rotation axis then maintained the same inclination with respect to the Sun as the sphere turned, eliminating the seasons. To counteract this and keep the Earth's axis pointing in the same direction in space, an ugly third motion was needed, which was an annual contrary conical sweep of the terrestrial axis; however, by making the period of this sweep slightly shorter than one year, it was possible to explain a subtle effect whereby the axis about which the stars rotate appears to slip westwards by a small but easily measurable 20 arcseconds per year. This effect, the precession of the equinoxes, was a long-standing unexplained problem, having been discovered in the second century BC by the Greek astronomer Hipparchus.

Copernicus's spheres and epicycles did not last long. Observations of the Great Comet of 1577 showed it had penetrated through the various planetary spheres, casting their existence into doubt. In 1609, Kepler published the first two of his laws of planetary motion. Kepler found that the planets travel in ellipses, with the Sun at one focus. As any given planet orbits, the line from the Sun to the planet sweeps out equal areas in equal times. Deferents and

The Earth in motion

epicycles were no longer needed. Kepler's two laws established the Sun as the main force governing the motion of the planets, and this was highly suggestive of the Sun being immobile at the centre of the planetary system.

At about the same time, geocentric cosmology was further undermined by Galileo, in Padua (Fig. 9.4). First, Galileo provided an alternative to Aristotle's dynamics. He disproved the Aristotelian contention that different weights fall at different speeds, and elaborated the law of uniformly accelerated motion. He defined the law of parabolic flight and grasped the idea of force as a mechanical agent, although he was unable to devise the mathematical relationship between force and motion. Second, his telescopic discovery of the moons of Jupiter in 1610 revealed a Copernican system in miniature, with satellites that were in bigger orbits moving more slowly. A heliocentric planetary system became exceedingly plausible by analogy. Galileo also discovered that Venus showed phases, and at the same time varied in angular size – firm evidence that at least that planet orbited the Sun.

Fig. 9.3. Nicolas Copernicus (1473–1543) made the Sun the centre of the solar system.

Galileo became a convinced Copernican. Partly, no doubt, as a skirmish in the wider religious battle between the Protestant and Catholic Churches, the Catholic Church decided to act against this new doctrine, and in 1616, and again in 1633, proscribed the teaching or belief that the Sun was at the centre of the universe and that the Earth moved around it. Galileo was ordered to recant, and spent the last years of his life under house arrest. Unjust as this seems today, the uncomfortable fact for the Copernicans was that there was no hard, incontrovertible evidence that the Earth moved around the Sun, and not vice versa. As Cardinal Bellarmine, one of Galileo's inquisitors, wrote:

> If there were a real proof...that the Sun does not go round the Earth but the Earth around the Sun, then we should have to proceed with great circumspection in explaining passages of Scripture which appear to teach the contrary, and rather admit that we did not understand them than declare an opinion to be false which is proved to be true.

Evidence suggestive of the Earth's rotation came with the discovery by Cassini in 1665 and 1666 of the rotation of Jupiter and Mars, but again, this was only evidence by analogy. Nevertheless, the Copernican world view was so persuasive that few students of natural science doubted that it was the Earth and not the Sun that moved – though surprisingly, no one in the seventeenth century seems to have considered the possibility that both might be in motion.

At this point it is worth stating explicity that different observational evidence is required to demonstrate the two principal motions of the Earth. For the annual orbit around the Sun, evidence of motion with respect to the distant stars is needed. The annual parallax is one such phenomenon and is purely *kinematic* – that is, it depends only on showing the orbital motion of the Earth relative to the distant stars taken as a reference frame. On the other hand, there was absolutely no doubt that there is a daily *relative* rotation between the Earth

Fig. 9.4. Galileo Galilei (1564–1642). He is reputed to have muttered 'Eppur si muove' ('Yet it does move') when ordered by the Church to recant his deduction that the Earth moves around the Sun.

and the stars – the stars rise and set each day. The question was, which was moving? Copernicus could argue:

> And since the sky is what contains and encompasses everything, the common stage for all that happens, it is not immediately clear why the rotation should be attributed to the container rather than the contents.

But again this argument was only plausible and not conclusive, even when dressed up in more concrete terms of a roast on a spit turning in front of a fire.[2] Kinematics – motion alone – cannot on its own prove the Earth's rotation, in any *absolute* sense; for that, a physical theory – dynamics – is required. Aristotle's dynamics had suggested rotation would deviate a vertically fired projectile towards the west. The translator of Galileo's works into French, Father Marin Mersenne (1588–1648), actually tried firing an arquebus vertically upwards, and later a cannon, but the shot apparently did not fall back to Earth (Fig. P.1). Searching for deviations in freely falling weights was a less hazardous procedure, though in practice just as inconclusive, not least because there was argument over what deviation to expect. Newton had made sense of the relation between force and motion after seeing the apple fall, and introduced dynamical laws based on the ideas of absolute time and an absolute space in which the centre of mass of the solar system was at rest. From these dynamical laws Newton predicted an eastwards deviation. His argument was that being farther from the Earth's rotation axis, the top of a tower would have a greater eastwards velocity than its base, and that this excess velocity would carry a weight dropped from the top eastwards during its fall to the ground. Hooke did not agree with Newton over the nature of light (page 60), and neither did he agree about this deviation. He saw, correctly, that the fall of a weight must be treated as an orbit and predicted an additional strong deflection towards the equator.

There was, however, a subtle flaw in this conclusion. In practice a deviation is measured with respect to a vertical established with a plumb line or level, but just as a conical pendulum needs a transverse force to maintain its swing, so too does every object on Earth in order to sweep out a circular path as the Earth rotates. The required force is effectively subtracted from the local gravity with the result that the local vertical does not pass through the centre of the Earth but leans slightly away from the equator by an angle which depends on the latitude. (In Paris, the angle is close to its maximum of one tenth of a degree.[3]) This deviation of the vertical exactly cancels Hooke's equatorwards deviation, which is thus unobservable in practice. It was not until the early nineteenth century that this error was recognized and correct derivations, which also accounted for air resistance, were produced by Laplace and the German mathematician Karl Gauss. The measurable deviation for Earthbound experimenters is purely eastwards and to reach a mere 10 mm requires a drop of almost 80 metres in Paris.[4] The deviation of the vertical is a centrifugal effect: Newton predicted that there should be others such as reduction of gravity at the Earth's equator along with an equatorial bulge and polar flattening (Fig. 9.5).

Bradley's discovery of the aberration of starlight was characterized by Jean d'Alembert, co-editor of the *Encyclopédie*, as 'The greatest discovery of the eighteenth century' because it established unambiguously and incontrovertibly that the Earth was moving in an orbit around the Sun. Proof of the Earth's rotation followed soon afterwards, with measurements of polar flattening in Lapland by Maupertuis in 1736–37 and of the equatorial bulge by La Condamine and Bouguer in Peru in the 1740s. As already noted, another century passed before the Earth's orbit was further indicated by the first observed stellar parallax in 1838.

There was then absolutely no doubt that the Earth rotated daily and that it orbited annually around the Sun. However, there was still the unfinished business of dropped weights. This was a clear dynamical question where the Earth's rotation should cause an eastwards deviation. Numerous philosophers had attempted the experiment: Hooke in 1679–80, but with a fall of only about 8 m; the Abbott Giambattista Guglielmini from the Asinelli Tower in Bologna in 1791 (78 m); J. F. Benzenberg from the tower of St Michael's church in Hamburg in 1802 (76 m) and in 1804 in a mine shaft in Schlebusch (85 m); and Ferdinand Reich in 1831 in a mine near Freiberg (158 m).

With hindsight, and correctly interpreted, the weights dropped down mineshafts by Benzenberg and Reich probably revealed the rotation of the Earth. But in 1851 *dynamical* proof of the Earth's rotation was still awaited. As Laplace had put it five decades earlier in his *Celestial Mechanics*:

> Although the rotation of the Earth is now established with all the certainty available in the physical sciences, a direct proof of this phenomenon would nevertheless be of interest to mathematicians and astronomers.

The Earth's motion betrayed

In 1849 June, Foucault reported on a theory developed by Babinet in which the rotation of the Earth drove the circulation of ocean currents just as it did the prevailing winds. Foucault concluded his report:

> This dual circulation of air and water plays an immense part in tempering the climate. We can hardly imagine what would become of our land if the Earth [stayed] still as in the Ptolemaic system ... The rotation of the Earth is therefore an important element in the harmony of the world in which we live; and it betrays its presence in phenomena without our needing to cast our eyes on the vault of the sky in order to find fixed reference points.[5]

Further, Foucault and Fizeau had worked together to detect the Earth's orbital motion optically. The underlying conception was that light waves were

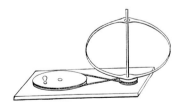

Fig. 9.5. When spun, the spring-metal band bulges at its equator and flattens at its poles in analogy with the Earth. The local vertical (perpendicular to the band) deviates away from the equator.

vibrations of a medium, the luminiferous ether, analogous to the way that sound waves are vibrations of air. The observed velocity and wavelength of light would be affected by motion of the source and observer through the ether in just the same way as for sound, as recently explained by Fizeau at the Société Philomathique. Most simply, the ether was presumed to be at rest, where 'rest' was that of Newton's absolute space. Annual variations were expected in terrestrial experiments because of the Earth's changing direction of motion through the ether as it orbited the Sun, but no such changes had been seen.

On May 27, just a few weeks after the success of his spinning mirror experiment, Foucault deposited a *pli cacheté* at the Academy describing trials made 'in my laboratory':

> The impossibility of noting any aberration phenomenon due to the translation of the Earth other than on the stars led M. Fizeau and myself to the idea that the *ether* is dragged along by ponderable matter ...[6]

The idea that moving matter might drag along the ether was not new. To eliminate any effects in terrestrial laboratories, the dragging needed to be complete, but Fresnel in his wave theory had predicted a *partial* drag depending on refractive index and the velocity with respect to absolute space. To search for their drag, Foucault and Fizeau adopted the 'double-tube' devised some decades earlier by Arago.[7] The apparatus was a simple application of Young's interference, but with the two light beams passing through separate tubes before they interfered. Arago had put humid air in one tube and dry air in the other. The resulting differences in wavelength due to the different refractive indices produced a slight shift of the fringe pattern. Foucault and Fizeau passed oppositely flowing air currents through the two parallel tubes so that the drags would oppose, but did not obtain a convincing fringe shift.[8]

With this background, Foucault's mind was prepared for the idea that the Earth's motion might be 'betrayed' in subtle phenomena.

The germ of the pendulum experiment

According to Louis Figuier, rough seas on a steamboat trip between Honfleur and Le Havre inspired the pendulum experiment when Foucault saw a cross spar remain in a fixed orientation despite the pitching of the ship.[9] This anecdote is further misinformation from Figuier's inaccurate pen, because Foucault himself said that what 'put me on the path' of the pendulum experiment was twanging a round steel rod held in the chuck of a lathe.[10] 'A rod that vibrates seems to expand and lets us see objects, as if by transparency, through a slight shadow,' Foucault noted (Fig. 9.6, *upper*).[11] He rotated the chuck and found that although the rod turned with the chuck, as it must, the plane of vibration did not. This effect still causes surprise (Fig. 9.6, *lower*). It occurs because in the absence of any force which might deviate the vibration, the rod's inertia, or resistance to change in its state of motion, keeps it vibrating in the same plane.

Fig. 9.6. (*Upper*) A uniform rod twanged in a stationary lathe chuck vibrates in a fixed plane. (*Lower*) When the chuck spins, no forces act which can deviate the vibration. The rod's inertia keeps it vibrating in the same plane.

The germ of the pendulum experiment

It is not recorded when Foucault drew the crucial analogy between the chuck and the Earth, and between the rod and a pendulum free to swing in any direction, but draw it he did. He realized that Earthbound observers will see the Earth's rotation reflected in an apparent slow contrary turn of the swing plane of a pendulum. 'The pendulum...', Foucault wrote, is 'one of physics' most precious instruments, one of Galileo's most beautiful conceptions',[12] and it is for this wonderful insight into its motion that posterity remembers him. To adopt nautical parlance usually reserved for winds, a pendulum in the northern hemisphere will *veer*, that is, it will appear to move clockwise (as seen from above) as the Earth turns anticlockwise beneath it. In the southern hemisphere, the situation is reversed and a pendulum will move anticlockwise, or in northern-hemisphere terms, *back*. In what follows, *veer* will be used to mean a clockwise deviation, and *back* an anticlockwise one, irrespective of hemisphere. (The veering of the swing plane is called *precession* by many authors, but this is a misnomer (cf. Fig. 9.33). The motion is technically a rotation of the line of apsides of the bob's orbit.)

Foucault's first pendulum swung in the cellar of the house on the Rue d'Assas (Fig. 5.15). Froment made this and all subsequent pendulums. A substantial piece of cast iron was fixed into the vaulting to provide a solid suspension for a 5-kg brass bob hung on a 2-m steel wire. Though convenient, Foucault's cellar was not the perfect site for this work, because in 1850 the rebuilding of Paris had not yet expelled the small industries and workshops which abounded throughout the city. Foucault explained:

> He who has tried to devote himself to precision experiments rapidly becomes aware of the troublesome, disturbing and provoking influence of vibrations caused by the passage of carriages, or by the hammer blows of a neighbouring blacksmith, or by the regular beat of a steam engine whose pulsing echoes pursue you to the depths of your hard-working retreat. Isolated and solid as it is, [even] our Observatoire de Paris does not escape completely from persecutions of this sort...[13]

We can imagine Foucault persecuted into working in the depths of the night to avoid vibrations, a pot of his beloved coffee at his elbow:

> Friday, [1851 January 3] 1–2 a.m.: first trial, encouraging result; the wire breaks.
> Wednesday, January 8, 2 a.m.: the pendulum turned in the direction of the diurnal motion of the celestial sphere.

Foucault told Arago of his discovery. Lukewarm as he may have been towards Foucault as a person, Arago recognized the demonstration's worth and authorized Foucault to swing his bob with an 11-m wire in the Meridian Room of the Observatory. With the north–south line set into the floor, where else could have been more appropriate?

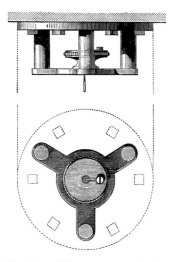

Fig. 9.7. 'The suspension is always a very delicate matter,' wrote Foucault.[14] To avoid fatigue fracture of the wire at its anchor point, the suspension of his first two pendulums probably laid the wire around a curve before mooring it under a screw head.

The suspension may be high up and out of sight, but it is one of the most troublesome issues in the construction of a Foucault pendulum and is never far from the physicist's mind. Access to the suspension is often tricky for a pendulum of substantial length, but in the Meridian Room easy-access locations were available either through the zenithal ceiling 'well' visible in Fig. 8.10, or through a weather-vane hole at the north end of the Room. The most obvious way to grip the wire is to thread it into a hole drilled in a metal block and clamp it with a grub screw. A wire in a close-fitting hole may move only slightly as the pendulum swings, but move it will; and this cyclic working causes metal fatigue and ultimately failure at the place where the stresses are highest, which is where the wire has been crushed by the end of the screw. Perhaps this is why Foucault's wire snapped on January 3. In his *Astronomie populaire* Arago printed an illustration of a better arrangement in which the drilled hole is curved open at the top like the mouth of a trumpet. Laid around this curve, the wire is better-isolated from cyclic flexure at its mooring point under a less-damaging screw head (Fig. 9.7). With this artifice, it seems likely that Foucault had no further problems with breaking wires in his cellar or the Meridian Room.

Foucault announced his discovery at the Academy on February 3 with a written memoir (Fig. 9.8), which was presented by Arago 'in a manner most flattering to the author', according to the reporter for *Le Siècle* newspaper. Complimentary remarks were also made by C. S. Pouillet, who was a physicist well known for his clearly written and widely read textbooks.

Foucault noted that though the pendulum swing plane would appear to veer by 360° per sidereal day at the Earth's poles; elsewhere, the rate is reduced by a factor of the sine of the latitude. (For those who have forgotten, Fig. 9.9 recalls the definition of this trigonometric function.) To derive the sine factor, he wrote, 'one must resort either to analysis or to considerations of mechanics and geometry which are outside the limits of this note...' He was wise to remain non-committal over this point, as we will see.

Fig. 9.8. Beginning of Foucault's memoir in the *Comptes rendus* announcing the pendulum experiment.[10]

> PHYSIQUE. — *Démonstration physique du mouvement de rotation de la terre au moyen du pendule;* par M. L. FOUCAULT.
>
> (Commissaires, MM. Arago, Pouillet, Binet.)
>
> « Les observations si nombreuses et si importantes dont le pendule a été jusqu'ici l'objet, sont surtout relatives à la durée des oscillations; celles que je me propose de faire connaître à l'Académie ont principalement porté sur la direction du plan d'oscillation qui, se déplaçant graduellement, d'orient en occident, fournit un signe sensible du mouvement diurne du globe terrestre.

Foucault went on to note that the veering of his pendulum accorded with the predictions of a memoir read at the Academy in 1837 in which the mathematical physicist Denis Poisson (1781–1840) had considered the rightward azimuthal deviation produced by the Earth's rotation on a horizontally moving projectile in the northern hemisphere.[15] We will discuss this deviation further, but it seemed to Foucault that the pendulum bob could be compared to Poisson's projectile. With a projectile or dropped weight there is but one, small, difficult-to-observe deviation. 'However the pendulum has the advantage of accumulating the effect' and the rightward deviations add up until they are big enough to produce an unquestionable leftward drift for an observer standing outside the pendulum (Fig. 9.10). This accumulation is why Foucault succeeded when so many in towers or mines had failed.

Arago and Foucault decided additional publicity was in order. ' "You are invited to see the Earth turn, in the Meridian Room of the Paris Observatory, tomorrow, from 2 to 3 pm." That is the substance of the note sent last week to several people by M. Léon Foucault,' reported a journalist called Terrien in *Le National* newspaper. He continued, 'At the appointed hour I was there, in the Meridian Room, and I saw the Earth turn.'[16]

The longer pendulum in the Observatory presented several advantages. A Foucault pendulum is subject to various perturbing effects, as will be discussed below, and these are reduced with a longer wire, yielding a purer effect. The veering is also easier to see because a longer pendulum swings more slowly, and during a single swing the Earth has turned more; and further, a given *angular* change results in a greater *linear* displacement with a longer pendulum. To make the veering even easier to appreciate, Foucault attached a downward-pointing stylus to the bottom of the bob and positioned an upward-pointing one on the floor beneath, with their tips initially almost touching. With the slower, more regular oscillation in the Meridian Room, the leftward drift of the bob was apparent in a separation of the tips after a single swing.[17]

The sine factor

Foucault's sine factor and over-elementary explanations have caused misunderstanding and confusion since 1851, so before attempting to understand *why* the Foucault pendulum moves as it does, it is important to develop a clear mental picture of *what* the motion is.

Figure 9.11 shows a table-top pendulum in which the bob is attached to its point of suspension by a wire which, because it is short, is torsionally rigid (i.e. it will not twist) but which can flex in any direction. A bearing underneath the base board allows the whole apparatus to be rotated azimuthally about a vertical axis. If this pendulum is set swinging, the orientation of its plane of oscillation remains fixed even when the apparatus as a whole is rotated. That the wire and bob turn along with the rest of the apparatus can be deduced from

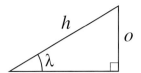

Fig. 9.9. The trigonometrical sine of angle λ (Greek lower-case lambda) is the ratio of the opposite side to the hypotenuse in a right-angle triangle containing λ: $\sin \lambda = o/h$. The value decreases smoothly from 1 for $\lambda = 90°$ to 0 for $\lambda = 0°$, and can be defined for negative angles so that, for example, $\sin(-90°) = -1$.

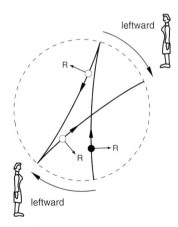

Fig. 9.10. In the northern hemisphere a succession of rightward deviations R of the bob produces a leftward or clockwise drift of the swing plane, as seen by an outside observer. In reality the deviation is much smaller than illustrated.

 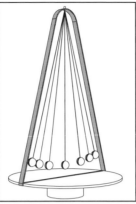 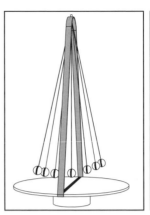 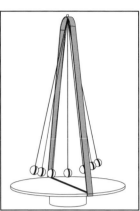

Fig. 9.11. The immaterial swing plane of a table-top pendulum stays fixed even though the material of the pendulum is rotated. (Note that the black lines on bob and baseboard remain parallel.)

Fig. 9.12. The first table-top pendulum. An engraving published by *L'Illustration* on 1851 April 5 to illustrate a reprint of Foucault's own article a week earlier in the *Journal des Débats*.

the black lines painted on the base board and bob, which remain parallel. There is no bearing at the point of suspension, and this is exactly like the vibrating rod, which is gripped firmly by the jaws of the lathe chuck. It is only the immaterial plane of oscillation or vibration that stays fixed. This immobility is only approximate, of course. The azimuthal rotation is quickly made and the table-top demonstration lasts for only a few seconds. If the demonstration was prolonged with scrupulously engineered apparatus, the swing plane would veer very slowly because of the Earth's rotation. It should be emphasized that for unadulterated veering the pendulum must be equally free to swing in any direction. Veering does not occur in a clock pendulum where the suspension constrains the oscillations to a single plane.

The table-top pendulum has been used in popular explanations of the pendulum since Foucault himself used the analogy in the *Journal des Débats* (Fig. 9.12), but it can mislead. The table-top analogy is sufficent to comprehend the motion of a Foucault pendulum at the Earth's poles where the vertical is coincident with the Earth's rotation axis, which, except for the minuscule precession of the equinoxes, maintains a fixed orientation in space. It is irrelevant that this axis slips sideways as the Earth orbits the Sun, or as the solar system moves through space. Its orientation is unaffected, as indicated by the reference frame of the distant stars, for example. The restoring force, which keeps the pendulum swinging, acts towards this unchanging polar vertical. It therefore acts in an unchanging plane despite the Earth's rotation, and this plane is the one in which the bob was set swinging. No force deviates the bob from this swing plane, which stays locked relative to the stars. To an Earthbound polar observer, however, the swing plane appears to veer, taking one sidereal day to complete a full circuit.

For locations away from the poles, the direction of the vertical changes to sweep out a cone in space as the Earth turns and the point of suspension is dragged around its parallel of latitude once per day. The forces on the pendulum are therefore more complicated away from the poles, and this is what gives rise to the sine-of-the-latitude slow-down announced by Foucault. A pendulum located at 30° N, for example, for which the sine of the latitude is one-half, takes not 23^h56^m to veer a complete 360°, but twice this, or 47^h52^m. Times for some other latitudes are listed in Table 9.1. The Foucault pendulum provides an example of what mathematicians call *anholonomy*, where although the environment returns to the initial state (after a 360° rotation), the system does not.[18] Other examples can be found in quantum mechanics.

The motion of a Foucault pendulum swinging at the latitude of Paris is pictured in additional ways in Fig. 9.13. Mathematicians often specify a plane by a line perpendicular to it; hence in Fig. 9.13 the long, light-grey arrow indicates the azimuth of the swing plane. For the Earthbound observer, the swing plane and its indicator arrow turn smoothly clockwise around the horizon at a fixed speed. Since the sine of the latitude of Paris is almost three-fourths, four-thirds of 24 sidereal hours, or almost 32 hours, must elapse for the pendulum to complete a full circuit around the horizon. As it happens, the pendulum's veering speed equals the component of the stars' motion around the horizon.[19] *The pendulum swing plane follows the stars as closely as it can consistent with the constraint of remaining vertical.*

It is also instructive to view the motion in inertial space, i.e. a frame of reference where there are no rotations or other accelerations and Newton's laws of motion hold in pristine simplicity. For practical purposes, the distant stars provide this reference frame for which the stationary celestial sphere is a geometrical representation. To visualize the motion of the swing plane, we can imagine the long, grey arrow as a paintbrush leaving behind a trail on the inside of the celestial sphere. However, this trace alone does not specify the orientation of the swing plane fully. The pendulum vertical might lie anywhere within the plane perpendicular to the paintbrush, so the orientation of the vertical must also be specified, as shown by the white arrow. The white circular trace in the right-hand part of Fig. 9.13 and the grey track between 0 and 1 together show how the orientation of a Foucault pendulum in Paris changes over a single sidereal day. The vertical paints out precisely one loop of a closed circular path, while the perpendicular draws only part of a complicated curlicue track. The points 2 3 4 indicate how the path develops over a further 3 days. If the sine of the latitude of Paris were exactly three-quarters, point 4 would coincide with the starting point 0, and the track would repeat; but as it is, the track scribbles over additional parts of the celestial sphere, like some heavenly spirograph.

The six cusps in the grey track, where the paintbrush has stopped to change direction, give the impression that the plane of vibration flutters and is occasionally stationary in space. This would be a misconception. For the swing plane to be stationary requires simultaneous cusps in *both* defining tracks, but there are never any cusps in the white track because the direction of the vertical never stops changing. The swing plane of Foucault's pendulum moves with-

Table 9.1. Times for a 360° circuit by Foucault's pendulum.

Veering
(Clockwise rotation)

North Pole	90° N	23^h56^m
Fairbanks	65° N	26^h24^m
Edinburgh	56° N	28^h53^m
Paris	49° N	31^h47^m
Los Angeles	34° N	42^h48^m
Bombay	19° N	73^h49^m
Lagos	6.4° N	214^h
Singapore	1.3° N	1070^h

Backing
(Anticlockwise rotation)

Rio de Janeiro	23° S	61^h15^m
Sydney/ Cape Town	34° S	42^h56^m
Christchurch	44° S	34^h45^m
Scott Base	78° S	24^h29^m
South Pole	90° S	23^h56^m

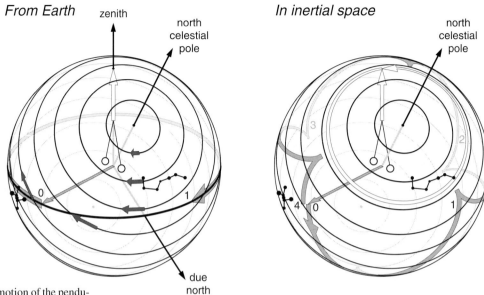

Fig. 9.13. The motion of the pendulum plotted on the sphere of the sky for the latitude of Paris. (*Left*) For the Earthbound observer, the plane of oscillation, indicated by the perpendicular, long, light-grey arrow, veers slowly clockwise. One sidereal day after a start at point 0, the swing plane has veered to point 1, only three-quarters of the full way around the horizon. It takes almost 32^h to complete a full circuit of the horizon. Dark grey arrows represent the motion of the stars and celestial sphere. (*Right*) Viewed from inertial space, the pendulum perpendicular arrow paints out a slow, never-ending curlicue path on the celestial sphere. The track over 4 days is indicated 0 1 2 3 4. To characterize the orientation of the pendulum completely, the zenith must also be specified (white arrow). This direction is fixed relative to an Earthbound observer, but draws out a circle on the celestial sphere of inertial space once per sidereal day.

out pause in its waltz through inertial space, as will have been quite clear to anyone who saw the didactic model displayed to the Academy and Société d'Encouragement in 1851 July by a certain Capitaine Édouard Silvestre.[20] This model (Fig. 9.14) could be adjusted for any northern latitude, and the sine law veering of its tiny pendulum swing plane was obtained via an inclined disc turned by the friction of a horizontal plate. 'By turning the handle,' wrote Foucault, 'one pretty much obtains the spectacle that would be reserved to an observer who, able to isolate himself from the motion of the Earth, set himself at some fixed point in space, and with an attentive gaze followed the absolute and relative motions of the meridian, the vertical and the plane of oscillation.'[21]

The noble walls of the Panthéon

'M. Foucault will not stop there, I hope,' continued Terrien after witnessing the pendulum in the Meridian Room:

> An experiment of this quality deserves to be made publicly, at a more grandiose scale, and with some pomp. It is under the dome of the Panthéon that it would be appropriate to suspend a wire some sixty metres long; those noble walls... are worthy of serving as the setting for one of the most beautiful experiments which has ever come out of the mind of a physicist.

In ancient Greece, pantheons were temples dedicated to all gods. The Panthéon in the heart of Paris's Latin Quarter was the former church dedicated to the city's patron saint, Saint Geneviève, whose prayers in the fifth century AD had reputedly saved Paris from Atilla the Hun (Fig. 9.15). A new church building in the classical style had replaced the old one in the late eighteenth century, and almost immediately, in 1791, it had been transformed by the Revolution's Constitutional Assembly into the *Panthéon français* destined 'to receive the remains of great men'.

The Panthéon was not neutral territory. It symbolized the struggle between anticlerical republicanism and the Catholic faithful. To appease supporters, Napoléon I had returned the edifice to the Church, while Louis-Philippe had restored it to secular use for the same reason. The Second Republic dreamed of making it a Temple of Humanity, and had appointed a painter, Paul Chenavard, to decorate the interior walls. During the June Days of 1848, insurgents made a stronghold there and had been bloodily removed. The building had since remained closed.[22] Only the Prince-President could give approval to install a pendulum in so potentially controversial a location and open it to the public. Foucault explained in the *Journal des Débats*:

> But under the lofty ceilings of certain buildings the phenomenon would take on a magnificent splendour. In the Panthéon we found a marvellously appropriate location for the installation of a gigantic pendulum ... A brief description of the experiment and the results that it would yield for science and put before the eyes of the world was enough for the President of the Republic to resolve that the thing should be done. At lightning speed his high influence flashed to the uppermost rungs of the administration ...[12]

'So that the apparatus would be worthy of the location,' Foucault later wrote, 'I was told that that the setting-up costs would not fall on me.'[23]

Chenavard's plans included a mural entitled 'Galileo imprisoned by the Holy See for having suggested that the Earth was not at the centre of the Universe'. The pendulum was palpable proof of Galileo's world view, and it would be possible to see its installation in the Panthéon as rationality flaunting its triumph over faith in a most provocative location. Terrien certainly made the connection in one of his columns. ('Between the words of the bishop and those of the astronomer, choose,' he wrote.[25] 'M. Foucault proclaims himself a heretic in Galileo's ilk,' wrote another.[26]) Chenavard's planned mural was certainly deliberately provocative, and was roundly condemned in the fervently Catholic newspaper, *L'Univers*: 'The Church of Saint Geneviève seems permanently consecrated to profanity and sacrilege.'[27] Foucault may or may not have been a believer, and he certainly made the odd mocking remark about church candles in the *Journal des Débats,* but he was not anticlerical: he was too familiar with the Church through his religious mother and the jovial Abbé Moigno. It seems the choice of the Panthéon was motivated only by the technical advantages of length and the splendour of the location, and Louis-Napoléon's approval for it derived simply from his affection for science and the majesty

Fig. 9.14. A model devised in 1851 by a Captain Silvestre which can be set to reproduce the veering of the pendulum swing plane at any northern latitude.

(©Musée des arts et métiers-CNAM, Paris/Photo Studio CNAM)

Fig. 9.15. The Panthéon. When this former church was converted to a national sepulchre in the 1790s, windows were blocked up to make the interior more solemn and the frieze above the portico was adorned with the motto *Aux grands hommes, la Patrie reconnaissante* ('To great men, the country's thanks').

Fig. 9.16. Foucault's pendulum in the Panthéon, 1851 March–April. 'As soon as it starts to move, this pendulum in a way belongs to celestial space,' Foucault wrote. The artist has exaggerated the bob for effect; its diameter was actually 17 cm. The engraving was published by *L'Illustration* on 1851 April 5, two or three weeks after public demonstrations began. The Panthéon is sketched schematically. The statue representing Immortality (long-since removed) was actually located in the apse, some 40 metres behind the pendulum.[24]

of the proposed installation. *L'Univers* took no offence: with the pendulum, it thought, the Panthéon 'must be very proud to have at last found an honest occupation'.[28]

The idea of using the Panthéon for what has become one of the world's most famous physics experiments was almost certainly not Terrien's, because on the same day as his article appeared, February 19, permission for Foucault's use was finalized.[29] Did the idea come from Foucault, or Arago, or yet someone else? We cannot say; but the Panthéon figured prominently on the Paris skyline in the 1850s, when the surrounding buildings were much lower, so the choice was obvious enough. In addition to the question of showmanship, a longer Foucault pendulum shows the Earth's rotation more clearly, as we have seen, and the Panthéon was lofty, centrally located and vacant. Nor was it new to use the Panthéon as an experimental site. Some fifteen years earlier, yet another doctor with an interest in physics had made measurements on the vertical from the gallery below the dome; and a few months later, a professor at the Medical School would be authorized to take air samples from the dome.[30]

Foucault's installation in the Panthéon can be judged from the one contemporary engraving of the experiment, published by the weekly *Illustration* (Fig. 9.16). Once more it was Froment who made the equipment, in less than two weeks.[12] Installation must have been similarly rapid, because by March 20 *Le Siècle* was reporting that the demonstation had been running 'for several days'.

The Panthéon offered an ideal location for the suspension. The dome has a triple-shelled structure. The innermost shell has an open oculus at its top to provide sightlines through to the inner face of the middle shell, which is adorned with a fresco by one of Napoléon I's favoured artists, Baron Antoine Gros. A plug at the apex of the middle shell was removed along with a painted cover representing the Sun in Gros' composition. Floorboards and a 40-cm square deal beam were laid across the opening, and from the beam the 1.4-mm diameter suspension wire was hung (Fig. 9.17). The wire was gripped in the same way as for the earlier pendulums.

The bob 67 metres below was a 17-cm diameter brass shell into which molten lead had been poured to produce a total mass of 28 kg (Plate VI).[31] A bob of this size brings dangers. A balustrade stopped the public from interfering with its advance, but equally importantly guarded the public from the bob, which will have had the momentum of a small demolition ball when swinging. The wire was under no more stress than the thinner wires used in the previous experiments, but Foucault clearly anticipated that the wire might break and whip down to lacerate spectators below. Further, the falling bob might shatter the mosaic floor and roll on to break toes and ankles. He spread a 20-cm layer of soil to provide protection. A 6-m diameter scale marked into quarter degrees allowed visitors to measure the pendulum's progress, but for those requiring a more immediate indication, a stylus screwed to the underside of the bob

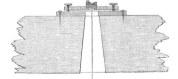

Fig. 9.17. The Panthéon pendulum was suspended from a 40-cm square deal beam laid across the opening at the top of the building's frescoed middle dome.

widened breaches in piles of damp sand by an easily perceptible 2.3 mm at the extremity of each swing (Fig. 9.16). Anyone who has ever scrutinized spectators around a Foucault pendulum will know that their imagination is very directly caught by ever-widening breaches or by the knocking over of pegs. The experiment was no less powerful in 1851, because Foucault commented:

> Every man, whether converted or not to prevailing ideas [about the Earth's rotation] remains thoughtful and silent for a few moments, and generally leaves carrying with him a more insistent and lively appreciation of our unceasing motion in space.

March was a miserable month in 1851: flu had broken out,[32] it snowed, and rain fell so heavily that the Seine rose. Nevertheless, crowds flocked to see the pendulum. Reports are unclear and possibly contradictory, but it would seem that although the first demonstration began at 11 a.m., hours were soon set between ten and noon on Thursdays.[33] Entry was free, and Foucault was present 'many, many times' to explain his experiment.[21] He could be proud and self-important with scientists, but surrounded by laymen (and laywomen) he was at ease: '...those who saw him in action under that magnificent dome were struck by the modesty and simplicity of his demeanour,' Donné remembered later. Amongst those reported to have viewed the pendulum were the British ambassador, pupils from the Lycée Henri IV, and Christian Œrsted, the Danish physicist.[34] This latter claim, at least, is suspect, since Œrsted had died at the beginning of March.

Pendulum mania

'Pendulum mania', as one magazine called it, spread like wildfire.[35] 'The newspapers...have flooded the provinces with details of this important discovery,' wrote a country doctor from near Bordeaux.[36] Savants and home philosophers alike tried to understand the veering theoretically, as will be discussed in the next section, and they also rushed to reproduce it experimentally. Reports appeared in England in the *Literary Gazette* dated March 22, and by April 9 a pendulum was swinging at a *soirée* at the Russell Institution in London's Great Coram Street using a 28-lb bob and a 30-foot wire. 'The experiment excited the astonishment of every beholder,' *The Times* reported, 'and many eminent scientific gentlemen who were present expressed their great delight in witnessing a phenomenon which they considered the most satisfactory they had witnessed in the whole course of their lives.'[37]

By early June a pendulum was under trial in the Radcliffe Library in Oxford.[38] Others swung in Bristol, Dublin, Liverpool and York. In France pendulums were set up in Rennes and in the cathedral in the royal city of Rheims, while others were erected in Rome by the Vatican astronomer Father Angelo Secchi, in Geneva by some of de la Rive's colleagues, in Cologne, in

Florence, in Ghent, in Brussels, in Colombo, in Rio de Janeiro and doubtless many other places. Pendulum mania even extended to the United States, where the experiment was repeated in at least twenty five cities and towns during the summer of 1851 (Fig. 9.18).[39]

Interpreting the motion

Public explanations of the sine factor began at the Academy on the Monday following presentation of Foucault's memoir. First to speak was Jacques Binet, who was professor of mechanics at the École Polytechnique and professor of astronomy at the Collège de France. Binet was an arch Catholic, but clearly felt no impiety was involved in admitting that the Earth turned. He made no lasting contribution to science and is completely unknown today, but along with Arago and Pouillet he formed the commission appointed by the Academy to deliberate on Foucault's discovery. (Cauchy and Babinet were later co-opted to help cope with the flood of papers.)

For Binet, the question was a matter of dynamics, a matter of considering the forces and therefore the accelerations acting on the pendulum, and the results of these accelerations on the bob's trajectory. Newton had understood the relationships between force, acceleration and motion over two centuries earlier, which, simply stated, are that forces cause accelerations, accelerations cause straight-line motions to *change*, and that a given force causes more acceleration in a smaller mass than a bigger one. However, Newton's Laws apply only when motions are measured with respect to an unaccelerated or *inertial* reference frame; that is, with respect to a defining set of axes which are stationary or moving uniformly through space in a straight line (Fig. 9.19). However, on the surface of the Earth we measure Foucault's pendulum and other motions with respect to local, Earth-based axes, such as the directions towards east, north and the zenith. Our axes move uniformly, but in a circle and not a straight line. A body at rest with respect to these non-inertial axes is therefore accelerated, since a centripetal acceleration is needed to make it follow the circular path of the axes. No new forces act if we choose to make dynamical computations with respect to our non-inertial, Earth-based axes, but we need to frame new laws of motion to allow for the fact that we are measuring motions with respect to a rotating reference frame. This is done by adding two extra acceleration terms into the laws of motion; they reflect the acceleration of the reference frame. It is common to represent these accelerations in terms of the forces that would be required to produce them if our non-inertial frame were in fact an inertial one. The introduction of these *fictitious* forces is fraught with potential misunderstandings, so in our non-inertial frame we will talk only of accelerations, which are perfectly real and measurable, being a property of only the motion and the axes against which it is measured.

Fig. 9.18. Pendulum mania in the United States. Location of thirty-nine Foucault pendulums set up between 1851 May and July. The grandest American pendulum was located in the Independence-War monument on Bunker Hill, near Boston, and was only about 3 metres shorter than the one in the Panthéon.

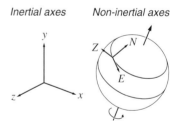

Fig. 9.19. Towards Binet's explanation, (i). The frame defined by unaccelerated axes xyz is inertial. (Such a frame is also called *Galilean*.) A frame defined by accelerating axes such as ENZ is non-inertial. Dynamics calculations in an accelerating frame require additional acceleration terms in Newton's Laws.

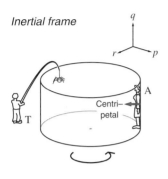

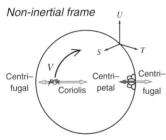

Fig. 9.20. Towards Binet's explanation, (ii). The same events analysed with respect to inertial and non-inertial frames of reference. (*Upper*) Pleasure-seeker A is inside a rotating fairground cage. As analysed in inertial frame pqr, the reaction of the cage on her back provides the centripetal acceleration that causes her to move in a circle with the cage. Zero net force acts on the stationary spider held by tease T. (*Lower*) Analysed in rotating frame STU, the pleasure-seeker is stationary and must experience no net acceleration. An outward centrifugal acceleration therefore balances the inward centripetal acceleration. The centrifugal acceleration also acts on the spider, but a larger inwards Coriolis acceleration acts to produce the net inward acceleration required for the spider to move in a circular path with speed V relative to STU.

The first additional acceleration in a rotating frame is the *centrifugal* acceleration. Viewed from the rotating frame, an object that is stationary with respect to the frame is unaccelerated. The inward centripetal acceleration must therefore be cancelled by an equal, outward, centrifugal acceleration. Thinking in these terms, it is centrifugal acceleration which causes the inclination of the vertical and equatorial bulge encountered earlier.

The second additional acceleration is more subtle and is now called the *Coriolis* acceleration after one of Poisson's pupils, Gustave Gaspard de Coriolis. The Coriolis acceleration accounts for the fact that bodies which are in motion with respect to the rotating axes are in inertial space left behind by the axes' movement. For an anticlockwise rotating frame like the Earth's northern hemisphere, the Coriolis acceleration veers a moving body rightwards, as in Fig. 9.10. For clockwise rotation, as in the southern hemisphere, the Coriolis acceleration acts to the left.

Because the Earth rotates slowly its centrifugal and Coriolis accelerations are small, and we are rarely aware of them. Figure 9.20 presents an example where these accelerations are bigger.

The dynamical equations incorporating the centrifugal and Coriolis accelerations were perfectly well known in 1851. They had been written down by Laplace in his *Mécanique céleste* at the beginning of the century and again by Poisson for his work on projectile motion in 1837. Poisson had even analyzed the effect of the Coriolis acceleration on a swinging pendulum, but failing to appreciate its cumulative nature, had decided that it was 'too small . . . to have a noticeable effect on its movement'. At the Academy, Binet wrote the equations down for a third time, adding rather snootily that he had no goal other 'than to show how M. Foucault's great experiment could have been indicated by the equations of dynamics if interpreted *without inadvertence*'. Through a series of approximations and some rather long-winded mathematics he deduced that the pendulum indeed veers according to a sine-of-the-latitude law.[40]

Next to speak was Joseph Liouville, who had just been elected to replace the exiled book-thief Libri as professor of mathematics at the Collège de France. Much of Liouville's research complemented or extended the results of others, so tackling the pendulum's motion ran true to form. Pendulum veering reflects the rotation of the local horizontal plane about its vertical (Fig. 9.21). Simple trigonometry shows that the component of the Earth's rotation in this direction is proportional to the sine of the latitude (this is the decomposition of polarizations again in different guise). 'The idea is very simple; it must have occurred to everyone,' said Liouville, 'but the developments which I have added form, I think, a mathematical demonstration which is sufficient in itself, and which yield everything that can be got through calculation.'[41]

A third point of view was propounded the following week by Louis Poinsot. Poinsot was a geometer who applied geometrical analyses to mechanics. In particular, he had applied geometry to rotary motion in his *New Theory of the*

Rotation of Bodies published in 1834,[42] producing the idea that the rotational equivalent of a body's mass (its *moment of inertia*) could be represented by an ellipsoid (a three-dimensional ellipse) and that in certain circumstances the rotational motion of the body could be represented by curves on this ellipsoid with such exotic names as polhode and herpolhode. He felt that this geometrical approach yielded more physical insight than the algebra developed by eighteenth century mathematicians such as Leonhard Euler and Louis Lagrange, whose book on mechanics contained not a single diagram. As such, Poinsot was like Foucault, following no master and no tradition.[43] Poinsot's view of the pendulum was as follows:

> ...the phenomenon... depends in its essence neither on gravity nor on any other force. The motion... I say, is a purely geometrical phenomenon and its explanation must be given by geometry alone, as M. Foucault has done, and not at all by the principles of dynamics, which are in no way relevant.[44]

Poinsot went on to propose another device for demonstrating the Earth's rotation involving the release of a folded spring. He claimed Foucault was taken with this proposal and planned to make one; but this was one of the many ideas that scientists toy with but never implement. In any case, we shall see that Poinsot prompted Foucault to build a far superior device to demonstrate the Earth's rotation.

It would be tedious to detail the heated arguments that developed over the ensuing months both in the Academy and around the globe as the learned world debated these different approaches towards explaining the pendulum's motions. Suffice it to say that analytical mechanics received a great boost.

Foucault wisely kept quiet as the world argued over his experiment, but what was this explanation of his 'by geometry alone' that Poinsot had mentioned? Foucault never presented it officially, but in the *Recueil* Gariel published the draft of a letter in which Foucault explained his idea. 'If I have not yet published it,' Foucault wrote, 'I have already talked about it a bit and have discovered that it does not go down well with everyone.' He continued:

> I begin by the brazen statement of the following postulate: when the vertical, which is always contained in the plane of oscillation, changes its direction in space, successive positions of the oscillation plane are determined by the requirement that the angle between them is minimum. Or to put it in common language: when the vertical moves out of the plane of the initial impulse, the plane of oscillation follows while staying as parallel as possible.

This brazen idea is easier to understand through Fig. 9.22, based on a drawing given by Foucault himself. He began by swinging his pendulum along the meridian; this plane of oscillation is shaded light grey. A little later the Earth has turned *anticlockwise* through a small angle, and the pendulum is

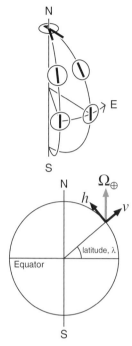

Fig. 9.21. Liouville's explanation. (*Upper*) The pendulum veers because of the rotation of the local horizontal plane about its vertical, represented by circles and direction-indicating thick lines for two successive times at three latitudes. This local rotation is maximal at the pole, but decreases to zero at the equator, where a Foucault pendulum does not veer, as indicated by the parallel thick lines there. The thick lines lie along the meridian, but any compass bearing could have been used. (*Lower*) The argument is expressed mathematically by decomposing the vector $\Omega_\oplus$ (grey arrow) which represents the Earth's total rotation rate into horizontal and vertical components, h and v. Pendulum veering occurs with respect to the component v, which is proportional to the sine of the latitude.

Foucault's explanation

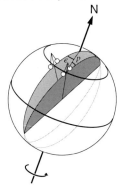

Bertrand's generalization

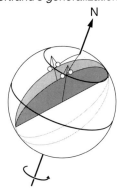

Fig. 9.22. (*Upper*) Foucault's explanation postulated that successive swing planes change their direction as little as possible while still containing the vertical. Simple trigonometry then shows that while the Earth rotates anticlockwise by angle e, the pendulum rotates clockwise by the angle p, which is smaller by a factor of the sine of the latitude. (*Lower*) Foucault only considered an initial swing plane in the meridian. Decades later his friend Joseph Bertrand generalized the geometry to arbitrary initial orientations.[46]

now swinging in the plane shaded darker grey. This new plane of oscillation must of course contain the new vertical, and Foucault's postulate is that this new plane makes the smallest possible angle with the previous plane. For this to be the case, the planes must be disposed symmetrically with respect to the two verticals and intersect 90° away from them. With any other intersection positions, for example with an intersection at the North Pole, the angle between the planes is greater. With respect to the new position of the meridian, the oscillation plane has rotated *clockwise*. Foucault needed the help of a friend out walking in the Luxembourg Gardens for the trigonometric proof that the clockwise rotation of the pendulum swing plane is smaller than the anticlockwise rotation of the Earth by a factor of the sine of the latitude.[45] It is reported that Foucault illustrated his explanations to the Panthéon crowds with a wooden ball: no doubt it was inked with lines similar to those in Fig. 9.22.

As is often the case, some tried to diminish Foucault's achievement. In England, the Reverend Baden Powell, Savilian Professor of Geometry in the University of Oxford and father of the future founder of the Scout movement, dismissively informed an evening meeting of the British Association that Foucault's work 'if fully verified... would however hardly amount to any more *palpable* proof to the *senses* than other astronomical phenomena afford'.[47] Old papers were scrutinized, and it was imagined that pendulum veering had previously been conceived or observed by Italian physicists, by a military engineer in Metz, and in a footnote by a translator of Pliny the Elder's *Natural History*.[48]

Happily, these incomplete, might-have-been discoveries made little impression. It was Foucault who had clearly conceived the effect that the terrestrial rotation would have on a pendulum and had demonstrated it experimentally. The honour was rightly his. 'One will ask, perhaps,' Moigno wrote sarcastically concerning the translator's footnote, 'how a fact of this importance was able to rest infertile for so long. It is because modern physicists only rarely look to Pliny for scientific observations.'[49]

Academicians were not the only ones who attempted to explain the motions of Foucault's pendulum. An artillery officer, a M. Joseph Tardieu, rushed out a brochure with ten pages of text in which, in Foucault's opinion, he 'very clearly' showed how various translational and rotational motions combine to produce veering.[21] Even Foucault's old friend Belfield-Lefèvre produced an explanation in 1852, or perhaps, with his English connections, adapted it from what he had read in a letter to the *The Times* the previous year by one 'J.M.H.' of Westminster.[50] The analysis was geometrical, and we can hardly doubt that it was thoroughly discussed between the two friends. The previous August (1851) Foucault had made a presentation to the Société Philomathique concerning the vibrating rod, which we will discuss shortly. In it he expressed Liouville's argument in a new way:

> The day when M. Foucault produced his pendulum experiment... he also formulated a new Principle of Mechanics which consists in the fix-

ity of the oscillation plane... which is but a consequence of the inertia of matter... When there is at the same time transport of the suspension point and change in the direction of the vertical, the Principle is still maintained, provided that the plane in which the pendulum was launched is taken to be carried along by the change in the direction of the vertical – which it must always contain – but without ever turning around this same vertical. Thus the fixity of the oscillation plane is not an absolute fixity, but a refusal to take any rotation about the vertical.[51]

One might think that a flag pole, for example, 'refuses to take any rotation about the vertical' since its orientation with respect to the Earth is always the same, but Belfield-Lefèvre's paper makes it clear that this is not what was meant. What Belfield-Lefèvre had realized or discovered was that if the tangent cone to the pendulum's parallel of latitude is flattened out (Fig. 9.23), the verticals around the latitude circle become parallel, whereas previously they were all inclined to one another. The swing plane does not rotate with respect to the plane of this *flattened* cone. It does however veer with respect to the meridians, which radiate out from the cone's apex to the latitude circle, and it is a matter of simple geometry to show that this veering follows the sine law.

At first glance, this 'refusal to take any rotation about the vertical' seems as shameless as Foucault's brazen postulate, but it is true that there is no need for a table-top pendulum to be to be centred on the rotation axis of the table (Figs. 9.11 and 9.12), and a pendulum set swinging with its suspension point above the circumference of a smoothly rotating turntable will maintain its plane of oscillation. As such, Belfield-Lefèvre's explanation has more than just mnemonic merit and continues to appear in the literature, in part because it is a simple example of a notion called *parallel transport* which is important in the geometry of curved surfaces and spaces, and plays a key role in General Relativity.[52]

Other accounts of the pendulum were less reliable. Newspaper reports ranged from the obscure to the ludicrous, especially concerning the sine factor,

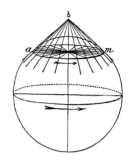

Fig. 9.23. Belfield-Lefèvre's explanation took the tangent cone *bamab* to the pendulum's parallel of latitude, flattened it into the sector *bama'b* of a circle, and then asserted that there is no rotation with respect to this flat sector by the swing plane (indicated by the succession of short lines). This produces the correct sine-law veering with respect to the meridian.

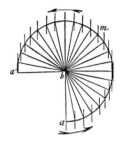

THE ROTATION OF THE EARTH.
To the Editor of "Punch."

"SIR,—Allow me to call your serious and polite attention to the extraordinary phenomenon, demonstrating the rotation of the Earth, which I at this present moment experience, and you yourself or anybody else, I have not the slightest doubt, would be satisfied of, under similar circumstances. Some sceptical and obstinate individuals may doubt that the Earth's motion is visible, but I say from personal observation it's a positive fact.

"I don't care about latitude or longitude, or a vibratory pendulum revolving round the sine of a tangent on a spherical surface, nor axes, nor apsides, nor anything of the sort. That is all rubbish. All I know is, I see the ceiling of this coffee-room going round. I perceive this distinctly with the naked eye—only my sight has been sharpened by a slight stimulant. I write after my sixth go of brandy-and-water, whereof witness my hand, "SWIGGINS."

"*Goose and Gridiron, May 5th*, 1851."

"P. S. Why do two waiters come when I only call one?"

*** We hope our correspondent did not conclude his evening in the station-house.—PUNCH.

Fig. 9.24. *Punch*'s explanation of pendulum veering.

which was sometimes completely omitted, as it continues to be to this day in over-simple accounts.[53] The satirical magazine *Punch* proposed an entirely different explanation involving brandy and water (Fig. 9.24).

To recap, the deviation of the pendulum's swing plane was seen as either a dynamical or a geometrical effect. We will assess the merits of these two viewpoints in the final section of this chapter.

Voting at the Academy

Just days before the pendulum began swinging in the Panthéon, the Académie des Sciences proceeded to fill the vacancy in its physics section created by the death the previous May of the physicist and chemist Louis Gay-Lussac. In its wisdom, the physics section proposed an astonishing *ten* candidates.[54] One was the siren maker, Cagniard-Latour, who at seventy-three was slightly older than the deceased Gay-Lussac; but the others were all much younger men, and for most of them this was their first attempt at election to the Academy. It was *de rigueur* for candidates to call on the Academicians to request their votes with humble words and a printed summary of their work. Foucault certainly prepared a seventeen-page publication list (Fig. 9.25) but we can sense that he would have hated soliciting votes, especially from Academicians who were, in his eyes, buffoons.

With an embarrassment of choice, the Academy took three rounds of voting to whittle the ten candidates down to four, with Cagniard-Latour at their head followed closely by Foucault.[55] In a final ballot, Academicians who had been voting for younger men reverted to what was becoming tradition and elected the senior candidate, Cagniard-Latour, for whom this was his fourth candidacy (Table 9.2).

It is difficult to gauge whether Foucault will have been heartened or dismayed by this result. Despite his solid credentials with the speed of light and the pendulum, there was realistically little chance in 1851 that the Academy would elect a thirty-year-old first-timer. The average age of election was drifting steadily upwards. From a low of thirty-six years in the 1760s, it had risen to forty-three by 1850, and would rise to well over sixty during the subsequent century.[56] 'The Academy has rewarded M. Foucault's zeal and skill by awarding him first place amongst the novices – if I may be pardoned this expression from racing,' Moigno wrote soothingly in *La Presse*,[57] and in the final ballot Foucault had polled very respectably with over a third of the votes. Terrien however was scathing in *Le National*:

> And then a young man appears... will he be given the cold shoulder on the pretext that his whiskers are still soft? Note that I am not talking about a candidate with no qualifications who is pleading only his as-yet unfruitful youth... No, I am speaking about a serious candidate, a young man such as were M. Biot, M. Arago, M. Regnault, M. Liouville, and so

Fig. 9.25. Aspiring Academicians handed out publication summaries as part of their campaign for votes.

many others, when the Academy made them one of its own ... So why change the system? Why not elect the savant who is most active and most skilful ... ?[58]

Foucault must have looked back wistfully on the earlier and more radical decades when youth was less of a handicap, but no doubt he was confident that his hour would soon arrive.

Practical difficulties

'The [pendulum] experiment is so simple that the least scientific of your readers can try it,' wrote a certain 'H.C.' in a letter to *The Times*,[59] but the simplicity is deceptive. At middle latitudes a single revolution of the swing plane takes over 100 000 s, during which time a pendulum of, say, 10-m length (6.4-s swing period) will make nearly 20 000 swings. The veering effect is very small – of the order of 1 part in 20 000 – so precision of construction is essential for it to be revealed.

How well did Foucault's pendulums perform? He left no clear statement, but we can deduce that all was not plain sailing when he wrote:

> Passing from theory to practice, the physicist must expect disappointments; and, in the present case, he must think himself very happy if with a real pendulum he is able to obtain an unequivocal deviation in the expected direction.[12]

From today's fuller understanding of physics we can infer his difficulties.

A first problem that Foucault will have encountered is damping of the swing due to air resistance. This has no direct effect on the veering, but it does reduce the drama of the effect and requires the spectator to sight along scales to appreciate that the pendulum has veered.

Scientists had long been studying the effect of damping on the period of the pendulum because of the period's importance in geodesy and timekeeping, but relatively little attention had focused on the decay of the amplitude. Newton had considered drag in terms of the momentum imparted by simple collision with the air, which results in resistance proportional to the density of the air and the square of its speed relative to the object. At lower speeds and for smaller sizes, however, drag results from the internal friction (or viscosity) of the air. In the year prior to Foucault's experiment, G. G. Stokes had published a seminal paper on the effect of viscosity on pendulums, but even if Foucault had been aware of this work, he would not have read the paper, which was highly mathematical.[60]

Figure 9.26 indicates how Foucault's oscillations died away. His bobs operated in the momentum-drag regime, but the drag on the wires resulted principally from viscosity. The momentum drag on the bob dominated the damping

Table 9.2. Votes cast in the election to the Academy's physics section, 1851 March 17. Foucault polled well, but ultimately the Academicians favoured tradition and time-serving and elected the aged Baron Charles Cagniard-Latour.

	Age	Votes
First round		
Cagniard-Latour	73	12
Ed. Becquerel	30	11
Foucault	31	11
Bravais	39	8
Fizeau	31	8
de la Provostaye	39	1
Wertheim	36	1
Jamin	32	0
Masson	44	0
Verdet	27	0
Second round		
Cagniard-Latour		18
Ed. Becquerel		11
Foucault		11
Bravais		10
Fizeau		4
Third round		
Cagniard-Latour		19
Foucault		14
Bravais		12
Ed. Becquerel		9
Ballot		
Cagniard-Latour		34
Foucault		19

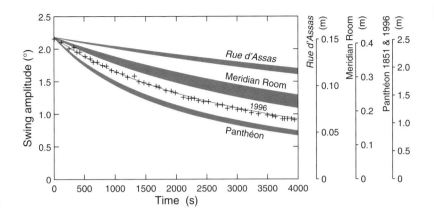

Fig. 9.26. Calculated amplitude decay for Foucault's pendulums in 1851, assuming they were all swung with the same initial 2.2° amplitude. Scales to the right convert angular amplitudes to metres. The widths of the traces indicate the ranges expected due to days of different air density and the uncertainty concerning the wire diameter in his cellar and Meridian Room, which Foucault specified only as varying from 0.6 to 1.1 mm. The reliability of the calculations can be judged from the line and crosses marked '1996' which respectively indicate the predicted and observed damping of the 47-kg bob in the most recent re-enactment of the pendulum experiment in the Panthéon. Amplitudes refer to the centre-to-extremity difference. Peak-to-peak differences are twice as big.

of the 2-m pendulum in Foucault's cellar, but longer wires increase the viscous loss, which dominated in the Meridian Room and the Panthéon. Foucault will have been unaware of these details, but he will surely have expected greater damping in the Panthéon, where his wire presented almost four times more cross section to the air than the bob. By his own account, the bob's stylus only cut breaches in the sand piles for the first quarter of an hour.

It has been claimed that a longer pendulum is less affected by damping, but as Fig. 9.26 shows, this is untrue.[61] Damping is the one issue for which a longer pendulum is not a better one. In early May Foucault sent a note to the Minister of Public Instruction asking for payment of Froment's bill of 450 francs, which had recently arrived.[62] He requested a further 500 francs to replace the 'provisional' 28-kg bob with one weighing 100 kg, which by providing over three times as much energy would have reduced the damping rate by a similar factor (Fig. 9.27). The letter languished unanswered.

'At first I had to let the crowd pass through', Foucault later wrote, and no doubt he frequently restarted the pendulum so that the public could see fresh breaches grow in the sand piles; but once the curious had abated he was able to let the pendulum swing for longer. He discovered that after a while the straight-line motion of the bob would sometimes deteriorate into a loop, or 'elliptical deformations', as he phrased it.

Looping arises if the pendulum is in any measure lopsided or if the point of suspension flexes more in one direction than another. To use the technical term, the pendulum is then *anisotropic*, and its effective length and therefore its period is different for different swing directions.

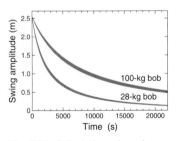

Fig. 9.27. Calculations show damping of Foucault's proposed 100-kg bob in the Panthéon would have been much slower than for the 28-kg bob actually used.

Foucault clearly suspected that anisotropy might be the cause of the looping because he investigated the problem, presenting his findings to the Société Philomathique at the start of August. He did not experiment with a pendulum, but rather reversed the logic of his initial discovery and used a stationary lathe chuck to grip a vibrating rod that was 'nearly round, such as is commercially

Practical difficulties

available.' Twanging 20-cm of this slightly anisotropic, 2-mm diameter rod, Foucault could not follow the instantaneous position of the rod with his eyes, but persistence of vision allowed him to perceive the overall path described by its tip (Fig. 9.28). He found 'ninety-nine times out of a hundred' that the rod did not stay fixed in the initial oscillation plane but swept out 'an incessantly changing elliptical curve'. The backing ellipse widened to a maximum and then narrowed down to a linear vibration in a new direction, from which the motion reversed to bring the vibration back to the initial direction, and so on. In one case in a hundred, however, Foucault found that the vibrations remained fixed in their initial plane, and that for a given rod there were always two such planes, at right angles, which he called 'planes of stable vibration'.

It may seem surprising that there should always be two perpendicular planes of stable vibration for an anisotropic rod or pendulum, whose lopsidedness can be completely arbitrary and possess neither 90° nor 180° symmetry. Foucault proffered no explanation for this, but it is true for the same reason that principal axes for rotation are perpendicular. The effect of pendulum anisotropy is analogous to putting signals with slightly different frequencies on the horizontal and vertical deflection plates of an oscilloscope. The resulting traces (Fig. 9.29) are now called Lissajous figures after Foucault's friend Jules Lissajous (Fig. 9.30), a physics teacher at the Lycée Saint Louis in Paris, who was to produce similar figures a few years later when studying sound by reflecting light off vibrating tuning forks.

A long pendulum reduces anisotropies of length. However accurately made, a real pendulum will *always* be anisotropic to some extent. It will therefore inevitably have two perpendicular planes of stable vibration and the bob's path will always tend to open up into a succession of ellipses which veer and back. However, three decades were to pass before this was clearly understood. The Dutch physicist Heike Kamerling Onnes (Fig. 9.31) is remembered for having liquefied helium and the discovery of superconductivity, but in his 1879 doctoral dissertation at Groningen, *Nieuwe bewijzen voor de aswenteling*

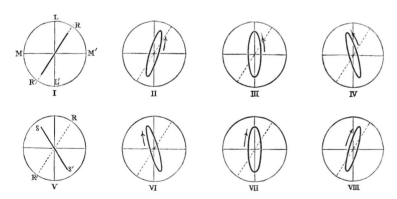

Fig. 9.28. Foucault attempted to understand the elliptical motion of his pendulum by studying the vibrations of the free end of a 'nearly round' rod that was clamped at the other end. Plane vibrations persist in only two mutually perpendicular directions, MM' and LL'. Plane vibrations initiated in any other direction such as RR' degrade into a moving ellipse, which widens to a maximum and then narrows to a plane vibration SS'. The motion then reverses. The planes SS' and RR' are always symmetrically positioned with respect to MM' and LL'. The arrows indicate the direction of motion of the ellipse. Comparison with the lower right-hand trace in Fig. 9.29 shows that the rod moves in the opposite sense.

20 per cent period difference

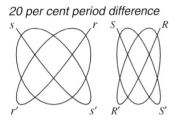

4 per cent period difference

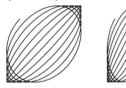

Fig. 9.29. Oscilloscope analogy of pendulum anisotropy. The upper traces illustrate the path of the oscilloscope spot (or pendulum bob) when there is a 20% difference between the oscillation periods of the horizontal and vertical deflections. Starting from r (or R) the spot loops clockwise to s (or S). It then returns anticlockwise back along the same path to r (or R). The cycle repeats. The lower traces show a smaller period difference for which the path is more obviously a succession of slowly varying near ellipses. If the two oscillation amplitudes are equal (left-hand traces), the ellipse axes remain fixed at 45° to the horizontal and vertical (or the planes of stable vibration) but their proportions vary over time. For unequal amplitudes (right-hand traces), the axes defining the ellipse back as the spot (or bob) moves from R to S, and then veer as it returns to R. The analogy is evident between $RR'SS'$ and $RR'SS'$ in Fig. 9.28. A Foucault pendulum must be made sufficiently isotropic that its veering or backing due to anisotropy is much slower than the Earth-rotation veering.

der aarde (New proofs for the axial rotation of the Earth) he was the first to produce a sophisticated theory of the Foucault pendulum which explained its deterioration into looping motion.[63] The topic had been suggested by Kirchhoff. Onnes built a pendulum out of a rigid rod supported on knife edges, just like Foucault's conical pendulum (Chapter 7). With clear mathematical insight into their cause, Onnes was able to study his pendulum's loops and from them deduce what adjusting weights were needed to eliminate the anisotropy.

Foucault would have been incapable of Kamerling Onnes' mathematical analysis, but despite his powerful mechanical intuition, it is unclear whether he understood the origin of pendulum looping even qualitatively. He certainly knew that he needed to make his pendulums as symmetrical as possible, hammering his bobs so that their centre of gravity coincided with their geometrical centre in just the same way as for his spinning mirror, and suspending his Panthéon pendulum from a stout beam that will have flexed by only a few tens of microns as the pendulum swung. However, he followed another tack in his presentation to the Société Philomathique. 'Let us now loosen the clamps which lock the lathe chuck,' he told the Société, 'let us give it back its mobility and, foot on the [lathe] pedal, let us set it turning.' That linear vibrations were unaffected by the lathe rotation was what had led Foucault to the pendulum experiment in the first place, of course, but he was surprised by another property of inertia. Whatever the rod's vibration — linear, elliptical or circular — it was frozen in this path and ceased to evolve when he turned the chuck at more than a couple of revolutions per second. He drew a moral concerning the suspension wire in the Panthéon. 'Supposing that this wire was in some way implicated in the elliptical deformations which were sometimes observed in the oscillations of this vast pendulum,' he continued, 'one could have eliminated it instantaneously by working in a plane of stable oscillation; even better, one could have eliminated it completely by rotating the suspension point, as in the vibrating rod experiment.'

Though theoretically correct, this was hardly a practical proposition, as Foucault himself must have realized in due course, because in his fourth and final pendulum four years later he did not rotate the suspension.

Anisotropy on its own causes veering and backing additional to the Earth-rotation veering, and no doubt was the cause behind the unexpected behaviour of pendulums such as the one set up by de la Rive's colleagues in Geneva, which veered at different rates in different compass directions.[64] Even if a pendulum is perfectly isotropic, however, its swing plane will veer if it is launched into an elliptical orbit because the swing period also depends — very slightly — on its amplitude. We have already met this anisochronism in Foucault's conical pendulum (Fig. 7.6), but here the ellipses in question are thin rather than fat (Fig. 9.32). The anisochronic deviation presents analogies with an effect called apsidal motion in binary stars, and the astronomer George Airy was one of several scientists who analysed the effect mathematically during 1851.[65]

An ellipsing pendulum is not stationary at the extremity of its path, but has a small tangential velocity there, and vice versa. To launch a bob into an ellipse-free trajectory it must therefore be set free from rest. To do this, Foucault tied the bob back with a loop of cotton, and when all twisting, rocking and bouncing of the bob had died away, he burnt through the thread (Plate VII). At the time, it seems many scientists had difficulty in making the distinction between rest with respect to Earth and with respect to inertial space, but with the cotton thread the bob was lauched from rest with respect to the non-inertial frame of the rotating Earth, and hence with some tangential motion in inertial space.[66] This results in some ellipsing and consequent backing, but happily the effect is negligible, in the Panthéon reducing the total veering by only about 0.05 per cent.

Unlike the anisotropic deviation, the anisochronic deviation has no particular relation to any particular mechanical axes of the apparatus and the ellipse always deviates in the same sense as the bob is orbiting (Fig. 9.32). The deviation rate is proportional to the area of the ellipse and inversely proportional to the five-halves power of the length of the pendulum. Whether ellipsing is caused by careless launch or anisotropies of construction, the anisochronism produces its own deviation. If anisotropic deviation is present, the evolution of the bob's orbit is not necessarily what would be expected from simple consideration of Lissajous figures alone because anisotropy and anisochronism couple together. The anisochronic deviation swings the bob path to new orientations, while the anisotropic deviation changes the dimensions of the ellipse, altering the rate of anisochronic change. In addition, the ellipse is fattened by air resistance, which damps the long axis proportionately more than the short one. The motion of a Foucault pendulum can thus be very complicated to understand.

Because of their dependence on length, the consequences of ellipsing and anisotropy are reduced for a long pendulum. For a pendulum swinging in the Panthéon with an amplitude of ±2.5 m, the ellipse minor axis can be as large as ±340 mm before the deviation due to ellipsing equals the veer due to the Earth's rotation. For six-times shorter pendulums swinging through the same angle, such as are more regularly seen in science museums, the constraint is some fifteen times more stringent for middle latitudes, and in practice it is easy for a real Foucault pendulum to deviate at any speed in either direction. As the colourful mathematician J. J. Sylvester wrote in a letter to *The Times* in 1851:

> It is perfectly absurd for persons unacquainted with mechanical and geometrical science to presume to make the experiment. Indeed, such efforts deserve rather the name of conjuring rather than experiment; but in this, as in many matters of life, it is true that 'fools rush in where angels fear to tread'.[67]

Readers are advised to scrutinize purported Foucault pendulums very carefully to determine whether their veering really results from the rotation of the Earth!

Fig. 9.30. Jules Lissajous (1822–80). Inspired by Foucault's spinning mirror experiment and aware of the patterns obtained with the vibrating rod, he studied acoustic vibrations through similar traces formed by light reflected off the prongs of vibrating tuning forks.

(Bibliothèque nationale de France)

Fig. 9.31. It was not until 1879 that the Dutch physicist Heike Kamerling Onnes (1853–1926) clearly stated why real Foucault pendulums deteriorate into elliptical orbits.

Fig. 9.32. Even if a pendulum is isotropic, an ellipsing bob deviates because the swing period is slightly greater for a large amplitude a than for a small amplitude b. The deviation Δ (upper-case Greek delta) occurs in the same sense as the ellipsing, which can be either clockwise or anticlockwise. The deviation is proportional to the product ab and can easily mask the slow clockwise veering due to the Earth's rotation.

Because length minimizes the consequences of anisotropy and ellipsing, Foucault's Panthéon pendulum must have performed reasonably well. Because of the rapid damping, the pendulum's longer-term progress had to be measured against the graduations on the central table (Fig. 9.16). 'By this means,' Foucault told his readers in the *Journal des Débats*, 'the plane of the oscillations, small as they are, can be followed in its apparent movement for five or six hours.' By this time, the oscillations would have had an amplitude of only about ±130 mm (Fig. 9.27).

Inextricable loops

In early May, Foucault's worries over the suspension proved justified:

> We know what accident came to interrupt the run of observations that were being made in the Panthéon. Through having had to operate for everyone from morning to evening, the immense pendulum ended up becoming detached; its suspension wire, 67 metres long, broke at its highest point, and coiled up over the bob in inextricable loops.[21]

Fatigue having been eliminated at the mooring screw, the wire had probably fatigued where it flexed against the lower edge of the suspension plate (Fig. 9.17). The wire will have bent in a small radius and the internal stresses will have been great. Metal fatigue can be very rapid, as anyone who has ever flexed a paper clip knows, and plausibly caused the wire to rupture after only a few thousand oscillations.

The rupture of the wire almost certainly curtailed the pendulum demonstration. Froment's bill was still unpaid, and the wire had not been replaced in early July when Foucault wrote again to the Minister of Public Instruction, this time at greater length and more forcefully, dropping the Prince-President's name several times, and pointing out that foreign vistors 'who, in this season, flock to our capital' were disappointed not to be able to see the famous experiment. His thinking had also evolved. Instead of a more-massive bob, he asked for 500 francs to build a motor to drive the pendulum and keep its oscillations at a constant size.[23] Meanwhile, he told his readers in the *Journal des Débats* that re-establishment of the pendulum awaited 'those who dispense the public purse' to 'deign' to act. This display of irritation was perhaps counter-productive, because the Minister finally only appropriated 300 francs towards Froment's bill, in November.

Any possibility of re-installing the pendulum was extinguished by Louis-Napoléon's *coup d'état*, on December 2. Because it was a potential focal point for dissent, the Panthéon was occupied by troops,[68] while on December 6 Louis-Napoléon decreed that it would revert to the Church as part of a package of measures designed to consolidate Catholic support for his new regime.[69]

A handsome gift

Precision equipment is costly, and Foucault had spent large on his pendulums and more especially his spinning mirror. His next step was undoubtedly facilitated by Louis-Napoléon who in 1852 February gave him the considerable sum of 10 000 francs.[70] Without doubt this largess derived in part from the Prince-President's love of science. (He was later to support Sainte-Claire Deville, Pasteur, and even impoverished geology students.[71]) The gift compensated generously for the unpaid pendulum expenses. Doubtless it also bought some positive publicity for the controversial new master of France.

Devising the gyroscope

Among the public, many had great difficulty comprehending the sine factor in the pendulum's motion. In *The Times*, J. J. Sylvester continued:

> [there are] many persons exceedingly perplexed to follow in their minds this relative motion at intermediate points, and who think it a hard tax upon their faith to believe that when the earth has gone once fairly round the *status quo* as between the horizon and the plane of vibration should not be restored.[72]

Foucault's pendulum had been a great novelty in both theory and practice, but the *idea* of his next device, the gyroscope, was not. For a fully direct demonstration of the terrestrial rotation, what was needed, as Louis Poinsot had pointed out, was some object or direction which 'stays fixed in absolute space'.[44] Poinsot and Foucault were kindred spirits, independent of mind and united by a love of clear exposition. Foucault had commented favourably when Poinsot had been one of the few Academicians to oppose study of an obviously fraudulent teenage girl who was supposedly endowed with electromagnetic powers,[73] and Foucault with his squint doubtless felt a bodily empathy for Poinsot, who had a glass eye. The two men discussed how to obtain the required fixed orientation,[74] but it needed no great leap of imagination to realize that it could be provided by the rotation axis of a mass spinning freely about a principal axis, because if no forces act, nothing can change. The Earth itself provided a well-known example. The orientation of its rotation axis is nearly stationary, and would continue to point indefinitely toward the Pole Star were it not for the slow, 26 000-year precession of the equinoxes discovered by the Greeks. However, the cause of this precession was understood; it had been explained by Newton. The Earth is not in fact completely free-spinning. The Moon and Sun produce twists, or torques, on its equatorial bulge, and these cause the precession. Newton further predicted an even more subtle variation in the Earth's orientation, a wobble called nutation, which was discovered observationally by Bradley in 1748, two decades after his discovery of stellar aberration. Precession and nutation are easily seen in a children's top for the

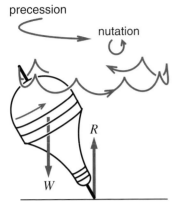

Fig. 9.33. The sweep of precession and the generally smaller, looping wobble of nutation arise when a spinning body is acted on by an external twisting force or torque. For a top, this torque is provided by the top's weight, W, and the ground's reaction, R. The direction of precession is perpendicular to the plane containing W and R, and is reversed if the top spins in the opposite sense. If W and R align there is no torque. There is then no precession or nutation, and the spin axis remains stationary in inertial space. Foucault's genius for experiment enabled him to achieve this alignment in his gyroscope. Note that because of friction with the ground, the long-term motion of a top is more complex than illustrated. The top will move into the vertical (or *sleep*) before slowing and toppling over.

reason that it too is not freely spinning, because its weight and the reaction force produce a torque which tends to make it topple over (Fig. 9.33).

In the eighteenth century, experiments had been made with spinning discs to provide artificial horizons at sea, while instrument makers had constructed various devices involving spinning discs to illustrate precession and nutation.[75] The best known of these had been devised in Tübingen in 1819 or so by Johann Bohnenberger, and introduced into France by Arago.[76] It consisted of a spinning sphere mounted in gimbals so that its axis could turn in any direction. A similar instrument called the rotascope had been produced in Philadelphia in the 1830s. The use of such devices to reveal the Earth's rotation had even been suggested in a lecture to the Royal Scottish Society of Arts in 1836, but had not been published.[77]

A freely mounted spinning rotor will indicate the terrestrial rotation through the same, slow motion of its spin axis as the exactly equivalent tracking of an equatorial telescope (Fig. 3.7), but none of these previous instruments was of sufficiently accurate construction. Their rotor's weight and the reaction force from the gimbals were always slightly offset, producing a torque which caused a precession that masked the Earth's rotation. The challenge that Foucault faced was to mount a rotor so accurately that there was neither torque nor precession, but only an apparent motion due to the Earth's rotation. 'But where to find an instrument maker, if you please, disposed to accept such a task?' Foucault asked. 'We would have despaired of success if we had not long-known M. Froment...'[78]

Plates VIII & IX and Fig. 9.34 show the finished device, which Foucault eruditely named the *gyroscope* from Greek words meaning 'to look at the rotation'. The heart of the device was a bronze torus mounted on a perpendicular steel axle. A rotor of this form places mass further from the rotation axis, resulting in greater stability. The torus was supplied with four balancing screws on each face and around its circumference. The experience gained with the spinning mirror will have helped Foucault adjust these screws for smooth running, along with the balancing weights on the inner gimbal. The torus axle rotated in two bearings set on the inner gimbal, which in its turn mated to an outer, vertical gimbal via two knife edges. The position of the outer gimbal was determined by locators, but its weight was taken by a long thread. Foucault called this a *fil sans torsion* or a torque-free thread, because it was sufficiently floppy that it provided negligible restoring torque when twisted. There was very little friction in any of the bearings, and properly balanced, the rotor and gimbals had no tendency to turn under their own weight.

The inner gimbal and torus were detachable and Foucault fastened them to a gear train and hand crank in order to spin up the torus. Replaced in the gyroscope, the torus was subject to negligible external forces, and stabilized by its spin, stayed locked to the inertial frame defined by the stars. This is illustrated in Fig. 9.35, which shows the great simplicity of the gyroscope mo-

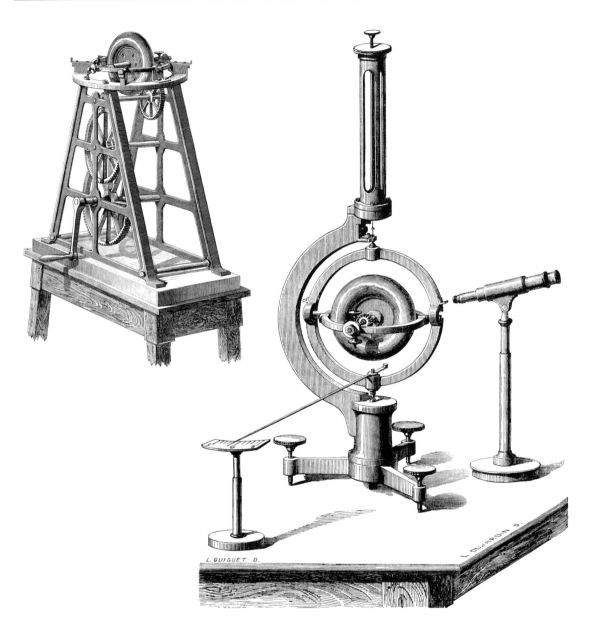

Fig. 9.34. Foucault's gyroscope. A pointer and a microscope could be used to view the slow rotation of the Earth. The torus was set spinning with a hand crank and gear train.

tion compared to that of the pendulum (cf. Fig. 9.13). Foucault could follow the motion of the outer gimbal via a pointer or with a microscope focused on a small graduated scale.

Of course no gimbal can be completely torque-free, but the effect of a given torque is less for a faster spinning rotor, which is why Foucault spun his rotor

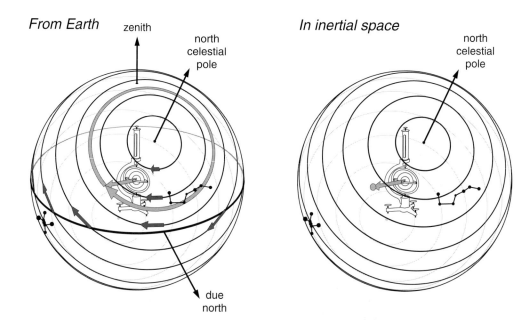

Fig. 9.35. The motion of Foucault's gyroscope for the latitude of Paris. For the Earthbound observer, the rotation axis of the torus, indicated by the long light-grey arrow, follows the stars, taking one sidereal day to complete a loop, just like an equatorial telescope (cf. Fig. 3.7). From inertial space the rotation axis is fixed (though the gimbals and stand of course move). The gyroscope motion is much simpler than that of the pendulum (cf. Fig. 9.13).

up to the 'great speed' of 150–200 r.p.s.[79] Besides being more stable, a faster rotor has more energy, and continues to turn for longer. Foucault reported that his gyroscope would perform for an astonishing eight to ten minutes. It also passed the crucial test: its drift was westwards whichever way the torus spun. The deviation was *not* precession due to poor balancing, but really was a reflection of the Earth's rotation.

Foucault later stated that the gyroscope worked and that his experiments with it were completed during 1852 May.[80] 'I had good fun this winter,' he told his friends.[45] But it was not until September that he made the device public. 'To inform the Académie des Sciences about this work,' he wrote, 'I was waiting peacefully for the end of the holidays and the return of a time more favourable for the presentation of quite a long body of work.' We must treat this assertion sceptically. Foucault no doubt saw the rotation of the Earth in his apparatus convincingly in May, but as Donné later wrote, 'nothing would leave his hands except in its final state'.[81] Tuning was needed to achieve the reliable performance essential for public presentation, and Foucault later admitted that it took 'eight months of unremitting supervision' to obtain 'an instrument so perfect that... the component parts move under the slightest breath'.

Nevertheless, Foucault was foolish to delay. Using a rotating mass to reveal the Earth's rotation was an idea whose time had come, and he should have realized that others would be working on it too. On Tuesday, September 21, most likely, he learned that there would be a communication on the subject at the Academy's next meeting. Was he going to be pipped to the post once more? He adopted the same expedient as with his spinning mirror. The following day he staked claim to his discovery in the *Journal des Débats*, in a letter which he admitted was 'unintelligible to people who have not made mechanics the object of their habitual preoccupations'.[82]

'The Lord be praised, the aim was accomplished,' he rejoiced. It turned out that *two* other scientists had presentations to make to the Academy. One memoir came from the Dean of the Faculty of Science at Besançon, Dr Charles Cléophas Person, but transpired to say little other than that Bohnenberger's apparatus would indicate the terrestrial rotation if 'constructed with precision'.[83] The other was from Person's assistant, a certain Georges Sire.[84] In Besançon, a town of ten thousand clockmakers, Sire had commissioned an actual instrument with a spinning wheel. 'Balancing such a wheel is very difficult to accomplish,' Sire reported, 'and I attribute the irregularities which I have observed to its imperfection.'[85]

Foucault, in contrast, presented the Academy with a working instrument, and repeated his experiments 'in front of competent spectators'.[86] In a second memoir immediately following, he discussed some additional properties of gyroscopic motion. Latches enabled him to lock the gyroscope's gimbals.[87] When he locked the inner gimbal, so that the torus axis could only move in the plane of the horizon, the axis moved through a series of decaying horizontal oscillations to settle in the meridian. The oscillations will have been relatively rapid, with periods of 10–20 seconds for Foucault's rotor speeds.[88] He then freed the inner gimbal and locked the outer one, so that the torus axis could only move in the plane of the meridian. The axis aligned with the celestial pole after a series of decaying oscillations in the meridian. This pole-seeking property is the basis of the gyrocompass, which aligns with the true pole (and not the magnetic one) because of torques exerted on the rotor by the locked gimbals. Further, the torus's sense of rotation was the same as the Earth's. A month later, Foucault summarized this as the tendency of one rotation to drive another into parallelism with it.[89]

Person and Sire were not Foucault's only competitors. An instrument builder on the Quai des Augustins in Paris, one E.-F. Hamann, had deposited a *paquet cacheté* with the Academy over a year previously, while Foucault was busy with preparations for the Panthéon pendulum. In it he outlined the basics of a gyrocompass, noting that when the inner gimbal did not move, the rotor axis had to be pointing at the pole. Hamann did not realize, however, that by locking the gimbal the rotor could be made to seek the pole automatically.[90] Five days after Hamann, a certain Ernest Lamarle had lodged another *paquet cacheté* with the Académie des Sciences in Brussels, in which he showed that a disc constrained to spin with its axis in the meridian would align on the pole.[91] Like Person and Sire, Hamann and Lamarle wanted their share of glory, and their *paquets* were promptly opened.

Foucault was magnanimous concerning the gyroscope:

> We make no claim other than priority of publication and of experimental realization. Now, for the last two months no one has been able to show any instrument able to function as the gyroscope does under the influence of the Earth alone. We say this once and for all, and will not speak of it again.[92]

Moigno was more forthright:

> ...the official glory, the scientific property... belong irrevocably to M. Léon Foucault... because, by everyone's consent, a sealed deposition can only have the effect of establishing a certain date for its author's work, but does not put him in solemn possession of the discovery which he consigns therein. In effect, a *paquet cacheté* is not a real publication, because, on the contrary, it expresses the intention and the will to keep the information it contains secret... Nevertheless, others may think differently, and for them, M. Lamarle will be the true inventor of the gyroscope, at least in the form which he gave.[93]

After some reflection, Person came to the eccentric conclusion that Foucault had misunderstood his gyroscope and the freely spinning rotor could not adopt an unchanging orientation in space.[94] Luckily, there was a physicist at the Collège Saint Louis with greater perspicacity. Antoine Quet had written a thesis on the oscillations of floating bodies in 1839 and had already sent the Academy a memoir on pendulum damping. Within little more than a month he explained the pole-seeking properties of the constrained gyroscope mathematically. Better, Quet made predictions, including that the gyroscope would find the pole if constrained to move in *any* plane passing through the pole, and not just the meridian; and that the frequency of the horizontal oscillations divided by that of those in the meridian should equal the cosine of the latitude.[95] Foucault tested Quet's predictions. 'We can assure him', Foucault wrote, 'that experiment shows that he is completely correct.'[92]

The gyroscope in England

Two years later Foucault went to England to attend the 'hurly-burly' of the annual meeting of the British Association for the Advancement of Science, which was being held in Liverpool. He travelled with Moigno, 'a M. Bernard whom the Abbé recruited just as we were leaving', and the gyroscope. The gyroscope made a striking demonstration. He reported back to Jules Regnauld from his billet at Chingwall Hall,[96] Broad Green, east of Liverpool:

> So far everything has gone for the best, good crossing, good bunks and so on. Only, to my taste I do not have enough time to myself. My host is a saviour who looks after me so closely that since my landing it has not been possible to roam as I would like. I think in short that he is quite

Foucault's final pendulum

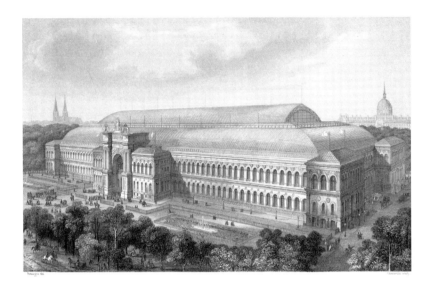

Fig. 9.36. The Palais de l'Industrie built for the Exposition Universelle in 1855. Foucault set up his final, motorized pendulum on the landing of the north-west staircase annexe (front-right corner). The Palais was demolished at the end of the century to make way for the 1900 Exposition Universelle. The sketch from which this engraving was made was drawn either by Foucault's sister's second husband Adolphe Rouargue, or his brother Émile, both of whom were artists.

proud of the curious animal who is staying with him. From yesterday his pride has been particularly flattered. After having waited quite some time for my turn to lecture, the decisive moment arrived, the gyroscope came out of its box and it showed all the expected virtues of a talisman.

You know that in the middle of tumult I still see things quite clearly, and I think I was the success of the day. Besides, I have positive proof. I wasn't *encored,* but it was immediately decided that a second showing would follow in an even vaster hall...I was able to judge the power of the gyroscope over these people who the previous day seemed disinclined to pay any attention to me.[97]

The official report was equally enthusiastic.[98]

Foucault's final pendulum

Two events dominated French life in 1855: the Crimean War, and the international Exposition Universelle, or Universal Exhibition, which ran from May to November in a specially constructed building nestled between the Place de la Concorde, the Champs Elysées and the River Seine (Fig. 9.36). The Exposition will reappear in Chapter 11 because Foucault was one of the judges, but he was also an exhibitor. Entry No. 1998 from the French Empire was Foucault's best and final pendulum, and profited from the lessons of the earlier ones.[100] It was installed in the north-west staircase annexe of the Palais de l'Industrie, on the first-floor landing, and once more was Froment's handiwork. Only 11 metres were available, the same as in the Meridian Room of the Observatory, where Foucault tested out his new pendulum beforehand.[101] He incorporated a

Fig. 9.37. A suspension with parachute on the wire and catch ring below designed to halt the fall of the pendulum when the wire breaks.[99]

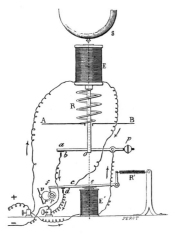

Fig. 9.38. Essentials of the electromagnetic drive for Foucault's pendulum at the Exposition Universelle in 1855. As the bob of soft iron swung in towards the vertical it was attracted and accelerated by the main electromagnet, E, so replacing the energy lost to air damping and maintaining a constant swing amplitude. The magnetic attraction was reciprocal, and the electromagnet rose on its spring, R, and through shaft o and the lever ap cut the current at contacts ab. With the current cut, there was no attraction while the bob swung away from the vertical. Gears, a rotating vane, v, and other parts restored the field once the bob had moved well away. Pulling in on the bob increases any ellipsing present. It would have been better to push out as the bob swung away, but passive detection of the approaching bob was beyond the technology of 1855.

'parachute' in the suspension to stop the bob from falling if the wire broke (Fig. 9.37). He gave effect to his earlier thought and built an electromagnetic drive to keep the pendulum swinging 'for as long as the Exposition Universelle will last', to borrow Moigno's words[102] (Plate X and Fig. 9.38). The motor seems quaint and complicated today when electronics can provide the same functionality without moving parts, but no doubt the design was as simple as possible at the time, and spectators were amused by its regular clicking.

It is recorded that this pendulum attracted 'the public's most lively curiosity',[103] while the reporter for *L'Illustration* wrote, 'It is the one item at the Exposition Universelle which is truly and purely scientific, it is the one contribution of the mind in this great congregation of material interests.'[104] The Emperor's cousin, Prince Jérôme-Napoléon, complimented Foucault over his pendulum during an official inspection.[105] Sir David Brewster and other members of the jury judging precision engineering were impressed too, reporting that Foucault would have been 'awarded the highest distiction... had he not been as a judge himself debarred'.[106]

Subsequent pendulums in the Panthéon

The Panthéon returned to civil use as a national sepulchre in 1885. The following year members of the Paris Municipal Council proposed that Foucault's pendulum should be set up afresh, and permanently. 'The apparatus which was used still exists in its entirety,' they claimed. 'It is thus just a matter of putting it back in place,' they continued, showing utter ignorance of the difficulties involved.[107] Nevertheless, a 39.5-m pendulum was set up in 1887 in the sixteenth-century Tour Saint Jacques.[108] The Brazilian Emperor Dom Pedro II was among the visitors. Another head of state with a keen interest in science, Dom Pedro no doubt appreciated this pendulum greatly, having earlier decorated Foucault for his discoveries and having received a gyroscope in return.[109]

It was in 1902 that a pendulum swung again in the Panthéon, set up in part by the astronomy writer Camille Flammarion (Plate VII).[110] A further pendulum did not swing until 1995 October, erected by the late Jacques Foiret, a retired engineer from the Conservatoire National des Arts et Métiers.[111] A 38-cm bob was used from the Conservatoire's collection under the mistaken impression that it was Foucault's original,[112] but with such a large bob and a 1.8-mm wire the damping was extreme, halving the amplitude within a quarter of an hour. It was suspected the bob might contain a clockwork-driven moving mass to maintain the oscillations, but a radiograph showed it was empty (Fig. 9.39). Foucault's pendulum in its most majestic location doubled the number of tourists visiting the Panthéon, so the demonstration was prolonged beyond the initially foreseen six months with a more satisfactory bob (Fig. 9.40).

Space and rotation, Mach and Einstein

No account of Foucault's pendulum and gyroscope would be complete without putting them in the modern context.

For Foucault, the veering of the pendulum was philosophically an easy matter. It was the consequence of the existence of an absolute space in which unaccelerated objects stay at rest or continue in their state of uniform motion in a straight line. 'The plane of oscillation is not a material object;' he wrote, 'it belongs to space, to absolute space.'[12] The inertia of bodies – their resistance to change in their state of motion – was a given, embodied in the dynamical laws formulated by Newton, which describe how forces cause accelerations in bodies of particular mass. Further, there was no distinction between this mass and the mass which enters into the formula describing the force that results from a body's gravitational attraction. The inertial and gravitational masses were one and the same thing.[113] The rising concept of the luminiferous ether was in accord with the existence of absolute space and an absolute standard of rest against which motions could be measured.

Matters were not this simple, however, for the Moravian physicist and polymath Ernst Mach (1838–1916). Taking up a theme from Descartes, Mach wished to eliminate what he called 'the conceptual monstrosity of absolute space', and a space that acts but is not acted upon.[115] In a positivist spirit, he theorized that space is unobservable and that positions and motions can only be measured relative to other bodies. In consequence, a single body such as a planet in an otherwise empty universe can have neither rotation nor inertia, and there is no inevitable equality between gravitational and inertial mass. The swing plane of a Foucault pendulum on Mach's solitary planet could never veer and the rotation axis of a gyroscope would stay fixed relative to the planet. According to Mach, it is the mass distribution of the universe that produces inertia, through some unspecified interaction. This notion came to be known as Mach's Principle, although it was never explicitly enunciated by him.[116] Since the observed equality between gravitational and inertial mass is independent of direction, and the distribution of material in the nearby universe is very lumpy, it follows that Machian inertia is overwhelmingly produced by material at the most distant reaches of the universe, which on this grand scale must be distributed equally in all directions.

Mach's ideas have an observable consequence for the pendulum and gyroscope. Consider the planet in the otherwise empty universe on which the Foucault pendulum does not veer. Imagine adding some matter to the far regions of this universe. It will now be possible to observe a rotation of the planet with respect to this distant matter, but if there is sufficiently little of it, hardly any inertia will result. The pendulum will veer only slightly with respect to the planet, but as increasing quantities of matter are added to the universe the swing plane will be increasingly affected. Ultimately the drift of the swing

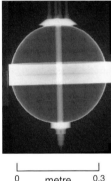

Fig. 9.39. (*Upper*) The nineteenth-century pendulum bob installed in the Panthéon in 1995. Oscillations quickly died out because of air resistance on the wire and enormous 38-cm diameter bob. (*Lower*) Radiograph confirming that the bob is hollow. Gamma rays from a cobalt-60 source were used because X-rays will not penetrate thick metal. The envelope containing the radiographic film can be seen taped behind the bob in the upper photograph.[114]

(*Bernard Rattoni, Saclay*)

Fig. 9.40. The 47-kg, 20-cm bob of the most recent Panthéon pendulum has given satisfactory performance since it began swinging from a 1.6-mm diameter wire in 1996 (cf. Fig. 9.26). It is usually restarted every hour or so.

plane will be dominated by the universe as a whole. However, the planet will still have some small effect, and the swing plane will not quite keep up with the distant galaxies. Similar reasoning applies to the gyroscope, and after a year's operation a pendulum or gyroscope might deviate by a few seconds of arc compared to the stars.[117]

For a while around 1900 it seemed that Foucault's pendulum might not prove the rotation of the Earth and that a new Machian dynamics could be formulated in which a relative rotation between Earth and the universe would be enough to make the pendulum and gyroscope veer. Bolstering this view was the famous Michelson–Morley experiment of 1887, which had failed to detect any ether wind as the Earth orbited the Sun, undermining the concept of an absolute standard of rest. To quote the talented French mathematician Henri Poincaré, writing in 1902:

> ...absolute space, that is to say the fiducial mark which would need to be used to determine whether the Earth really turns, has no objective existence... these two propositions: 'the Earth turns' and 'it is more convenient to suppose that the Earth turns' have one and the same meaning; there is nothing more in one than in the other.[118]

The modern view is different, however. Mach's ideas greatly influenced Einstein in his development of relativity; indeed, it was Einstein who coined the term 'Mach's Principle' in 1918, a couple of years after Mach's death.[119] Einstein made two postulates: that no frame of reference is more fundamental than any other for the formulation of physical law; and that gravitational and inertial mass are proportional through his famous Principle of Equivalence. General Relativity restores space in the new form of spacetime, which exists independently of matter, and both acts and is acted upon, because its geometry is affected by the masses within it, and the masses move under the influence of its geometry. Within spacetime inertial frames exist, though they extend through only small volumes of spacetime, but within them inertia, acceleration and rotation are absolutes which can be identified from observable dynamical effects, just as in the Newtonian conception. The pendulum and gyroscope track the stars because the stars define a frame which is inertial, or very close to it. From the viewpoint of General Relativity, which is the current best theory of space and dynamics (having passed all experimental tests), Foucault is indeed the man who proved the Earth rotates.

However, a spinning planet drags the surrounding inertial frame with it a little, and General Relativity predicts that a pendulum or gyroscope on the planetary surface should not quite follow the stars. This is the Lense–Thirring effect. Unlike Mach's additional veering, the Lense–Thirring frame drag differs with latitude and even reverses near the equator. At the terrestrial poles it amounts to about 0.2 arcsecond/year, much bigger than any rotation of the fixed stars or the universe itself (Table 9.3).[120]

A proposal has been made to search for the Lense–Thirring effect with a scrupulously engineered Foucault pendulum at the South Pole and claims have been made that it has been revealed through laser ranging measurements of the orbits of the LAser GEOdynamics Satellites (LAGEOS).[121] A long-standing and technically brilliant NASA project, Gravity Probe B, aims to detect the effect from the motions of four orbiting gyroscopes in which the golf-ball sized rotors are electrically levitated, niobium-covered, quartz spheres of unprecedented roundness, which spin in a magnetic field that is so low and a vacuum that is so tight that they would slow down by less than one per cent over a thousand years. Scientists eagerly await incontrovertible evidence of the Lense–Thirring effect; it will be a major prize.

A geometrical or dynamical effect?

When Poinsot said that the motion of the pendulum depends 'neither on gravity nor on any other force' he made it clear what is meant by a geometrical phenomenon. It is one where there is no need to invoke any of the laws of dynamics other than the inertia of bodies. In this respect, the rising and setting of the stars, and their tracking by a perfectly frictionless gyroscope are indeed geometrical phenomena, resulting only from the relative motion of the sky and Earth. Inspired by the pendulum, Wheatstone presented a device to the Royal Society of London on 1851 May 15 which exhibited sine law veering but for which gravity was clearly irrelevant.[122] The instrument is sketched in Fig. 9.41. It was composed of a turntable, which for convenience was horizontal, but could have been at any angle, with a perpendicular metal hoop carrying an adjustable clip. A helical brass spring was attached between the clip and the centre of the turntable. By moving the clip, the spring could be inclined at any angle with respect to the turntable. When the spring was twanged sideways and the turntable rotated, the vibration plane was found to veer with respect to the turntable at a rate which depended on its inclination in just the same way that the pendulum veering law depends on the latitude. Gravity enters nowhere into the vibration of Wheatstone's spring, for which the restoring force comes only from the springiness of the brass. In Foucault's pendulum, gravity provides the restoring force, but Wheatstone's apparatus shows, as asserted by Poinsot, that gravity is not fundamental to pendulum veering.

Yet the veering of gyroscope and pendulum cannot be entirely geometrical, because veering does not occur if the gyroscope does not spin or the pendulum does not swing. As the Earth turns, the spinless torus and swingless bob are carried around the parallel of latitude by their inertia and remain fixed relative to the Earth. Both pendulum and gyroscope are set going from rest *with respect to Earth*, with consequences for their motion. The pendulum veers slightly slowly, as has already been explained, while the gyroscope axis exhibits a very slight wobble because it is not rotating precisely around a principal axis.

Table 9.3. Rotation rates compared to the Lense–Thirring effect. The Lense–Thirring effect is an example of what is called *gravitomagnetism* because it arises from a flow of mass, just as magnetism arises from a flow of charge.

Earth	361 °/day
Local ('fixed') star	1×10^{-6} °/yr
Universe	$\sim 10^{-11}$ °/yr
Lense–Thirring effect (at poles)	60×10^{-6} °/yr

Fig. 9.41. Wheatstone's apparatus. When the turntable is rotated, the spring's plane of sideways vibrations follows an $\Omega \sin \lambda$ veering law relative to the turntable, in the same way that Foucault's pendulum follows an $\Omega_\oplus \sin \lambda$ veering law relative to the Earth. Gravity plays no part in the spring's vibrations, showing that gravity is not fundamental to the veering of the Foucault pendulum.

To understand these real as opposed to idealized motions, it is essential to consider forces, accelerations and the laws of dynamics. This conclusion is reinforced when other difficulties inherent in real pendulums and gyroscopes are considered, such as the anisotropic and anisochronic perturbations of the pendulum, friction in the gyroscope gimbals, and air damping.

It is true the mathematics of motion may be representable in geometrical terms, and since motion takes place in space, geometrical representations have mnemonic value and frequently provide physical insight. However, these are matters of analogy and fashion. The equations of spacetime can be seen as describing the geometry of a four-dimensional space, but we inhabit only three spatial dimensions and the fourth dimension, time, cannot be measured with a ruler. Newton presented the analyses and proofs in his *Principia* in geometrical terms because that was what the learned world expected, even though his discoveries were made using fluxions (calculus). The current fashion is away from geometry, and Poinsot's polhodes and herpolhodes are largely forgotten. The mathematics of physics is an analogy to reality, not reality itself. When the mathematics can be expressed geometrically, it does not necessarily follow that the physics is geometrical.

* * *

The pendulum experiment continues to capture the imagination and has been replicated – with greater or less success – many thousands of times. The gyroscope and gyrocompass were of immense practical importance throughout the twentieth century for their rôle in inertial guidance and navigation, though navigation is once again evolving through the radio signals of Global Positioning System satellites, while mechanical gyroscopes are being superseded by optical ones in which two beams of light interfere after having passed in contrary senses around the same closed path. Relativistic contractions render the paths unequal as the apparatus turns, and the interference pattern shifts, revealing the gyroscope's rotation speed. Vibrating micro-rods of silicon are providing inertial guidance.

In 1852 Figuier published a volume entitled *Description and History of the Principal Discoveries of Modern Science*.[123] It contained no reference to the pendulum experiment. Piqued, Foucault commented that Figuier '...really must add a further tiny little chapter to say that it is not only the wheels of locomotives and steam boats that turn.'[124] The pendulum and gyroscope are Foucault's most famous legacies to science, not because they indicated any new laws of physics, but because they showed how rich the existing ones were, when interpreted without inadvertence.

Chapter 10

Biding time

Foucault now entered a phase where he must have felt he was drifting. His name was made with the pendulum, gyroscope and speed-of-light experiment, but he held no official position and had no employment except as a hack-writer for the *Journal des Débats*. He set about looking for a job.

Passed over

After the *coup d'état*, all public servants were required to swear allegiance to the new government. C. S. Pouillet, the textbook writer, was a confirmed supporter of Louis-Philippe and refused. He was stripped of his professorship in applied physics at the Conservatoire des Arts et Métiers. Foucault, the impatient Edmond Becquerel and the elderly César Despretz proffered themselves in replacement.

Appointment was formally by the Minister of Public Instruction, but both the Academy and the Conservatoire's Council made recommendations. The Council eliminated Despretz on the grounds that he already held a professorship. Becquerel too held a professorship, at the National Agricultural Institute in Versailles, but the Institute had just been closed by the new regime. Wishing to compensate, the Council unanimously recommended Becquerel, noting that he had proved his teaching mettle, even though his research showed none of Foucault's originality. The fact that Becquerel's father was a well-known physicist was no doubt influential too.[1] The Academy voted on only the young men, and also preferred Becquerel, who was duly appointed.[2] This must have galled Foucault considerably.

Foucault's doctoral thesis a few months later was probably an attempt to strengthen his credentials for future posts, but it had not been examined (perhaps luckily) when the chair of experimental physics at the Sorbonne became available for a six-month term. Temporary positions are often poisoned chalices. Foucault decided to go to the top, and wrote to the Emperor, making it clear that he wanted a permanent job.[3] But before his letters had been forwarded to the Minister of Public Instruction with a cover note emphasizing the Emperor's 'benevolent inclinations', the Minister had decided there would be no permanent appointment and Foucault had withdrawn.

'My hour has not yet come,' Foucault wrote to Dumas, who was advising Foucault in his quest for employment. 'You will not blame me for having avoided a real danger and returned to the sideline until the day comes when the occasion is better.'[4] In the meantime, Foucault returned to an earlier occupation – the science of electricity.

An act of courage

We have mentioned that Foucault completed an auxiliary regulator for the arc lamp in 1849 September (page 114). This device consisted of platinum plates dipped by an electromagnet to variable depth in acid. It was needed because the brightness of the arc depends not only on the current through it, which was maintained constant by the regulator in the lamp, but also on the electrical tension across it. The auxiliary regulator compensated for the decline with use of the Bunsen cells' tension – or electromotive force, as it is called in analogy with mechanics. The auxiliary regulator also offered a solution to the problem of what Foucault called *distribution*, or the prospect of being able to drive several arcs from a common battery.[5] Deleuil, the supplier of Bunsen cells in Paris, 'energetically solicited' Foucault for his solution to this important practical problem, and Foucault consented; but Deleuil then procrastinated. Meanwhile, Foucault devised a simpler regulator. He put the platinum plates in an inverted jar which collected the hydrogen and oxygen evolved by the passage of the current through the acid. The gas reduced the immersed length of the plates, choking off the current. When the tension across the arc fell below the desired value, an electromagnet opened a valve to release gas, thereby reimmersing the plates.

Among the chemistry topics set by the Faculty for Foucault's doctoral examination in the spring of 1853 had been the subject of electrochemical equivalents. Perhaps it was his lacklustre performance in that examination combined with reflections on the evolved gases in his auxiliary regulator that led Foucault to experiment on the law of electrochemical decomposition developed during the 1830s by his hero Faraday (Fig. 10.1). These experiments are interesting because they are one of the few cases where Foucault went up a blind alley, or at least one where it was too dark to see.

Faraday had found that electricity was conducted by some liquids, such as water, dilute sulphuric acid and molten lead chloride, but not by others, such as molten glass or fused glacial acetic acid. Conduction, when it occurred, was accompanied by decomposition of the electrolyte (as Faraday termed it) into simpler species at the electrodes (another Faraday term). This is just what Foucault was witnessing in his auxiliary regulator. Further, Faraday had con-

Fig. 10.1. The English chemist and physicist Michael Faraday (1791–1867). 'If you were to see him today,' wrote Foucault in 1854, 'you would be unable to prevent yourself from *cracking jokes* with him.'

cluded, 'The chemical power of a current of electricity [i.e. the amount of each decomposition product] is in direct proportion to the absolute quantity of electricity which passes.'[6]

Faraday had measured some weak conduction by solid electrolytes such as salt, saltpetre and ice. He also found that feeble currents could pass through liquid electrolytes without any apparent decomposition. He concluded that electrolytes possess a small intrinsic conductivity, in slight violation of his law of electrochemical decomposition.[7]

In 1853 October Foucault sent a paper to the Academy reporting thoughts and experiments which he believed provided evidence of this additional 'physical' conduction, as he dubbed it.[8] In Moigno's opinion, it was 'an act of courage' to question Faraday's Law.[9]

Foucault first presented theoretical arguments for physical conduction. Although incorrect, the reasoning was ingenious. The preferred conception of electricity was the two-fluid theory first elaborated in the eighteenth century in which electricity was composed of two weightless fluids. One was positive and the other negative. Normally they were combined in a single, neutral fluid; but they could be separated, for example by friction, or when two liquids mixed and reacted. Foucault reasoned that in a reaction either the fluids would separate to build up a tension which would stop further action, or they would flow through the mixture and neutralize. Since Faraday-type conduction through the mixture would decompose the reaction product, the net effect in either case would be that the reaction would not proceed. But reactions do proceed, and so, concluded Foucault, electrolytes must conduct – at some level – without decomposition.

Foucault then reported some experiments. In the simplest, he connected two identical electric cells together through a very sensitive galvanometer (an instrument that indicates whether a current is flowing). Each cell comprised zinc and platinum plates immersed in dilute acid. No current flowed when the spacing of the plates in the cells was identical, but current did pass when the spacing was different (Fig. 10.2). According to Foucault, the electrical resistance due to physical conduction was reduced when the plates were closer. The corresponding internal currents then no longer balanced, so giving rise to an external current. 'The experiment... is tricky to repeat and requires a lot of care,' Foucault noted, so he devised a more robust arrangement akin to the Volta pile, with a column of zinc and copper discs separated by acid-soaked rounds of cloth. The pile began and ended with the same metal and being symmetrical produced no current. However, current did flow when double cloth circles between alternate plates destroyed the symmetry. Irregularities in the individual plates were averaged out by increasing their number, which also produced a higher tension and a clearer effect.

The notion that liquids were physically conductive led Foucault to realize that batteries, like electric eels, can be made without metals; though in fact

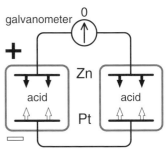
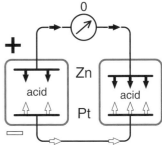

Fig. 10.2. (*Upper*) The galvanometer measured zero current when Foucault connected two identical cells together. (*Lower*) Current did flow when the symmetry was disturbed. In the two-fluid theory of electricity, it was conjectured that there were both positive and negative flows (filled and open arrows respectively) which neutralized in the acid.

this is possible because electrolytes are *chemically* conductive. He made piles with rectangles of sailcloth soaked with various liquids. Currents were drawn through wetted end layers backed by platinum plates.

Besides publication in the Academy's *Comptes rendus*, Foucault sent his paper to Auguste de la Rive in Geneva. De la Rive was an old hand in electricity, notably having invented a method for galvanic gilding in 1840, but he was also one of the editors of the *Archives des sciences physiques et naturelles* which appeared monthly as a supplement to the more literary *Bibliothèque universelle de Genève*. Like many journals of the time, the *Archives* reprinted interesting papers that had appeared elsewhere, translating them into French as necessary; but the *Archives* were unusual in adding editorial comment, which Moigno was emulating in his lighter-weight *Cosmos*. Foucault approved:

> The editors of the *Archives* have understood perfectly that a scientific journal lacks interest and excitement if it is composed uniquely of disparate papers slapped down one after the other. Such is the need in the commerce of the sciences for judgment and assessment... that a simple summary written by another often presents more interest than the original memoir... Thus there is no astronomer, no physicist, no chemist who is not anxious to know what opinion is held of him by the *Bibliothèque universelle*.[10]

Four years earlier de la Rive had found Foucault's arc light 'a bit complicated'.[11] What was his opinion concerning physical conduction?

De la Rive was sceptical. 'The results... are very unexpected,' he wrote, and he was not convinced by Foucault's conclusion.[12] Others weighed in. Heinrich Buff, a physicist in Giessen, thought the currents between the platinum–zinc cells might result from sloshing, and devised a slosh-free experiment where no current flowed whatever the separation of the electrodes.[13] Foucault meanwhile produced another experiment involving the electrolysis of water. He passed the current from three Grove cells through two voltameters in series (Fig. 10.3, cf. Fig. 1.9). One contained dilute sulphuric acid and the other distilled water. Believing that the physical conduction was proportionately greater in pure than acidulated water, he expected less gas to be evolved from the distilled water, which is what he observed; but he was obviously less confident about this work, which did not make it to the Academy and was published only in the *Archives* and *Cosmos*.[14]

Foucault was right to be cautious. Had he studied Faraday's papers he would have known that Faraday himself had found that, like the veering of the pendulum, electrolysis in practice is rarely as simple as elementary theory suggests. Microscopic gas bubbles can dissolve in the electrolyte and additional or different reactions may be occuring at the electrodes from what the experimenter supposes. The electrolysis of water rarely produces the theoretical 2:1 volumes of hydrogen and oxygen. Temperature, added electrolytes and the size, geometry and material of the electrodes can markedly affect the evolution

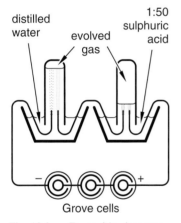

Fig. 10.3. Foucault's decomposition of water. He passed the same current through two voltameters (or *volta-electrometers* as Faraday called them) and mistakenly attributed the unequal gas production in them to conduction without decomposition.

of gas, especially oxygen at the positive electrode (or anode – another Faraday term). Faraday had noted the bleaching action of the water near the anode, and at about the same time as Foucault was setting up his voltameters, Heinrich Meidinger in Heidelberg showed that electrolysis of aqueous sulphuric acid can produce hydrogen peroxide instead of gaseous oxygen, as de la Rive was quick to point out.[15] However *less* gas should then be evolved from acidulated water, in contradiction to Foucault's observation.

Others entered the debate. Matteucci replicated Foucault's experiment in Pisa and wrote in good French to confirm the result, though his volume difference was smaller.[16] At the École Polytechnique, Jules Jamin was able to obtain hydrogen-to-oxygen ratios between 0.55 and 9.3 and even stop the evolution of one or other gas by using different electrodes and other conditions.[17]

In contrast to de la Rive's scepticism in the *Annales*, Moigno ran a campaign in favour of physical conduction in *Cosmos*. He wrote to Faraday at the Royal Institution, and under the emphatic headline 'Solemn and comforting support', announced that Faraday's reply 'will greatly delight the young French physicist.' Faraday had written, 'M. Foucault's results have impressed me very greatly, and I am happy to see that they confirm my own opinions; opinions which, since you ask, no reason tends to make me abandon.'[18] Despretz was on Foucault's side this time too, concluding two years later that water could conduct electricity without decomposition after examining electrodes at high magnification and seeing no bubbles.[19]

No decisive experiment disproved physical conduction. The debate petered out. Because of their complexity and inherent sub-microscopic nature, chemical phenomena are far more difficult to fathom than the macroscopic world of most of Foucault's physics. Despretz's microscope and Foucault's simple arguments were inadequate for what was really chemical investigation at the ionic scale in an era before the basic nature of electricity was understood, let alone the mechanism of electrolytic conduction. Indeed, even today the exact reactions at electrodes are not always clear. It is difficult to ascribe an origin to the currents produced by Foucault's asymmetrical batteries,[20] and to his strange gas ratios, but their cause was certainly not physical conduction. 'The science of electricity is certainly not complete,' Foucault wrote,[21] and the additional reactions that manifestly occurred may plausibly have been influenced by factors such as impurities, adsorption, concentration gradients and electrokinetic effects (potentials generated by electrolyte flows).

England

Britain was the richest and most powerful country in Europe. Although British scientists were in despair about the decline of their science, British science was influential and it was sensible for Foucault to promote his cause across the Channel. In the autumn of 1854 he attended the British Association meeting in Liverpool, where his gyroscope was so favourably received. 'To sum up, I think the campaign has gone well and will bear fruit,' he wrote home to his friend Regnauld. Here it seems Foucault was angling for the Royal Society's Copley Medal. This annual prize was, as one commentator remarked, 'the greatest scientific honour to which a savant can aspire',[22] on a par with today's Nobel Prizes, although no money was attached. Foucault's hopes proved justified, because having seen him in the flesh, the Royal Society awarded him the Medal the following year 'in testimony of our admiration of the skill, ingenuity, and talent displayed in your very remarkable experimental researches'.[23] Foucault will not have known it, but he was chosen in preference to Pasteur – who was still working in physical chemistry at that time.[24] Foucault was not the first Frenchman to win the Medal – it had already been awarded to Arago, A. C. Becquerel, Dumas and Le Verrier.

In Liverpool Foucault met his idol Faraday again. Faraday unfortunately had to return to London just before the gyroscope demonstration, but Foucault told Jules Regnauld that he was not too disappointed:

> I should be wrong to complain, because two days previously, invited to dinner at the Mayor's, I found myself seated between Richard Owen [the zoologist] and Michael Faraday. I was thus able to enjoy the good man at my ease... Michael Faraday is not at all as we imagine him, and singular fact, he does not conform to our memories of him. He's a lively little man with grey hair and a sharp eye... So give up thinking of Faraday as an old patriarch... The gent is not without mischievousness. During dinner we were talking about the personalities that death has taken from science lately, and without changing his tone, he added, 'And you have also lost M. Dumas.' I though he was confused and meant Arago [who had died the year before], but not a bit of it. He is fully aligned with our way of seeing things; according to him, Dumas is lost to science...[25]

No doubt the dinner-table conversation also touched on physical conduction.

In Liverpool Foucault saw 'the infamous Powell who makes it his job to teach the gyroscope without mentioning me', a comment prompted by a further Friday-evening discourse at the Royal Institution by Baden Powell in which he had mentioned the gyroscope but omitted Foucault.[26] (Powell must have been chastened, because the following spring he fulsomely acknowledged Foucault's 'remarkable experiments'.[27])

Foucault found he was on 'very good terms' with three English physicists 'about my age and who appear to be quite good chaps': someone called Hockes, who left no mark on science; William Thomson (later Lord Kelvin); and Faraday's assistant John Tyndall. With Moigno, he also enjoyed the 'learned and hospitable solitude' offered at nearby Bradstones by William Lassell, a wealthy brewer and passionate amateur astronomer.[28] Lassell had built a telescope with a giant 24-inch (0.6-m) metal mirror with which he had discovered Neptune's moon Triton scarcely a fortnight after discovery of the planet itself. This was a time when reflecting telescopes (those which gather light with a mirror rather than a lens) incorporated heavy metal mirrors, and while Lassell's mirror was not the biggest in the world, as will be discussed further in Chapter 12, it was the first large mirror to be placed in a successful equatorial mount. With employment at the Paris Observatory in view (see next chapter) Foucault must have studied the mechanical and optical details of Lassell's telescope with interest while enjoying the views it gave of Triton, asteroids, star clusters, a spiral nebula and the double, double star Epsilon Lyrae.

To see more of Faraday, and meet Wheatstone, who had not been in Liverpool, Foucault stopped off in London on his way home. He stayed with the Orleans' family physician, Henri Guéneau de Mussy, who had come to London with Louis-Philippe in 1848.[29] Perhaps de Mussy was an acquaintance from medical school days or a friend of a friend; certainly the relationship was not intimate, for Foucault made fun of him in another letter to Regnauld:

> I ended up by falling into the arms of Guisneau de Mussy; it is in his office, at his desk and on his paper that I am writing to you at present, taking advantage of the few free moments he leaves me while he goes piously to mass!!! Apart from that, he is perfect, he is charming, he is as good-natured as you could want, and he wants to introduce me to the whole world.[30]

Foucault was a trifle homesick. He continued:

> ...it seems to me that you could very well... tell me a little about what is happening in Paris. One is not removed from one's home for a month without developing some vague disquiet and without wondering if one's house is still standing. You know this half-stupid feeling...

He went on to question his friend as to why he had received so few letters, and to ask for intelligence on what was happening on the Rue de Seine (i.e. the Academy) 'so as I'll know on which foot to dance when I make my return'. He also asked Regnauld to go and see his brother-in-law and report back on his sister and her children.

Foucault reported that he had managed to see Wheatstone:

> I even spent a whole day with him and tried out all his gadgets. I didn't learn much, but at last I made acquaintance with this man who is so interesting and who seems disposed to support my rights on all occasions.

There were further meetings with Faraday. Faraday knew that Prince Albert, Queen Victoria's husband, was curious about rotating discs as possible stabilizers for the royal yacht, so he wrote to the Prince's private secretary suggesting the gyroscope might be of interest. 'M. Foucault is a perfect gentleman,' Faraday announced. 'His apparatus is portable.'[31] Six days later Foucault demonstrated the gyroscope to the Prince at Windsor Castle.[32]

Fig. 10.4. Three chemists. (*Centre*) The vivacious A. J. Balard (1802–76), discoverer of bromine, who was one of Foucault's doctoral examiners in 1853. (*Left*) The bright-eyed Henri Sainte-Claire Deville (1818–81), who replaced Balard at the École Normale Supérieure and was a loyal friend to Foucault. (*Right*) Adolphe Wurtz (1817–84), another of Foucault's friends.

(*Académie de Médecine*)

Henri Sainte-Claire Deville

It was about this time that Foucault may have made another important friendship. It was with the kindly chemist Henri Sainte-Claire Deville (Fig. 10.4). Henri and his older brother Charles, who became a geologist, had been born of French parents on the tiny Danish island of Saint Thomas in the Virgin Islands. As adolescents they had been sent to Paris for schooling. Henri entered the Paris medical school at about the same time as Foucault, and perhaps they met then. Unlike Foucault, Deville qualified (in 1843); but like Foucault, he found science more engaging than medicine and set up a personal laboratory in the garret of a house on the Rue de la Harpe. Almost immediately after graduation, Deville was appointed Professor of Chemistry and Dean of the newly created Faculty of Science in Besançon, where he stayed until appointed to the École Normale Supérieure in Paris early in 1851. It will be recalled that the École Normale trained teachers, but Sainte-Claire Deville set up a laboratory and began research. He discovered methods for producing more than microscopic quantities of aluminium metal. Napoléon III dreamed that this 'silver from clay' might provide lightweight armour for his armies and financed research towards large-scale production. Aluminium elicited universal admiration – 'it was welcome, one made it ring, one lavished brilliant horoscopes on it', wrote Foucault[33] – and we shall encounter it in some of Foucault's instruments, but the metal was curiously slow to find industrial applications, and it was not until after World War II that its use became widespread.[34]

Napoléon III detested pedants, but the open-faced Deville was not one. With his bubbly demeanour and soft, enchanting smile he obtained a lasting influence with the Emperor. In Chapter 1 we have already encountered the loyalty that he showed to Foucault's memory after his friend's premature death.

Gaslight

One job that came Foucault's way, and no doubt earned him some money, involved the illuminating power of peat gas.

By the early 1850s gas flames had replaced oil lamps for street lighting and in commercial premises. Several companies were supplying gas – though they were soon to disappear in a forced amalgamation in 1855 – and gas use was spreading rapidly for lighting in homes. Most Parisian gas was produced by heating bituminous coal, but there was always the possibility that some other source might be cheaper.[35]

A company had been formed at 53, Boulevard de Strasbourg, in northern Paris, with the intention of making gas from peat. Foucault's task was to see if the gas burned brightly enough to be economically competitive. 'As far as we know, this is the first time that work of this sort has been done by an eminently capable physicist,' Moigno declared.[36]

Foucault made his measurements at the end of 1854. Brightnesses were commonly measured with the Rumford photometer (Fig. 10.5). In this device two sources were compared through the shadows they cast of a black rod. The sources were moved closer to or further from the rod until the two shadows appeared equally dense. The illumination from the two sources was then identical. Flux decreases according to the inverse square of the distance, and from this law the relative strength of the two sources was deduced.[37]

Gas lights were simple butterfly flames (it would be forty years before incandescent mantles became common) and Foucault rejected the Rumford photometer on the grounds that the flames' large extent produced penumbra that were too wide. He experimented with a photometer devised by Babinet, but it needed a reference source.[38]

Abandoning these earlier photometers, Foucault conceived his *compartment photometer*, which compared fluxes directly (Fig. 10.5, Plate XI). The underlying principle is shown in Fig. 10.6. A central, dividing vane cast shadows so that each half of the diffusing viewing plate was illuminated by only one source. The trick for good sensitivity was to adjust the dividing vane so that the intersection of the shadows fell precisely on the viewing plate. As

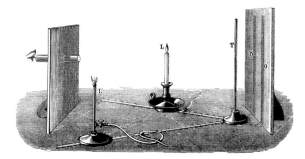 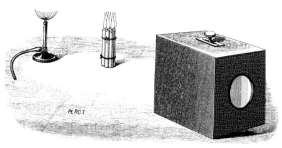

Fig. 10.5. (*Left*) Rumford's photometer. The screen might be opaque, as shown, or made of ground glass for viewing in transmission. (*Right*) Foucault's compartment photometer. He found that a bundle of candles provided a good reference source because of the number of flames and the stable air flow they produced.

Starched glass plate

Fig. 10.6. In Foucault's compartment photometer, the light sources S and S' cast separate shadows of the movable, central vane onto the viewing plate. For precise comparison of the two sources it was necessary to adjust the position of the vane so that the intersection point, P, of the two shadows fell on the viewing screen (lower diagram). Foucault found that a ground-glass viewing screen was too transparent and looked differently bright from different angles, while paper was too irregular. Starch deposited from a water mixture proved ideal and the null setting could be found using both eyes.

the distances to the sources were adjusted, the vane needed to be repositioned. Equality of the two fluxes could then be judged by the disappearance of any intensity step.

The peat gas was denser than coal gas and for the same pressure difference flowed more slowly. Since gas was sold by volume, Foucault reasoned that he should make comparisons at equal flow rates. 'We have found...,' he reported, 'that during a fortnight the mixture of ... gases, as produced by the gasworks on the Boulevard de Strasbourg, was always more luminous than coal gas, and that its lighting power compared to that of ordinary gas, represented by 100, was always contained within limits between 150 and 300.'[39]

Dumas and Victor Regnault used Foucault's photometer a few years later when it was discovered that the gas produced by the amalgamated company did not meet the Municipal Council's specifications.[40] Broadly similar photometers were used for gas testing well into the twentieth century.[41]

Chapter 11

The Observatory physicist

Arago's death

After the *coup d'état*, C. S. Pouillet was not the only scientist who refused to swear allegiance to the new regime. Arago, sick with diabetes and Bright's disease, wrote to the Minister of Public Instruction saying that as a member of the provisional government in 1848 he had contributed to the abolition of political oaths, and asked the Minister to:

> appoint a day on which I shall have to quit an establishment which I have been inhabiting now for near half a century. That establishment...has risen from its ruins and its insignificance, and can now be offered to foreigners as a model. It is not without a profound sentiment of grief that I shall separate from so many fine instruments...but my conscience has spoken...[1]

The Minister consulted the Prince-President. Louis-Napoléon repeated the tolerance shown by his uncle fifty years earlier to the student Arago and exempted him from the oath. In the following autumn, Arago died (Fig. 11.1).

Sensitive to Arago's reputation, the Emperor (as Louis-Napoléon had now made himself) ordered a state funeral (Fig. 11.2). Foucault reported for the *Journal des Débats*:

> Arago was an orator, and in addition he had memory and imagination; he discovered great things, and in the development of his career, the glory of the inventor was consummately underpinned by a power of observation and a wealth of erudition.... he put more new ideas into circulation as teacher at the Observatory and as Permanent Secretary at the Académie des Sciences than a whole generation of others...[2]

It is said that Arago's final words were 'Work – work diligently!' but according to Foucault, they were a more believable and prosaic 'I'm suffocating!'

Fig. 11.1. Arago photographed a fortnight before his death in 1853 October.

(©*Observatoire de Paris*)

Fig. 11.2. Arago's funeral procession leaving the Observatory in driving rain. 'All levels of society were present in the crowd which grew as it followed the cortège,' Foucault reported, 'and one saw with interest that Arago had numerous followers even in the poor parts of town.'

Urbain Le Verrier

Arago may have thought that the Observatory could be offered as an example to foreigners, but this was not the general opinion. Like many directors at the end of their mandate, he had let things slip. The instruments were all old. The most recent were Gambey's equatorial telescope from the 1830s and a special telescope for measuring celestial positions called a meridian circle, completed by the same constructor in 1843. The enormous dome on the west wing contained its equatorial mount, but still awaited a telescope.[3] George Airy would later assess the Observatory as 'in a very meagre state'.[4]

The Observatory needed reinvigorating in order to regain a position at the forefront of international astronomy. Within a month of Arago's death, a commission was instituted to propose reforms. Decrees in 1854 January split the Observatory from the Bureau des Longitudes and appointed Urbain Le Verrier as director (Fig. 11.3).[5]

Fig. 11.3. Urbain Le Verrier (1811–77) was appointed Director of the Paris Observatory in 1854. He had a mean spirit: one of his early acts as Director was to tear down Arago's lecture theatre.

This choice of director was unsurprising. Now that Arago was dead, Le Verrier was France's best known astronomer, and that was because of his prediction which had led to the discovery of Neptune. Like Arago, Le Verrier was from a modest background and had risen in France's meritocratic institutions because of his abilities; but unlike the southern Arago, Le Verrier was from the north, having been born in St Lô, in Normandy. Le Verrier graduated from the École Polytechnique in 1834 and then worked for the state tobacco monopoly for a few years, which enabled him to pursue his interest in chemistry. In 1837 he applied for a vacant position of *répétiteur* (assistant) in chemistry at the École Polytechnique, but Victor Regnault was also a candidate. Luckily a second post of *répétiteur* fell vacant. Regnault filled the position in chemistry while Le Verrier was appointed in 'geodesy, astronomy and machines'.

Le Verrier took this change of subject philosophically. He began by working on a question that continues to be of interest in celestial mechanics, that of whether the planets' mutual gravitational interactions may eventually disrupt the solar system. Le Verrier soon found that planetary masses were not known with sufficient accuracy for his analysis.

At Arago's suggestion, Le Verrier had analysed the orbit of the seventh planet, Uranus. Uranus had been discovered in 1781 by William Herschel and was not following the expected path. One hypothesis was that the unexplained swerves were due to the gravitational influence of an unknown planet. Le Verrier's genius was to realize how to formulate this hypothesis mathematically and to work back from the observed irregularities to the location of the putative planet. As already stated in Chapter 6, astronomers at the Berlin Observatory found the planet within a degree of Le Verrier's predicted position on their first night of searching.[6]

It was, Foucault wrote soon afterwards, 'such a great discovery', and as for Le Verrier, 'the glory of this man is imperishable and nothing will be able to tarnish its brilliance'.[7]

Le Verrier's quiver held more than a single brilliant arrow, however. In 1849 he had been appointed Professor of Astronomy at the Sorbonne. He was involved in politics, having been elected a deputy for La Manche (his natal region) as a member of the authoritarian Friends of Order party. In the National Assembly he had written reports on railways, the electric telegraph and the École Polytechnique. He was no republican: after the *coup d'état* he had been appointed to the imperial Senate. He had sat on the commission considering the Observatory's future. He thus had the insider knowledge, and scientific, administrative and political clout to be an effective Observatory director.

Fig. 11.4. Planet Le Verrier, discovered through mathematical analysis in 1846. Ever rooted in experimental reality, Foucault commented, 'In the resulting enthusiasm, one has hardly considered anything other than these beautiful calculations forthwith confirmed by such a shining verification. One has, so to speak, lost sight of the eighty years of observations which are their foundation.'

Employment at last

The task at the Observatory was a big one, and Le Verrier hesitated before accepting the directorship, but accept it he did. His first occupation was to consider more carefully what work the Observatory should do. There was also the question of Léon Foucault. The Emperor had taken a personal interest in both the Observatory and Foucault. This could not be ignored.

'We had long discussions over this', Le Verriers later recalled,[8] and he advised Foucault to set down on paper what he might do if appointed to the Observatory. Foucault detailed possibilities in 1854 August in a letter written from a hotel in Dieppe, where no doubt he and his gyroscope were waiting to board the *Shamrock* steamer *en route* to the British Association meeting in Liverpool.[9] He envisaged two broad themes. The first was to apply the methods of experimental physics to improve astronomical observations. He imagined application of the electric telegraph, improvements in optics, photography of the Sun and Moon, and determination of the vertical, most particularly with respect to variations resulting from the gravitational attraction of the Sun and Moon, which could provide a further means of determining their masses.

The second theme was an attack on various problems of physics that required the use of astronomical instruments and was more in the line of what he had already accomplished concerning the speed of light and the Earth's motion. He particularly proposed an experimental study of whether the refraction of starlight is affected by the Earth's motion. The received opinion was that it was not, but 'I have good reasons', wrote Foucault, 'to doubt that this is so...' Here he was obviously thinking of the ether-drag experiment begun with Fizeau, and the erroneous belief that it would reveal the orbital motion of the Earth with respect to Newtonian absolute space.

In Liverpool, Foucault grew impatient. What progress was there over his appointment? Jules Regnauld consented to act as a go-between and a month later reported on an audience with Le Verrier:

> ...and thanks to your name I was able to enter into the temple and approach the God...Neptune did not fear to initiate me into his mysteries and I was able to enjoy this divine commerce at my ease... at all costs he wanted to make it clear that he was not neglecting your interests. 'Be sure to tell M. Foucault that I absolutely want him to join us. But in order that his position here is worthy of him in every way, it is necessary that the post which is created for him is part of the vast reform plan which I am pondering and directing. I want him to come here with all the equipment and money necessary to show that it is not a sinecure that he is being given. But you understand that a programme of improvement ends up as a request for funds. I have only 30 000 francs for all my people, all my instruments. What do you expect me to do with this bagatelle? I am amongst ruins, amongst rubble, in a hovel; it is disgraceful, disgusting, hideous. I took the key to the shop and I shall only enter as victor, or I shall depart; yes Monsieur, I will get what I want or I shall deparrrrt!! ...But, to sum up, M. Foucault will join the Observatory, not immediately, it is true. Because to be successful one must choose one's moments; and the moment is not right.[10]

Indeed, the moment was not right. In the spring of 1854, the French and British had allied together and joined the Ottomans in war against the Tsar. This was part of Le Verrier's troubles. The money he needed and had been promised was now delayed until 1856, when it was expected the war would have long been won. In mid-September, when Regnauld saw Le Verrier, Anglo-French troops had just disembarked in the Crimea and were preparing to attack the principal city. 'If I went to see the Emperor here and now,' Le Verrier had continued, 'he would no doubt reply: Will we take Sebastopol? and send me packing. Patience!'

Already we can see a divergence between Foucault's and Le Verrier's ideas that was soon to cause trouble. Foucault had suggested a number of separate projects that he would undertake personally, but Le Verrier was recasting him as a leader who would direct assistants as part of a grand plan of research.

By December the attack on Sebastopol had degenerated into siege. Le Verrier judged that the Emperor would be less preoccupied, and forwarded his plan through the Minister of Public Instruction. It was an extensive document, totalling eighty-eight pages, and comprised seven themes including improvements to equipment and publication of results; research into newly discovered objects such as nebulae and variable and double stars; and work for other branches of the state, such as better verification of chronometers for shipping, distribution of accurate time by telegraph, and the setting up of a meteorological service.[11] Le Verrier also proposed that the resources of experimental physics could render great services to astronomy. He sharpened up some of Foucault's ideas, noting that there were a large number of scientific instrument-makers keen to attempt the manufacture of large telescope lenses, but that they needed support and advice. The speeds of light and electricity also merited further study.

Le Verrier's proposals were not without their detractors. Even before they had been fully formulated, the Swiss astronomer Émile Plantamour, who later became director of the Geneva Observatory, objected that the Paris Observatory had a scientific vocation as well as a utilitarian one, and that it was wrong to reform the Observatory on the lines of the Greenwich Observatory, whose mission was the improvement of navigation and where 'scientific research is almost entirely absent'. The Paris Observatory had employed many scientists of high standing as well as 'simple observers . . . What savant of merit will consent to become the subordinate of M. Le Verrier? . . . a man whose character is not such as to allow colleagues around him, but only underlings, machines.'[12]

Whatever Plantamour may have felt, people without jobs cannot always choose their bosses. Le Verrier wrote to the Minister proposing Foucault as the candidate of first choice if a post of physicist were to be created, with a certain Emmanuel Liais from Cherbourg as second preference.[13]

Two months later, on 1855 February 20, the Emperor approved Foucault's appointment as *physicien*, or physicist, at the Observatory on a salary of 5000 francs per year.[14] 'We may be permitted to say that it was we who first seriously expressed the wish that this eminent position should be created, and that M. Foucault should be appointed to it,' claimed a self-satisfied Moigno.[15]

Foucault's pay started immediately. One of his early tasks was to report on a recording instrument from Froment that was misbehaving,[16] but he had commitments besides the Observatory. The report on the luminous power of peat gas needed finishing, and duly appeared proudly headed *Par Léon Foucault, Physicien de l'Observatoire de Paris*. Electricity was still occupying his thoughts, and he was busy with the forthcoming Exposition Universelle.

New doctrines and 'Foucault currents'

There was undoubtedly a symbiosis between Foucault's science and his reporting which caused him to reflect on emerging concepts. A particular example concerns the 'new idea', as he put it, of the mechanical equivalent of heat, which he introduced to his readers during a series of articles in 1853 and 1854 which described Victor Regnault's research on gases:[17]

> This idea, we have already said, consists of seeing in heat one of the forms of the dynamical principle which animates the universe. This principle being, like matter, indestructible, it was a question of extreme embarrassment for the old physics to explain the apparent annihilation of all the movements which we see decay away around us. Furthermore, motion being unable to come from nothing, one had no more idea as to how animals frisked and why steam engines produced force. The recognized and well-proven existence of a mechanical equivalent of heat has made everything fall into place.[18]

Some explanatory remarks are needed. The nineteenth century was an era of mechanization, and even though rivers provided an important source of energy in France, the steam engine was of great significance. Searching to make this motor more efficient, a young *polytechnicien* and army engineer called Sadi Carnot published a booklet in 1824 entitled *Réflexions sur la puissance motrice du feu* (Reflections on the Motive Power of Heat). Carnot's theories were the 'ancient errors', which, Foucault told his readers, 'one will have a lot of difficulty in dislodging from more than one head.... From no more than the tedium of it,' he continued, 'one judged that it was wrong.'

These dismissive opinions will surprise the reader with knowledge of thermodynamics (the physics of heat and its relation to mechanical work) because Carnot has been enshrined as one of the fathers of thermodynamics, but in his *Réflexions* Carnot assumed that heat was an indestructible fluid, *caloric*, which produced mechanical work when it flowed from a higher temperature to a lower one, similar to the way in which water produces work when it flows down from a reservoir to turn a mill-wheel. The idea that heat and mechanical work might be manifestations of the same thing – energy – was an idea that was not quite as novel as Foucault had implied, however, having been proposed in various forms by scientists such as Newton and Laplace, but the name that history retains is that of Julius Robert Mayer, a doctor from the town of Heilbronn in Württemberg. In 1842 Mayer explicitly proposed the 'fertile principle', as Foucault judged it, that heat and mechanical work were but different manifestations of the same thing, and that an equivalent quantity of heat was lost when work was produced, and vice versa. This remains the modern view, but Carnot's adoption of indestructible caloric was not so eccentric because the heat loss in early nineteenth century steam engines was quite unmeasurable, being only a small percentage. Carnot died in the cholera epidemic in

1832, but is remembered because of his approach, which was developed and extended during the 1850s by William Thomson, one of the physicists of about his age that Foucault had found so congenial in Liverpool.

In 1824 Gambey had reputedly noticed that a compass needle mounted in a metal box moved lazily, as though the metal somehow acted as a brake on its oscillations. Inverting the circumstances, Arago had found that a rotating copper disc placed underneath a magnet dragged the magnet with it. This discovery had won Arago the Copley Medal, but had remained unexplained until it led Faraday to discover induction currents in 1831. The magnetic field induces electrical currents in the moving copper (this is the principle of the electric generator), and these currents produce a contrary magnetic field which acts to slow the disc. 'One can generalize', wrote Foucault, 'and say that through a mutual influence the magnet and the conducting body tend towards relative rest.'

Foucault was led to speculate on this effect when he visited the Paris workshop of the instrument-maker Heinrich Daniel Ruhmkorff (Fig. 11.5). Ruhmkorff was one of the numerous children of a postillion in Hanover. He had led a peripatetic life, working in various workshops around Germany, in Paris, and in London; but finally, wishing to improve his scientific knowledge, he had returned to Paris where he had become foreman in Charles Chevalier's optical shop. In 1839 he founded his own workshop, specializing in electrical apparatus.[19]

It was in this workshop that Foucault witnessed the dramatically rapid deceleration of a metal block or plate dropped into the field of a powerful electromagnet.

Fig. 11.5. H. D. Ruhmkorff (1803–77) specialized in making induction coils and other apparatus for the study of electricity.

(Bibliothèque nationale de France)

> 'So what happens to this motion which dies into nothing?' we inquired with insistence of the pensive spectators who were sharing our admiration. And everyone stayed mute as though floored by this mystery.[20]

In due course Foucault found the key to the conundrum:

> If... one wants the motion to persist, one must provide a continuous amount of work... which, according to the new doctrines, I judged should reappear as heat.[21]

Mayer had obtained a numerical value for the conversion rate between heat and mechanical energy, and using this figure, Foucault calculated that significant temperature rises should be achievable in practice.[22] Luckily, the necessary equipment was at hand for a prompt verification. He placed the spinning torus of his gyroscope between the poles of a strong electromagnet and found that within a few seconds the torus stopped. When he used the hand-crank to keep the torus spinning, the torus temperature rose from the ambient value of 16 °C to 20, 25, 30 and then 34 °C, and later was even 'burning'.

Babinet presented Foucault's experiment to the Academy on 1855 September 17 with what Moigno described as 'remarkable delight and liveliness'.[23] Foucault's memoir concluded:

> If the experiment appears worthy of interest, it would be easy to arrange an apparatus to reproduce and increase the phenomenon... so as to place a curious example of the conversion of work into heat before the eyes of the public assembled in lecture theatres.

Engravings in contemporary textbooks show that instrument-makers produced several easy arrangements for the demonstration. Figure 11.6 sketches the principal parts of the apparatus illustrated in Plate XII and made by Ruhmkorff. In it the temperature rise is increased by spinning a copper disc. In other apparatus Ruhmkorff put the copper in the form of a bottle-shaped slug spinning around a vertical axis; a hole was bored into the slug for a thermometer.[24] Other arrangements placed the electromagnet and crank horizontally, or even caused a cork to pop when water within a spinning copper tank boiled.[25]

None of the physics was new, and Foucault made no claim other than that his experiment might be interesting in lectures. Nevertheless Moigno was later to note, 'At that time, the fundamental question of the conversion from heat into force [work] and force into heat was hardly raised in France.'[27] It was no doubt for this reason that Foucault's demonstration 'rapidly acquired a just renown', as another reviewer put it.[28] The fame extended to the Palais des Tuileries. No doubt most of Louis-Napoléon's 10 000 francs had been spent on the gyroscope. A fortnight after the meeting of the Academy, Foucault received welcome news from Ildephonse Favé, the Emperor's aide-de-camp:

> I had the honour of informing the Emperor of your new experiments... His Majesty was deeply interested by this discovery, and wishing to facilitate and favour your work, He has decided that all the costs of your future experiments will met by the privy purse.[29]

The response was more measured in Britain, however.[30] Six weeks later Foucault was in London to receive the Copley Medal, where we have seen that he dined with the British physicist G. G. Stokes. The sodium D lines were not the only topic of dinner conversation, because a couple of days later Stokes wrote to William Thomson in Glasgow:

> Foucault has lately brought forward an experiment which seems to have created quite a sensation in Paris... A very pretty result certainly but I apprehend nothing new in principle, at least new in England.

Thomson replied:

> It is only from ignorance of what Joule taught twelve years ago, that Foucault's experiment... has caused anything like a 'sensation.' I saw the account of it in the *Comptes Rendus*, and *admired* but did not wonder at

Fig. 11.6. Foucault's demonstration of the equivalence of mechanical energy and heat. The 8-cm diameter, 7-mm thick copper disc heated up when cranked between the poles of the electromagnet.[26]

it. I knew well from Faraday how instantaneously a mass of continuous copper comes to rest between the poles of a strong magnet... and had frequently shown the phenomenon in my lectures. I do not think I ever showed it without saying at the same time that if the copper were kept rotating uniformly, an equivalent to the work spent would be produced in heat generated in the copper.

Indeed, in 1843 James Prescott Joule in Manchester had already reported induction experiments in which, like Mayer, he measured the equivalence between mechanical work and heat, but his results had been very coldly received because the caloric theory was still too strongly engrained and his measured temperature differences were small.[31]

Induction currents where energy deposition is the major effect have numerous practical applications ranging from the testing of pipes to cooking, while applications of induction in general are even broader, including electrical transformers and metal detection. In France induction currents of all hues have come to be known as *courants de Foucault*, or Foucault currents, though the honour should more fairly be Faraday's, as Foucault himself would have readily acknowledged. Nevertheless, the appellation is not completely unmerited, because Foucault's experiment was new and clear and, unlike Thomson's, published, and it was influential in advancing the cause of the conservation and convertibility of energy, which proved slow to lodge in many heads on both sides of the Channel.[32]

An understanding of the conservation of energy was one of the keys to Kirchhoff's explanation of the reversal of the D lines in 1859. Since Foucault was an apostle for this principle, his failure to explain the D lines' reversal is all the more poignant.

The induction coil

There were more than falling plates in Ruhmkorff's workshop. Foucault encountered a fearsome and captivating device, which hummed and sparked and gave off smells like a living creature. It was the induction coil (Plate XIII) and it was making Ruhmkorff famous.[33]

Faraday was not the only one to discover induction currents, which were discovered independently in America by Joseph Henry. News of Henry's work led to the simultaneous development of the induction coil on both sides of the Atlantic. Figure 11.7 presents the fundamentals of the mature *inductorium*, as it was sometimes called.

For induction the magnetic field must change with respect to the conductor, but in the induction coil it is the field's *strength* which alters, rather than its *location*. Two independent insulated conductors are wound around a common iron core, which serves to increase the field. A battery and contact breaker cause pulses of current to flow in the primary winding (as it is now called)

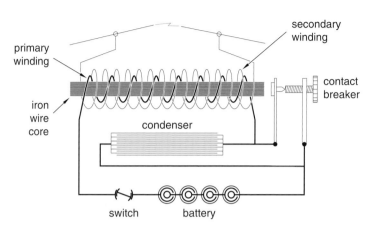

Fig. 11.7. Elements of a typical induction coil. Many of the important features were devised independently in the late 1830s by Charles Grafton Page, a physician in Salem, Massachussetts, and by Father Nicholas Callan at Maynooth College, near Dublin.

which is made of relatively few turns of thick wire. The current pulses generate a pulsing magnetic field, which as it grows and withers induces pulses of electromotive force in the turns of the secondary winding. With sufficiently numerous turns in this winding, the total induced tension in the secondary can be high enough to produce long sparks.

During the 1840s induction coils were produced in large numbers for doctors who found it fashionable, and no doubt profitable, to apply electricity for the supposed relief of diverse ailments. The contact breaker could take various forms, from a wire run along a rasp, through rotating toothed wheels, to a platinum anvil and hammer attracted by the magnetism of the core of the induction coil itself, as sketched in Fig. 11.7, and still usual in electric bells.

Ruhmkorff had made an induction coil for Antoine Masson, one of Foucault's co-candidates for election to the Academy in 1851. Ruhmkorff decided to add the device to his catalogue and made great efforts to increase the spark length, which in due course made the instrument useful for serious physics research. One expedient was to put more turns in the secondary (he used 8–10 km of wire in his biggest coils), but Ruhmkorff also sought advice. The hammer-and-anvil contact breaker would spark as it cut the current in the primary winding, and the sparks eroded the contacts, even though they were made of platinum. The sparking was due to self-induction, because the decaying magnetic field induced high tensions in both windings, but the sparking also slowed the decay of the primary current and the collapse of the magnetic field. Since induced tensions are proportional to the rate of change of the magnetic field, this in turn diminished the tension induced in the secondary. In 1853 Fizeau suggested connecting a capacitor (or condenser) across the contacts (Fig. 11.8). At contact break, current flows into the condenser, reducing the tension and sparks between the hammer and anvil. With less sparking, the magnetic field decays more rapidly, inducing larger tensions in the secondary winding and, for Ruhmkorff's coil, sparks that were 8–10 mm long.

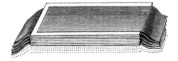

Fig. 11.8. Fizeau's condenser of insulated, interleaved tin sheets increased the length of the secondary spark in Ruhmkorff's induction coil.

The induction coil

The higher tensions required improved insulation with glass bobbins and supports. To separate parts that were at very different tensions, the secondary was wound in sections rather than layers.

Foucault worked to increase the spark length further. Even with Fizeau's capacitor, sparking and breakdown of the insulation on the secondary winding imposed a limit on the coil's output. Foucault therefore decided to double the output by connecting the secondaries of two Ruhmkorff coils in series, but for the induced currents to be in phase, he needed to synchronize the two circuit breakers. This was easy enough to achieve by wiring them in *parallel* and supplying both primary windings from a common battery. The output sparks became 'noisy, sinuous, and 16 to 18 mm long'.[34]

Fig. 11.9. Foucault's first mercury contact breaker. Such devices were sometimes called *electrotomes* in the nineteenth century.

His next improvement was to the circuit breaker. It was obviously redundant to use two in parallel, nor is there any particular need for the breaker to be driven by the magnetism of the induction coil. Like many others before him, as Foucault acknowledged, he thought to replace the unreliable solid contacts with ones where a platinum finger dipped in and out of mercury. Figure 11.9 shows Foucault's first mercury breaker, in which an electromagnet drives a strip of spring steel to which the platinum finger is attached. From his very first trials, Foucault found that the mercury splashed and vaporized when the circuit broke, so to quench this and prevent oxidation, he covered the mercury with a thick layer of distilled water, and later alcohol. The current broke more quickly, accompanied by a muffled click; the tension induced in two Ruhmkorff coils in series gave sparks 30–35 mm long.[35] This was about the limit of what the insulation between the primary and secondary windings could stand, so Ruhmkorff inserted a glass tube between the windings. With this improved insulation, Foucault was able to gang four coils together, and obtain sparks 7 or 8 cm long, which will have corresponded to a tension of some 150 000 volts.

The intermittent light of the spark acted as a stroboscope, and thus immobilized the circuit-breaker finger, which appeared well-retracted from the mercury, indicating a significant interval between rupture of the primary current and formation of the discharge.[36] One might have expected that experience with the electric arc would have taught Foucault to be cautious, but prolonged scrutiny of the most energetic sparks provoked painful ophthalmia some hours later.[37] No doubt a foot bath provided relief.

Plate XIII shows an induction coil commercialized by Ruhmkorff incorporating a Foucault circuit breaker powered by the coil itself. Similar breakers were also available separately.[38] The movable weight on the vertical rod could be adjusted to alter the spark rate.

Foucault made further improvements. He increased the condenser's plate area to 6 m^2, and found that it then hummed loudly.[37] He duplicated fingers and mercury cups so that he could run the breaker and primary windings from separate batteries (Fig. 11.10). He produced a symmetrical breaker with dou-

Fig. 11.10. Foucault double-cup circuit breaker. The breaker and primary winding of the induction coil were powered by separate batteries.

(Museo per la Storia dell'Università di Pavia)

Table 11.1. Exhibition categories in 1855 were divided into eight groups (Roman numerals) and subdivided into 30 classes. The classes were divided into a further 251 sections and some 3000 subsections. The classes within Group III are shown (Arabic numbers). Foucault sat on the prize jury for Class 9.

I. Extraction or production of raw materials
II. Use of mechanical force
III. Use of physical or chemical agents, or relating to science or teaching
    8. Precision engineering, science or teaching
    9. Economic use of heat, light or electricity
    10. Chemistry, dyestuffs, paper, leather, rubber
    11. Food preparation and preservation
IV. Industries relating to learned professions
V. Mineral manufactures
VI. Cloth making
VII. Furnishing, decoration, fashion, industrial design, printing, music
VIII. Beaux-arts

---

ble the number of fingers and cups and an oscillating armature that broke the current *twice* per oscillation.[39] Towards the end of 1857 he found that adjustment of the breaker fingers was easier if the mercury was stiffened with a little silver.[40]

Improving the spark length seemed to satisfy Foucault. With four coils, he told the Société Philomathique, he was able to obtain a discharge in a vacuum tube 2 metres long, with all the pretty colours and stratifications that were already known in gas discharges, but it was others who went on to use the Ruhmkorff coil to study the conduction of electricity through gases and, decades later, profit from it to discover radio waves, X-rays and the electron. Gone were the days when Foucault would have blushed to present an instrument before it had 'served to elucidate some new problem'. Foucault was turning from science to technology. He would still complete one purely scientific piece of work, the measurement of the speed of light (Chapter 13), but the rest of his career would be devoted to applications.

The Exposition Universelle, 1855

Let us now step back three years to the summer of 1854.

The Revolution had instituted annual national industrial exhibitions 'in order to strike a most deadly blow to English industry'. British industry had none the less survived, and the exhibitions had become five-yearly. Further, the British had held their enormously successful international exhibition in London in 1851. A few months after the *coup d'état*, Louis-Napoléon decided to postpone the next national exhibition for a year, and open it to international exhibitors in order that it should be a better showcase for his new regime.

There had been criticism that the choice of exhibition categories in 1851 had reflected British rather than European industrial production and that this had allowed the British to win too many prizes. There had been too much emphasis on cutlery, for example, which was a British speciality. The 'arid and difficult' task of devising a different classification scheme fell to Frédéric Le Play, a pioneer of quantitative methods in social economics and a paternalistic social reformer.[41] In the event, Le Play's classification favoured the host nation just as much as those of 1851, with subcategories like clothing and fashion, for which Paris was as pre-eminent then as now (Table 11.1).

In the summer of 1854, Foucault was appointed to a first exhibition committee. It was a national one, which checked that entries from the Département de la Seine (which included Paris) were of a sufficient quality to merit exhibition. Foucault was the vice-secretary of this committee.[42] Later, four hundred judges were recruited for the trickier task of awarding prizes. Foucault was appointed as one of the nine members of the international jury that would examine entries in Class 9, *Industries concerned with the economic use of heat, light and electricity*. Other French members included Babinet from the

Observatory, and the ubiquitous Edmond Becquerel. The jury's president was Charles Wheatstone. Foucault acted as the jury's secretary.

The Exhibition was opened with great pomp by Napoléon III in mid May. 'I delight in opening this temple of peace,' he declared.[43] There were some twenty-five thousand exhibitors. Over one hundred thousand people visited on May 17, a day on which entry was free; and by the close of the Exhibition in November, over five million had passed through the turnstiles. In the Palais de l'Industrie (Fig. 9.36) and other buildings the public saw Isaac Singer's sewing machines; the first oil-immersion microscopes; sparks from Ruhmkorff's induction coil; the spinning mirror and gyroscope on Froment's stand; a coffee percolator; ingots of aluminium; paintings by Ingres, Delacroix and Winterhalter; and a fabulous diamond from Brazil; as well as a gallery organized by Le Play of cheap, domestic objects aimed at improving the lot of the common people. 'The Exhibition here is *magnificent*,' reported Sir David Brewster, who was vice-president of the jury for Class 8.

It must have been fascinating and time-consuming work examining the vast array of devices and products on show. Jury members concentrated on their specialities. Foucault had recently completed his study of peat gas, and was assigned to report on oil and gas lamps. All but one of the thirteen medals were awarded to French manufacturers: along with fashion, lamp making was, as Foucault noted, 'an eminently Parisian industry'.[44] With Babinet, he also reported on another particularly French industry, the manufacture of 'lenticular' lighthouse optics.

In the times before radio beacons, lighthouses were crucial to the safety of coastal navigation; and Arago and Fresnel had improved lighthouse technology. To be visible at great distances a lighthouse beam needs to be parallel, or *collimated*, and this can done by placing the lamp at the focus of a lens or mirror (Fig. 11.11). Lighthouse optics do not need to be of the precision required in a telescope, but glassworks could not make large blanks for monolithic lenses. Working with the optician Soleil, Fresnel was able to construct a composite lens out of separate pieces of glass (Fig. 11.12). Fresnel's polyzonal lenses and reflecting prisms not only captured more light from the lamp, but eliminated spherical aberration. The result was a more parallel and therefore more penetrating beam. Owing to their manifest superiority over ones with metal reflectors, dioptric lighthouses (ones with refractive optics) were rapidly established around France's coasts. Brewster – who was pathologically afraid of dying in a shipwreck – immediately advocated the same system for British lights, without success.[45] The importance accorded to safe navigation was epitomized by a mock lighthouse tower that dominated the central nave of the Palais de l'Industrie, and every day its real optical parts were set working.

Had Brewster been heeded, British industry might have been a more competitive manufacturer of lighthouse optics in 1855. *Médailles d'honneur* were awarded to the French lighthouse administration and the two French manufac-

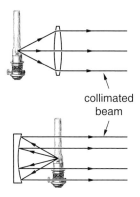

Fig. 11.11. The beam from a lighthouse is collimated (made parallel) by placing the lamp at the focal point of a lens or mirror.

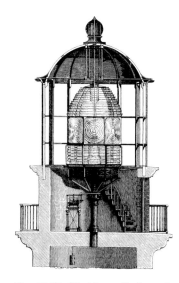

Fig. 11.12. Working with the optician Soleil, Fresnel had been able to construct the first polyzonal lighthouse lenses in the early 1820s. By 1855 French industry was predominant in the manufacture of lighthouse optics. With Arago, Fresnel had also improved lighthouse lamps, devising multiple, concentric, oil-cooled wicks.

turers who had exhibited, one of whom – L. Sautter & Co. – we shall meet again. The only other lighthouse exhibitor was Messrs Chance Brothers of Oldbury, near Birmingham. This British glassworks had displayed a Fresnel-style lighthouse head, but it was for a fixed light with only a few panels and none of the moving parts necessary for it to flash. There were no longer any thoughts of striking deadly blows to British industry when the judges rewarded Chance Brothers with a minor medal 'in recognition of their efforts to import the manufacture of lenticular optics into Britain'.[46]

The Chance discs

There was a Gallic side to Chance Brothers, however, because their master glassmaker was Georges Bontemps, a Frenchman. This *polytechnicien* had been director of the glassworks at Choisy-le-Roi, south of Paris, where he had won prizes for the production of flint glass, but the revolution in 1848 had caused him to seek sanctuary in England, where he obtained employment with Chance Brothers. At the Exposition, Chances displayed a flint and a crown glass blank with the staggering diameter of 29 inches or 75 cm. With them there was the prospect of fashioning a giant achromatic telescope lens. The flint disc had been shown at the Great Exhibition in London where the judges had been unable to decide on its quality in its rough form; but once the two faces had been polished, they found that the disc merited a medal, despite several internal striations. The crown glass was new, but in a rough state. The judges in France made no comment on the discs.

Before freighting the discs back to Birmingham, Bontemps thought to offer them to the Observatory.[47] Le Verrier was attracted: the judges in London had probably been right to think the flint's striations of no great importance, and the discs offered the prospect of equipping the Imperial Observatory with a world-beating telescope. With government authorization, Le Verrier entered into an imaginative purchase arrangement whereby the Observatory would examine the discs more closely and have first refusal on them. If the Observatory decided to attempt making an achromatic objective lens from them, Chances would receive an initial payment of 25 000 francs, and a further like sum if the enterprise succeeded.

Foucault began examining the discs early in 1856. He and Le Verrier discussed how to proceed. The great size of the discs made even the simplest test long and laborious. A very large dark room was needed and no doubt was set up specially, perhaps in the Meridian Room. The discs were weighty, over 200 kg for the flint, and probably about two-thirds that figure for the crown.[48] Substantial tables and frames would have been required to support them, and moving them would have needed several people.

Foucault first checked the discs for homogeneity. The method was simple: divergent sunlight from a small aperture was passed through each disc and

The Chance discs

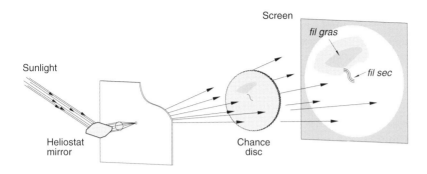

Fig. 11.13. To inspect the homogeneity of the glass in the Chance discs, Foucault shone light from a small aperture through them. Sharp density gradients in the discs (*fils secs*) cast sharp lines flanked by diffraction bands onto a distant screen. More gentle changes (*fils gras*) produced smoother variations of light on the screen.

allowed to fall on a screen of white paper (Fig. 11.13). Striations within the glass caused, and were characterized by, streaks on the screen. At one extreme, *fils secs* (thin threads) were sharp obstacles or rapid density changes within the glass and on the screen produced elongated shadows flanked by diffraction bands. *Fils gras* (thick threads) were more gradual changes of refractive index, and produced smoother, more fluid patches on the screen. Inspection of the screen thus gave a rapid overview of the quality of the glass, but did not provide any quantitative information concerning how annoying the defects might be in a lens. It was found that about half of the flint disc was affected to varying degrees by *fils gras*, with some *fils secs* at the centre, but that the rest was perfect. The crown blank had been exhibited rough cut. Once its faces were polished, it was found to be beyond reproach. 'Of all the glass discs that we have examined, large or small, this crown is the most beautiful that we have seen,' Foucault's report concluded.[49]

The discs' refractive indices were measured next. This was important because the discs would be worthless if their dispersions were inadequate for the fabrication of a chromatically corrected lens. Windows were polished into the edges of the discs so as to give three paths through the discs each about 600 mm long, which averaged the local variations in refractive index (Fig. 11.14). The refractive indices were determined 'using two theodolites'. The underlying theory is simple: it is only necessary to measure the three angles shown in Fig. 11.14, and Foucault will have adopted a procedure devised by Newton in which the total deviation of the light is minimized; but with large discs and theodolites the practice will have been complicated. He will certainly have used a heliostat and one theodolite in reverse to feed a collimated beam of sunlight onto one of the polished windows from a fixed direction. It will have been tedious to adjust the beam and discs so that the angles all lay in the same plane. Most probably he will have set up reference marks around his laboratory which he surveyed in order to determine the location of the theodolites in the various positions from which he measured the various beams. Whatever he did, it will have been the patient and insipid drudgery that he detested. Sunlight was used

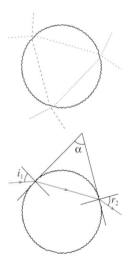

Fig. 11.14. (*Upper*) Windows were polished into the rough edges of the Chance discs so that their refractive indices could be measured over three different, long paths (broken lines). (*Lower*) Although the determination of refractive index in principle required measurement of only three angles, i_1, r_2 and α, in practice the experiment will have been complex and time-consuming.

partly because it was bright enough to be visible through 600 mm of glass but also because of the crucial wavelength references afforded by the Fraunhofer lines. The flint disc was too absorptive for Foucault to be able to see the most violet part of the spectrum, and the paths through the disc were affected by inhomogeneities, but he was able to make measurements along the clearest path at the wavelengths of the B, C, D, E, F and G absorptions. His conclusion was that the discs could be used to make a satisfactory achromatic doublet.[50]

Conflict and despair

Foucault's examination of the Chance discs took most of 1856, and the long wait can only have annoyed the impatient Le Verrier. As has been made clear, inspection of the discs was no trivial task, but Le Verrier's annoyance must have been exacerbated by the fact that during the previous year Foucault had done little for the Observatory, having been occupied with the conservation of energy and the Exposition Universelle. Further, the 1854 reforms decreed quite explicitly, as Plantamour had decried, that 'the Director alone ... controls ... all the scientific work undertaken at the Observatory', but in 1856 Foucault was diverting himself with unauthorized attention to the induction coil.

It was not in Le Verrier's personality to keep his displeasure private. Torn between his Director's imperious demands and his work with Ruhmkorff, Foucault despaired. Concerned, Donné wrote from Montpellier in 1856 August to Le Verrier, outlining his friend's fragile state. Parts of this letter have already been quoted in Chapter 7. Donné continued:

> *Order* him to take rest. Tell him to come and spend the month of September with me. I will send him back to you with new strength and new ardour to assist you in your great works.

It seems likely that Le Verrier saw the wisdom of this proposal; in any event, Foucault's report was completed in November. The Emperor had seen the discs when he had made a visit to the Observatory in the spring, and purchase of the Chance discs was rapidly approved.[51] Foucault began to devote more of his efforts to Observatory projects, but as we shall discover, this was insufficient to satisfy Le Verrier.

Chapter 12

Perfecting the telescope

To discover how to shape the Chance discs, Foucault went to the workshop belonging to Marc Secrétan (Fig. 12.1).[1] Secrétan had been born in Lausanne and began a career in law, but his interests turned to mathematics. In 1844 he moved to Paris where he took up partnership in N. M. P. Lerebours' daguerreotype and instrument-making company. The firm prospered, and when Lerebours retired in 1855, Secrétan took over as sole owner.[2] The company had a long association with the Observatory and Bureau des Longitudes, so it was the obvious place for Foucault to study optical fabrication.

Foucault knew that lens polishing was art rather than science. 'In the space of a few months,' he wrote, 'as a result of my daily contact with proprietor and workers, I was able to discover how things were done from start to finish.'[3]

As we shall see, Foucault soon abandoned the Chance discs in favour of making telescopes with mirrors rather than lenses. Anticipating this development, we will describe the manufacture of a spherical, concave mirror, because spherical surfaces form naturally and opticians could make them with some success.

A matched convex–concave pair of copper discs were turned to approximately the desired mirror curvature on a lathe. The *ball* and *basin*, as they were called, were ground against each other using successively finer emery grits to speed the abrasion until they slid over each other with equal ease in every direction, indicating matched, spherical surfaces (Fig. 12.2). The glass mirror blank was then ground against the ball (for a concave mirror) with finer emeries until it had hollowed out to match and its surface had developed a dull, uniform sheen. For polishing, a sheet of paper was glued to the ball. The paper acted as a matrix to hold the finer, softer rouge that brought the glass to a shiny finish, and it was in this step that the image-forming quality of the surface was determined. A good quality lens or mirror must be shaped with an accuracy similar to a wavelength of light, and it was distressingly easy to push too heavily or too lightly during the polishing strokes and end up with a misformed surface. Obtaining a good optical surface was a lengthy, hit-or-miss process.

Fig. 12.1. Marc Secrétan (1804–67). Foucault built all his telescopes in collaboration with this Swiss-born instrument maker.

(Bibliothèque nationale de France)

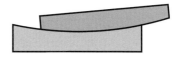

Fig. 12.2. The ball and basin are spherical when they move against each other with equal smoothness in all directions.

Fig. 12.3. Because celestial objects are very distant, only a very narrow spread of rays from each point reaches the telescope. The objective focuses these essentially parallel ray bundles to produce an image in its focal plane (dashed). In visual use, this image is examined with an eyepiece.

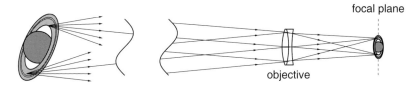

A telescope is designed to focus parallel rays, because the rays coming from individual points in distant objects are essentially parallel (Fig. 12.3). Checking a telescope lens therefore required a very distant light or test card.

Foucault realized that testing would be easier if parallel test rays were provided by a collimator, like in a lighthouse beam, but with greater precision (Fig. 12.4a, cf. Fig. 11.11). But this, said Foucault, is a vicious circle, because the collimator cannot be tested without a perfect lens, which cannot be tested without a perfect collimator. He realized a concave mirror would break the circle. It too can act as a collimator (Fig. 12.4b). Unlike a lens, however, a concave mirror can be tested directly. If the mirror is properly spherical, a point-like light close to its centre of curvature will form an equally tight image nearby since the rays are almost radial (Fig. 12.4c). The optician can use a microscope to examine the quality of this image, and hence that of the mirror.

The collimator must have a diameter at least as big as the lens to be tested. Foucault was experimenting at reduced scale, but his ultimate goal was a 75-cm diameter to test the Chance discs. He had seen a mirror of almost this size in Lassell's telescope near Liverpool, and was aware that Lord Rosse in Ireland had built even bigger mirrors in the 1840s. These mirrors, or *specula*, were cast in a shiny, metal alloy composed mostly of copper and tin. This alloy, called speculum metal, was very reflective; indeed, the first reflecting telescope, invented by Newton in 1668, incorporated a metal mirror (Fig. 12.5).

Foucault found that metal mirrors gave unsatisfactory images under the microscope. He therefore turned to glass, which Newton had known was easier to work than metal, and was able to obtain quality images that indicated a good spherical surface.[4] But, as with a window pane, polished glass only reflects a small fraction of the light that falls on it. This was enough for testing, but woefully inadequate for the stars. For Foucault, the remedy was obvious, because he had already used it with his spinning mirror. He silvered the glass.

Speculum metal and silvering

Reflecting telescopes are attractive because they are exempt from chromatic aberration: a mirror focuses all colours of light in the same way. The problem that had hindered their development was making good mirrors of adequate reflectivity. Looking glasses in drawing rooms were reflective, of course. They were made by smoothing a thin foil of tin against one side of the glass and immersing the whole in quicksilver (mercury) for a day or two. The result-

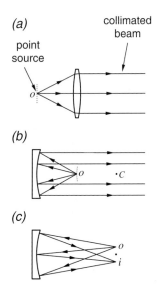

Fig. 12.4. (a) A parallel or *collimated* beam is produced when a small light, *o*, is placed in the focal plane of a lens (dotted). (b) A concave, spherical mirror can also produce a collimated beam. The focal surface (dotted) lies halfway between the mirror and its centre of curvature, C (the centre of the sphere of which the reflective surface forms part). (c) A concave mirror can be tested by examining the quality of the image, *i*, of a point-like object, *o*, placed adjacent to the centre of curvature.

ing amalgam was thick, rough and fragile, and only gave a practical reflecting surface when seen through the protective thickness of the glass. Newton, Airy and others had tried making telescopes with glass mirrors quicksilvered on their back, but had been thwarted by crystallization of the reflective amalgam and distortions induced by their imperfectly aligned faces.[5] For this reason, Lord Rosse and Lassell were using speculum metal, where only a single, intrinsically reflective surface of the mirror needed to be shaped.

Silver is more reflective than speculum metal, and Lord Rosse had tried making flat mirrors out of solid silver or by preserving a silver precipitate in shellac. However, it was not the needs of astronomy which led to techniques for making glass reflective through silvering, but the appalling effects of quicksilver on looking-glass workers, who soon developed tremor, delirium and other symptoms of mercury erethism. The silvering process used less toxic substances: silver was deposited by the chemical reduction of silver nitrate solution. The reaction had been noted in 1835 during an investigation of aldehydes by the German organic chemist Justus von Liebig and is still used in school laboratories as a test for aldehydes.[6] Liebig's reaction involved boiling and was hardly appropriate for precision optical components. In 1843 a certain Thomas Drayton, 'gentleman of Brighton', took out a patent in London involving reagents that did not require heating.[7] The details of the silvering process were to undergo numerous modifications as they were refined for optical and industrial use, but their essence was as follows. An alkaline, ammoniacal solution of silver nitrate was prepared. A reducing agent was mixed in, and the cleaned, wetted glass surface immersed in the solution. Numerous reducing agents were in vogue at one time or another, such as oil of cloves; grape, milk and invert sugar; aldehydes; and tartaric, saccharic and glyceric acids. Concentrations were critical in ensuring the deposition of a rugged, reflective, well-attached and uniform layer of silver.

In his patent, Drayton claimed that 'eighteen grains of nitrate of silver are used for each square foot of glass'. This corresponds to a silver layer on average 760 nm thick. The variations in thickness are smaller, and less than the wavelength of light. Used on the *front* side of a glass mirror, a chemically deposited silver layer enhances reflectivity without affecting the focusing quality of the underlying surface.

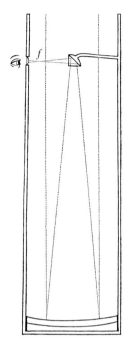

Fig. 12.5. The key components of a reflecting telescope are shown in this illustration adapted from Newton's *Optics*. A concave mirror reflects parallel rays (dotted) from its front surface to a focus, f, which, in order that it may be examined with an eyepiece, has been reflected into a convenient position by the total internal reflection of a prism. A flat mirror can replace the prism.

The polish of glass and the shine of silver

A silvered collimator mirror can just as well serve as the mirror in a reflecting telescope. Foucault's first silvered-glass mirror was completed around the very beginning of 1857, at about the same time as Flaubert was being prosecuted – unsuccessfully – for his story of the bored and profligate *Madame Bovary*. Foucault described his invention to the Société Philomathique at the end of January, and Le Verrier presented it to the Academy soon after.[8]

The mirror had been made with one of Secrétan's 'skilful workmen'; its useful diameter was 9 or 10 cm.[9] Its focal length (distance to the focus) was 50 cm. The silvering solution had been obtained from Drayton's French agents. Freshly deposited, the silver was dark and lustreless, but was easily buffed up with a chamois and a little rouge. Set up with a prism and eyepiece in the same way as Newton's telescope, magnifications up to 200 were satisfactory. Foucault enumerated the advantages of his instrument compared to a refractor. It was cheaper, it gave sharper images, and being based on reflection, it was of course free from chromatic aberration. He used his compartment photometer and found the light reflected was 90 per cent of what was transmitted by an achromatic doublet of the same diameter. Since the glass only served to support the silver, there was no need to search for clear, striation-free pieces as when making a lens. Speculum metal tarnished rapidly because of its high copper content. Foucault's mirror was too recently silvered for it to have tarnished, but he noted that when necessary, it could easily be resilvered. (Eighteen months later, the silver was still bright.[10]) If the telescope shown in Plate XIV is not Foucault's first telescope, it is certainly very similar. The mirror, Foucault later wrote, had 'the polish of glass and the shine of silver'.[11]

Babinet had begun to write on astronomy and meteorology for the *Journal des Débats* and waxed eloquent over Foucault's invention:

> ...[the] effect has been above everything obtained previously with similar apertures...with M. Foucault's invention, there is no doubt that the *popularization* of the telescope and astronomical notions will make great progress. Before long there there will be no château or manor where, for the pleasure and edification of its inhabitants, there is not a spyglass or telescope and tripod with which one can count the legs of a fly from one end of the garden to the other...With his telescope I have seen stellar images of the first quality, as well as very clear phases of Venus, a planet for which, according to the astronomical dictum, there is no good telescope...[12]

The advantages of the silvered-glass telescope were indeed great, and had been appreciated a year previously by Karl Steinheil in Munich. Steinheil had made a similar-sized mirror to Foucault's, as he wrote to the Academy to point out.[13] Moigno jumped to Foucault's defence: 'M. Foucault certainly did not know of M. Steinheil's trials...we ourself did not know of them, despite our attentive following of the scientific movement in Germany.'[14] In his letter, Steinheil made explict an advantage that Foucault had only hinted at. Repolishing destroyed the optical quality of a speculum mirror, which therefore had to be reworked and retested when tarnished. Tarnished silver, however, could be cleaned off glass without affecting the underlying surface, and a glass mirror could be resilvered without optical repolishing.[15]

Despite his predictions of a great future for the reflecting telescope, Steinheil subsequently devoted most of his energies to refractors.[16] It was Léon Foucault who was to turn mirror-making from an art to a science, but for the time being, he had other plans.

An improved polarizer

The fact that Foucault was discovering how to grind and polish optical surfaces undoubtedly contributed to an invention which would otherwise appear as a complete surprise, since it has no antecedents in his work, apart from the fact that he had made use of Nicol prisms with Fizeau over a decade earlier (Fig. 5.11). His discovery was a new way to cut a polarizer out of Iceland spar, and the device, fashioned by Foucault himself, was shown to the Société Philomathique at the end of July and soon afterwards presented by Babinet to the Academy on Foucault's behalf.[17]

Nicol's prism had the disadvantage of requiring a long crystal. The impossibility of procuring big, clear crystals meant that Nicol prisms could not be used with wide beams. 'Reflecting on the constraints of the matter', Foucault realized that much stubbier rhombs could be used, provided the Canada balsam glue was replaced by an air gap to change the angle at which total internal reflection occurred (Fig. 12.6). It was true that the new way of cutting the crystal could only polarize rays within a narrower cone of angles, but this rarely mattered in practice. Absorption of the unwanted ordinary rays was improved, but interference fringes could form because there was more reflection at the cut, where the refractive-index difference was greater than with Canada balsam. Because it contains no glue, the Foucault polarizer can handle high power densities and similar designs are now employed with lasers.

Foucault made no use of his polarizer in any subsequent research. What led him to devise it? Was he finally putting into practice a conception that he had had in the 1840s when he was actively working with polarized light? It seems likely, but we do not know for sure.

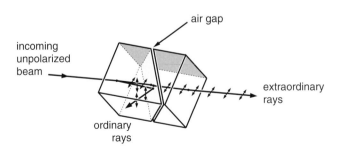

Fig. 12.6. Foucault's polarizing prism used less crystal than Nicol's prism (cf. Fig. 5.11). It could only polarize a narrower cone of angles ($\pm 4°$ compared to $\pm 11°$), but the rejected ordinary rays struck the crystal side at a steeper angle and were absorbed better by the external blackening. However, internal reflection between the crystal halves was strengthened by the larger refractive-index difference of the air gap, and extraordinary-ray interference fringes could form.

Ireland

Foucault's polarizer was completed with hardly a moment to spare, for he immediately took it and his telescope across the sea to show off in Dublin where the British Association was holding its annual meeting.[18] As with the meeting in Liverpool, he travelled with Moigno, and also two young travelling companions 'from a very good family'.[19] 'The Abbé was gay and good-natured, we teased him all the time,' Foucault wrote in a letter to Jules Regnauld, 'I wanted to convert him to the Devil and he me to God.'[20] The visit to Ireland was short, unlike the trip to England three years earlier, and within ten days of the meeting's end, Foucault was back in Paris to celebrate the *fête* (saint's day) of his mother, who had paid for the trip. In Dublin, Trinity College awarded him an honorary doctorate (Fig. 12.7). He travelled the 130 km west from Dublin to visit Lord Rosse's seat at Birr Castle, where he proudly signed the visitors' book 'Léon Foucault LL.D., T.C.D.' and inspected the 'Leviathan' telescope with its 6-foot, 4-tonne metal mirror (Fig. 12.8).[21] Foucault was not impressed with the world's largest reflecting telescope. 'Lord Rosse's telescope is a joke,' he wrote to Regnauld; but next to a 6-foot mirror, his 4-inch telescope had not evoked the same admiration as the gyroscope three years earlier, and Lord Rosse's telescope had had no shortage of defenders in Dublin.[22] 'For the British, mine does not exist,' Foucault sighed, 'It has been, it is, and for some time to come it will be, as nothing.' Nevertheless, on his way back to Paris Foucault visited the astronomer John Herschel at his home in Kent, and gave him two small glass mirrors. Herschel's household evidently found Foucault very exotic, and one of the womenfolk made a 'clever caricature of the Frenchman's face', which was remembered forty years later.[23]

Fig. 12.7. Foucault was one of many scientists awarded honorary doctorates by Trinity College Dublin during the British Association meeting in 1857. The degree was awarded *by diploma* because as a foreigner Foucault was unable to take an oath of allegience.

(Family collection)

The physicist cannot retain his position

Within a fortnight of his return to Paris, Foucault was making the first tests from the Rue d'Assas of a new silvered-glass mirror with an effective diameter of 18 cm, twice as big as his first one. Moigno reported:

> Seen in this second mirror, Jupiter presented a magnificent spectacle. One saw five distinct equatorial belts with different widths and perfectly outlined. Cloud masses varying from one day to the next were clearly portrayed near the poles ...[24]

Le Verrier, however, was losing patience. He was not interested in induction coils and polarizing prisms, or trips to Dublin. Foucault's mirrors were interesting, but were not much to show after two years as Observatory physicist. What was on Le Verrier's mind was a physics service led by Foucault with assistants dealing with weights and measures, terrestrial magnetism and meteorology, but what he wanted most immediately was to improve the Observatory's instruments, and for Foucault to fashion the Chance discs. Laboratory

Fig. 12.8. Lord Rosse's 'Leviathan' reflecting telescope. The 6-ft (1.8-m) diameter mirror (hidden in the pit) was spherical and polished to an enormous 56-ft (17-m) focal length to keep aberrations low.

space had been renovated for him at the Observatory, but Foucault had moved nothing in. Le Verrier decided to send a warning letter:

> I recognize that you work in your private laboratory on the Rue d'Assas; and no doubt your instruments there receive the necessary attention. But the laboratory at the Observatory has been left in the greatest disorder; you appear there for hardly more than a few moments each week.
>
> The Minister of Public Instruction has informed me that he wishes to honour us with a visit... If nothing changes, His Excellency will meet the Physicist through only the slimmest chance. He will see the most disorganized laboratory in Paris. As to the great glass discs, one cannot be shown, and the other is languishing in dust. There is no doubt His Excellency will say to us that this is not what His Majesty the Emperor wanted when he founded the position of Physicist at the Imperial Observatory...
>
> I therefore cannot allow this state of affairs to continue, nor the example which it sets.[25]

Le Verrier went on to order Foucault to start installing his laboratory the following Monday.

Foucault did nothing except stiffen the mercury in his circuit breaker with a little silver. Worse, absenteeism was spreading. Emmanuel Liais, who had been second ranked when Foucault was appointed physicist, had later been employed as the Observatory's meteorologist, but he too was staying away from work, distracted by what Le Verrier called 'dissipations' and outspoken claims for promotion.[26] Three weeks later an increasingly exasperated Le Verrier wrote to the Minister saying that his orders to Foucault had lain:

> *without response, without effect,* and as if *null and void*...
>
> ...in the present state of things, we will never succeed in figuring the great glass discs purchased already one year ago...
>
> In consequence, it is my honour, M. le Ministre, to ask for your instructions.
>
> Having as a painful necessity to give my advice, my opinion is that, by continuing to ignore his duties and placing himself above all authority, the Physicist cannot retain his position at the Observatory.[27]

Nevertheless, Le Verrier continued to make overtures towards Foucault. 'You know very well that though I cannot sacrifice the Observatory by ceding the principle [of personal work] to you,' he wrote in early January, '*I shall do whatever you want* as far as the execution is concerned.'[28] But Foucault had no desire to be cast as the leader of a great department, as he reiterated in a conciliatory reply:

> It really is impossible for me to change my mind over what I have told you on different occasions. I have no ambition to occupy an eminent

position; I only wish to conserve the facility of working and of modestly fulfilling my destiny.

Please do not hold this against me because I have not for one moment forgotten how much I owe you.[29]

Le Verrier seized on this letter which he forwarded to the Minister, adding:

Such a clear refusal to take part in the organization of the department to which one has been appointed must be considered, it seems to me, to be a resignation. I regret to say that it appears to me imperative that this resignation be accepted.[30]

Liais was in due course relieved of his duties, but Foucault enjoyed the Emperor's esteem and could not be fired so easily. Further, the Emperor was keen that Foucault should retain the title of Observatory Physicist.[31] Le Verrier changed tack. Was it 'uniquely because of ill will' that Foucault was refusing to head a physics service, he asked the Minister. 'Or is it beyond his powers?' he continued, forwarding Donné's letter from eighteen months previously which outlined Foucault's fragile mental state.[32] Six weeks later a compromise was in place. There had been lengthy discussions, a report by Foucault on how he saw his duties,[33] and no doubt Ministerial delays caused by the emotion surrounding the attempted assassination of the Emperor by an Italian nationalist called Orsini. The Observatory's needs in physics were to be catered for by Paul Desains from the Sorbonne while Foucault was 'left to his personal work, rather like M. Chacornac is in astronomy'.[34]

Evaluating telescope performance

The irony is that freed from the yoke of compulsion, Foucault devoted a large fraction of the next years to the Observatory's interests. He returned his attention to mirror making. Progress was rapid, and at one stage he was reporting advances almost weekly to the Société Philomathique.[35] In little more than a year, he developed definitive procedures for the construction of large, optically perfect, silvered-glass telescopes, which he then detailed in a comprehensive memoir published in the Observatory's *Annales*.[36]

Foucault had not gained a very high opinion of of the opticians' craft in Secrétan's workshop. 'When a mirror does not give good images,' he wrote, 'opticians are normally content enough to reject it without seeking in what way it sins; they redo the surface afresh, and they repeat this until they judge that they have succeeded.' In his memoir, Foucault claimed that it was only after the workman polishing a 42-cm mirror had failed five times to obtain a satisfactory result that he felt 'the imperious necessity' of studying the actual shape of the surface being produced. This is a simplification contradicted by earlier publications. However, let us simplify like Foucault, and trace the development of his ideas without seeking strict chronological accuracy.

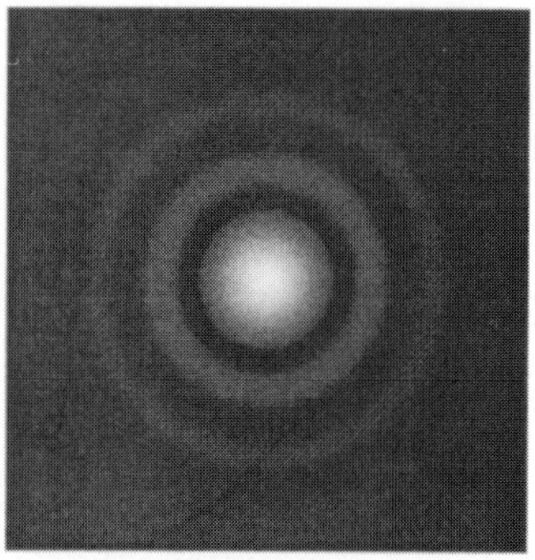

Fig. 12.9. Diffraction broadens the image of a point. The best, unaberrated image consists of a central bright spot surrounded by weak rings whose intensity is exaggerated in this representation. The spot and rings are smaller if they are formed by a larger lens or mirror.

Having produced 9- and 18-cm mirrors, Foucault next succeeded with 32- and 36-cm ones which produced images at a focal length of 3.5 m. 'I qualify these images as good,' wrote Foucault, 'because they present the same character of sharpness and clearness... as the smaller instruments founded on the same principles.' But Foucault was aware that it had become necessary 'to escape from the vague characterizations which leave room for delusions' and to specify his mirrors' capabilities through some numerical value derived from a standardized procedure.

An important indicator of the quality of any optical instrument is the extent to which it can resolve fine detail. Aberrations degrade performance, but even a perfectly shaped lens or mirror cannot resolve infinitely fine detail because of diffraction. The edges of a lens or mirror cause this spreading out of light waves, and in 1835 Airy had shown theoretically that the resulting fuzziness of the focus is less when the aperture is bigger. The best focus that can be obtained of a point-like object such as a star is not a simple blob of light but consists of a central bright spot surrounded by weaker rings of light (Fig. 12.9). A bigger lens or mirror, besides collecting more light, can resolve finer detail and support higher magnifications provided it is aberration free.

Foucault asked Froment to cut away strips of metal from a piece of silvered glass so as to make a test object composed of equally spaced opaque and transparent stripes (Fig. 12.10). Illuminating them from behind, he set the stripes face-on to the mirror under test and examined the image of them cast by the mirror. He then inclined the stripes to reduce their apparent separation. The separation at which they just ceased to be distinct gave a measure of the mirror's ability to resolve fine detail, which Foucault called the *pouvoir optique* or optical power, though the term has not stuck. Foucault initially expressed this quantity as the minimum angle that could be resolved, or as the separation of the just-unresolved stripes divided by the distance to them (which is just the angle expressed in radian measure), but he finally settled on the inverse of this latter quantity, which produced impressively large numbers. His 32-cm mirror, which could resolve 0.5 arcsecond, or some one four-thousandth of the angular extent of the Sun or Moon, had an optical power of 400 000 when expressed this way. 'The sharpness thus defined makes it possible to compare instruments without the necessity of testing them side by side', he wrote.

Examining the mirror shape

As he used his microscope to examine the focus produced near the centre of curvature of his mirrors, Foucault will have realized that a deformed image provides clues concerning how the mirror is misshaped. To examine the mirror and its defects Foucault developed three tests. The first we have already met: it was the examination with a microscope of the quality of the image of a point-like source close to the centre of curvature. For his point-like source,

Fig. 12.10. (*Left*) Target made by Froment for testing Foucault's mirrors. (*Right*) The apparent separation of the stripes was altered by inclining the target.

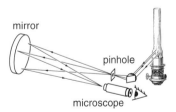

Fig. 12.11. Experimental set-up for Foucault's first test. The image of an illuminated pinhole was examined with a microscope for aberrations. A similar procedure had already been used by the English astronomer John Hadley in the early eighteenth century.

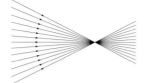

(a) Perfect geometrical focus

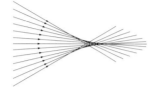

(b) Positive aberration

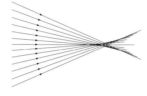

(c) Negative aberration

Fig. 12.12. (*a*) Geometrically perfect focus. Because of the wave nature of light, diffraction rings appear. (*b*) What Foucault called *positive* aberration, where the outer rays deviate too strongly. (*c*) Negative aberration, where the outer rays deviate too weakly.

Fig. 12.13. Foucault's second test. (*Left*) Set-up. (*Right*) Principle illustrated for a single straight wire of the grid. The image of the wire is examined through an aperture so that individual parts of the image are formed by rays reflected from only a limited region of the mirror surface. The curvature of the image then maps mirror-surface errors. To avoid confusion, the object wire and rays from it to the mirror are omitted; only reflected rays are shown.

Foucault illuminated a tiny pinhole in a screen with light brought in by a prism and lens from a lamp to the side (Fig. 12.11). If the image produced by the mirror was round, the mirror was rotationally symmetric – it had the same cross section along all diameters. If the mirror was correctly shaped, the rays would come to a perfect focus (Fig. 12.12*a*), where diffraction rings would be visible, but the focus would be poor if the mirror was incorrectly shaped. If the mirror was too strongly curved near its edge, for example, rays reflected from its periphery would focus closer to the mirror than those reflected near its centre (Fig. 12.12*b*) and vice versa if the mirror edge was too weakly curved (Fig. 12.12*c*). Foucault called these defects *positive* and *negative* aberration, respectively, though these are also terms that have not endured.

Careful examination of the aberrated image moving the microscope backwards and forwards could therefore indicate faults in the mirror shape.

A second test was easier to interpret. This time the test object was an illuminated square grid of wires, again placed close to the mirror's centre of curvature. To understand the principle of the test, we need consider only a single wire. Rays from every point on this wire reflect from every point on the mirror to produce an image whose quality of course depends on the shape of the mirror. Foucault's innovation was to observe this image through a small aperture so that the rays forming a given part of the image came only from a limited area of the mirror, not from the entire surface. This is illustrated in Fig. 12.13. Here the mirror edges have been represented as forming a focus too close in, and in consequence the image is curved. Each part of the image samples a different part of the mirror surface. The image of the grid as a whole maps the errors of the mirror shape, as illustrated in Fig. 12.14. Perhaps the idea of this test was prompted by Foucault's experience with the wire grid in his speed-of-light experiment.

The first two tests were sufficient to characterize a mirror's faults, but they were not very sensitive. The third test was exquisitely so, and has become known as Foucault's shadow or knife-edge test. The set-up is as for the first test, with a pinhole object at the mirror's centre of curvature. However, the optician's gaze is fixed on the mirror, which because it is reflecting light seems bright. The image of the pinhole is cut through with a sharp edge (the 'knife edge', Fig. 12.15). If the image is perfect, rays coming from all parts of the mirror will be cut simultaneously and the optician will see the mirror dim

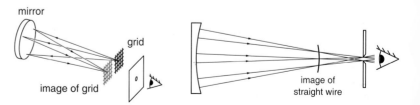

Examining the mirror shape

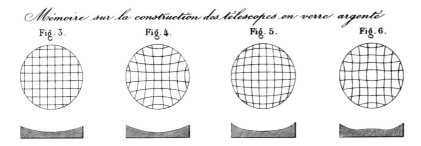

Fig. 12.14. The distortions of the image of a rectangular grid of wires indicate the shape of the mirror surface. In Fig. 3, the mirror is correctly shaped, whereas in Figs. 4 and 5 its edges and centre respectively have been polished too strongly. In Fig. 6 a ring has been over-polished.

uniformly (Fig. 12.16a). If the mirror is misformed, however, there will be 'a luminous aureole', as Foucault put it, of aberrated rays around the focus point, and once the rays falling close to this focus have been blocked by the advancing knife edge, the remaining aberrated rays will 'denounce' those parts of the mirror surface which are not reflecting at the correct angle. Figure 12.16b shows an example for positive aberration. After most of the light has been obstructed, the optician will still perceive rays coming from the upper part of the mirror. In terms of how this mirror 'sins', its surface is too forward and too inclined near its edges. Compared to the surface that would produce a perfect focus, the *difference* is like the inside of a shallow bowl (Fig. 12.16c). The mirror therefore gives an effect of light and shadow as if this difference surface were being illuminated obliquely by a light shining from the direction opposite to the knife's advance. This conclusion is generally true (Fig. 12.16d). The defects of the mirror surface are revealed in exaggerated relief.

Figure 12.17 shows some examples. As Foucault himself noted, these profiles are subject to a visual illusion whereby the experimenter may, for example, perceive a valley illuminated from a particular side at one instant, and a moment later a hill lighted from the opposite side. The correct interpretation must be sought from consideration of the knife-edge motion.

Where did Foucault find inspiration for his knife-edge test? Perhaps it arose from the combination of two ideas tucked away in his mind. One concerned magic mirrors. These polished metal mirrors from China had elaborate designs cast or engraved on their backs, but somehow the design was imprinted on the reflected beam. Magic mirrors had been discussed at the Academy ten years earlier, and Foucault had reported Babinet's suggestion that the designs on their backs produced minute corrugations on the polished fronts that deviated the reflected rays sufficiently to produce the effect.[37] Six years later, Lerebours confirmed this supposition experimentally.[38] The second idea comes from Fizeau's toothed-wheel experiment. Foucault had reported that Fizeau used 'a screen with a well-defined edge, the blade of a razor' in the focal plane of the Suresnes telescope to show that the returned image was focused to a small point. The knife-edge test combines both these ideas.

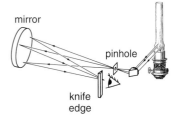

Fig. 12.15. Experimental set-up for Foucault's third or 'knife-edge' test. As the knife-edge cuts across the image of the pinhole, the faults in the mirror's shape are seen in exaggerated relief.

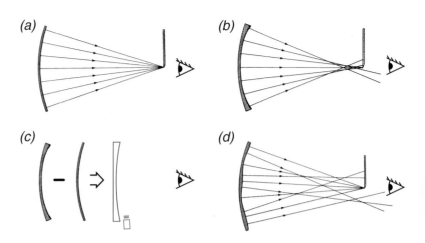

Fig. 12.16. (*a*) When the focus is perfect, the knife edge cuts all rays simultaneously and the mirror dims uniformly. (*b*) If the mirror is misshapen, the rays from some parts of the mirror still reach the eye. (*c*) The effect is as if the *difference surface* is illuminated obliquely from the opposite direction to the knife. (*d*) The test works for arbitrarily misformed mirrors.

Retouches locales

Although Foucault initially reported good results with his 36-cm mirror, Moigno was more reserved. In early May he observed Saturn. 'The image of the planet did not satisfy us fully,' he wrote; but when the mirror was diaphragmed down so light reflected only off its central part, the image was 'very pretty'.[39] Here Moigno was carried away by his unbridled enthusiasm for Foucault, because this praise was actually damning. Reducing the aperture is a way of improving the resolution of a *bad* lens or mirror because it reduces the relative importance of the aberrations, which arise most strongly from peripheral rays. It is the antithesis of an indication of superior quality.

With his tests Foucault found that the central part of the 36-cm mirror stood proud. It was then but a small step to conclude that opticians were wrong to repolish the whole surface of a bad mirror. Only the defective regions needed attention. This is the principle of *retouches locales*, or local corrections. The idea 'revolted the practitioners' in Secrétan's shop, and this scepticism clearly wounded Foucault, because he harked back to it again much later; but the operation 'succeeded far beyond expectations', and within a few hours the mirror surface was essentially spherical.[40] 'The establishment of a spherical surface by local corrections has for me become an acquired fact,' Foucault noted with satisfaction.

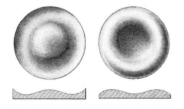

Fig. 12.17. The knife edge test reveals mirror errors in exaggerated relief. Plate XVII shows wooden models of these surfaces, which Foucault called *differential solids*.

Spherical mirrors, however, are not ideal for telescopes: their images suffer from spherical aberration if their diameter is too big compared to their focal length (Fig. 3.17), and it was to minimize spherical aberration that Lord Rosse had built his Leviathan so long. Foucault realized, of course, that his tests were not actually tests for a spherical surface; they were tests for whatever surface was required to produce a perfect focus. With pinhole adjacent to image, the tests do indeed check for a spherical reflecting surface. To obtain a perfect

Fig. 12.18. Foucault's ellipsoidal mirror with its complementary focal points, f, 9 m apart. Vertical dimensions have been doubled for clarity.

image that lies at some different distance an ellipsoidal mirror is required; and as the complementary focal points at which the pinhole and its image are situated become further apart, the ellipsoidal form tends towards a paraboloid, which is the ideal mirror shape for objects that are essentially infinitely distant, such as stars. Most importantly, with a paraboloidal mirror the focal length can be very short compared to its diameter, leading to very compact telescopes which are much less expensive to build and house.

Foucault started from a spherical mirror. He edged his test object towards the mirror and the image test station away from it, correcting the mirror with *retouches locales* as he went. In this way he produced an ellipsoidal 24-cm mirror with one focus at 1.10 m and the complementary one 9 m further out (Fig. 12.18). This mirror was just small enough to test within the calm of his house on the Rue d'Assas – presumably with the mirror in one room and Foucault in another (cf. Fig. 5.15).[3] With Froment's test stripes, the optical power was measured to be 300 000. This showed the mirror was diffraction-limited, where every millimetre of aperture contributes about 1200 to the optical power. The quality was further revealed when Foucault masked the mirror down to smaller apertures. The resolution got poorer, as predicted by Airy's calculations. The image quality was limited by diffraction, not aberrations.

New obstacles

The next step was to make a paraboloidal mirror, but at this juncture, Le Verrier threw a new obstacle in Foucault's path: he cut off his salary. Just at this time, a sixteen-year old called Camille Flammarion began work at the Observatory as one of the clerks employed to do arithmetic. Flammarion later became France's premier astronomy popularizer (Plate VII), but at his recruitment interview, he later recalled, there was talk that Foucault was threatened with madness.[41] Over the years, Le Verrier cut off the pay of many of his subordinates, but this action can only have worsened Foucault's mental state, which was perhaps the intention.

At the beginning of each month, Observatory employees signed a common pay sheet. This list was approved by Le Verrier and the Ministry, and at the beginning of the following month each employee could draw his salary from the Observatory bursar. Foucault was surprised on July 1 to be refused payment as a result of what appeared to be delays in forwarding the pay sheet. The following day Le Verrier refused to let him sign the next pay sheet. Foucault wrote to the Minister asking what to do; but then, obviously better informed about what was happening, he wrote another, much stronger letter, invoking the Emperor's name:

> Now is a moment when I am making the greatest sacrifices to bring a successful conclusion to an enterprise which has just recently once more been approved by the Emperor. This delay, if it were to continue, could have the most bitter consequences for me.[42]

Le Verrier had also contacted the Minister, asking that Foucault should be paid directly, rather than through the Observatory, stating:

> For the Observatory it is a matter of discipline that this employee with a special status does not appear each month to sign the pay sheet without having taken any part in the Observatory's work. It sets the most deplorable example for his former colleagues...
>
> What is no less grave is that M. Foucault adopts a provocative attitude and goes so far as to insult his colleagues.[43]

The Minister hesitated, and it took the intervention of Ildephonse Favé, the Emperor's aide-de-camp, to resolve the situation. It seems likely that the support from the Emperor's privy purse initiated three years earlier had dried up, if indeed it had ever materialized, since Louis-Napoléon was prone to flamboyant but unkept promises, and Favé was probably aware that Foucault was financing his telescope development personally. In these circumstances, every franc counted. Favé saw that Le Verrier was attempting to drive Foucault from the Observatory; indeed, although it had been agreed earlier in the year that Foucault could keep his laboratory there, Le Verrier had written requiring him to turn the keys over to Paul Desains, and entry to the laboratory was now 'rigorously forbidden' to Foucault.[44] 'And this when the Emperor had wanted to see M. Foucault's new telescopes for himself before he left for Plombières,' added Favé in his letter to the Minister.

The pressing issue was thus the money. The Minister chose the fastest solution, which was that Foucault would receive his salary at the Ministry of Finance, as Le Verrier wanted, but after signing an individual pay sheet at the Observatory.

Parabolization

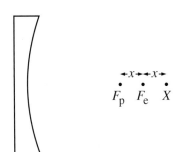

Fig. 12.19. Parabolization of an ellipsoidal mirror with near focus F_e located a distance x from the desired parabolic focus F_p. With the pinhole at F_e, there are no aberrations at the distant conjugate focus. The pinhole is moved to X and the resulting aberrations are noted. The pinhole is returned to F_e and *retouches locales* are applied to create identical aberrations. The mirror is then parabolized.

Despite these distractions, Foucault moved ahead to make paraboloidal mirrors. Initially he was vague about how he proceeded, but it cannot be done by removing the test station to infinity! Later he revealed his secret. When the mirror had the form of the most cigar-shaped ellipsoid that could be tested in the available space, the position of the pinhole was still a certain distance – let us call it x – from the position of the focus desired for the final paraboloid. In general, the image aberrations caused by *withdrawing* the pinhole by a small amount are approximately the same, but of the opposite sign, as those caused by *advancing* the pinhole by the same small amount. So Foucault *withdrew* the pinhole source by an amount *x* and examined the aberrations that had appeared at the knife edge or microscope. Returning the pinhole to its previous position, Foucault applied *retouches locales* until these aberrations were mimicked. The mirror then approximated the desired paraboloid (Fig. 12.19).

Foucault made two paraboloidal mirrors, one with a 25-cm diameter and a 1-m focal length, and the other with 33-cm diameter and a focal length of 2.25 m. Foucault tried this second mirror on the sky. Perhaps it was boxed up in a temporary wooden tube similar to his first telescope, but in any case it was soon set more properly in a wooden equatorial mount made by the Prussian-born Wilhelm Eichens, who was head of Secrétan's workshop (Figs 12.20 and 12.21). On the night of 1858 July 21/22 Foucault scrutinized a difficult test star, Gamma Andromedae. It was well known that this star appeared double in a small telescope, consisting of a bright orangish star with a fainter, blue-green companion, but in 1842 the Estonian-born Otto Wilhelm Struve found that the companion was itself double, comprising two blue stars separated by about 0.4 arcsecond when observed with the giant 15-inch (38-cm) lens of the new telescope at the Pulkova Observatory, near St Petersburg.[45] Would Foucault's parabolic reflector show these two components? It is not only aberrations and diffraction which hinder observations; turbulence in the Earth's atmosphere also smears the view, and to see detail finer than about 1 arcsecond requires calm conditions. Foucault and two companions waited patiently, fortified no doubt by coffee. As the first rays of dawn were beginning to appear, the atmosphere fell calm and the blue-green star was seen to separate into two components.[46] The parabolic telescope had proved itself.

Nevertheless, it soon became apparrent that the mirror, which was relatively thin, sagged under its own weight as the telescope was moved to different places in the sky. Foucault devised an expedient which was later adopted by others. He inserted a rubber air bag behind the mirror, linked by a tube to a stopcock near the eyepiece (Fig. 12.22). The observer blew air into the cushion, or released pressure, until the flexure was counteracted and the image was perfect.[47] 'It is one of these little strokes of genius to which we have become accustomed from this skilful physicist,' remarked Moigno, laudatory as usual.[48]

Nature decided to send the nineteenth century's brightest comet to greet Foucault's telescope. Discovered from Florence on 1858 June 2 by G. B. Donati, its tail grew to over 40° in length after it had passed closest to the Sun on September 30. Among those invited by Foucault to view the comet's head through his parabolic telescope were Moigno, Babinet, Faye and his antiquary neighbour, Honoré d'Albert, Duc de Luynes.[49] 'The success of his invention has exceeded all my hopes,' Faye recorded, 'and everything that I have seen of comets, with the instruments to which I have so far had access, seems to me to have been inferior.'[50] On October 5 the comet's magnificent tail swept in front of the bright star Arcturus. Newspapers made great play of the fact,

Fig. 12.20. F. Wilhelm Eichens (1818–84), head of Secrétan's workshop until he set up on his own account in 1866.

(Bibliothèque nationale de France)

Fig. 12.21. Eichen's equatorial mount for Foucault's larger telescopes.

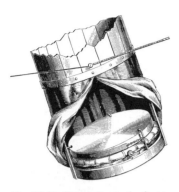

Fig. 12.22. Foucault was the first to counteract gravitational flexure by mounting his larger mirrors on an inflatable air bag. Here the idea has been taken up for a $15\frac{1}{2}$-inch mirror by Henry Draper in the United States.[53]

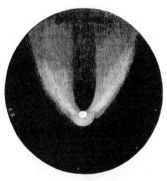

Fig. 12.23. The head of Donati's Comet observed with Foucault's 33-cm parabolic mirror on 1858 October 5. Moigno likened the bright nucleus to a carbon electrode from which an electric arc is emanating, and was struck by the dark cone separating the tail into two. The Duc de Luynes noted a dark splotch close to the centre of the nucleus.

and around the world especial attention was paid to the comet.[51] Figure 12.23 shows a woodcut of a drawing of its head made that night with Foucault's telescope by a certain C. Bulard, who had a talent for drawing and who was soon to depart for Algeria, sent by Napoléon III to set up the newly founded Algiers Observatory.[52]

Stalemate

In reporting his achievements to the Academy in 1858 August, Foucault ended with the remark that the expenses had been sufficiently moderate to be borne by a private individual (i.e. himself).[54] The fact that he had paid meant that he was free to exploit his invention commercially, and in fact 9- and 18-cm telescopes had been available for purchase from Secrétan's shop as early as May.[55]

Figure 12.24 illustrates one model, while Plates XV and XVI picture an equatorially-mounted version for easier tracking of the sky. Babinet remarked how little neck strain was involved compared to a refracting telescope.[56] The telescopes came complete with instructions that must have caused amusement in the châteaus and manors: 'When looking at your mirror, dread touching it with your fingers & splattering it with the tiniest drop of saliva ... '[57]

Foucault now had all the key elements to attempt a bigger mirror, but for this the resources of a private individual were inadequate: he needed the support of the Observatory, but he did not even have the keys. There was little he could do. Instead he devised another use for his silvered-glass mirrors, one which was sufficiently small-scale that the work could be done at home. It was a catadioptric microscope, that is, one involving both refraction and reflection in its optics. The idea was simple and no doubt in part it resulted from deficiencies that Foucault, the careful observer, had noticed when working with Donné fifteen years earlier, but it also took account of the limits diffraction places on resolution, which Foucault had so recently encountered with his telescopes.

In essence a microscope consists of an objective lens of short focal length which casts a magnified image of the specimen. This in turn is examined with an eyepiece lens. The problem with microscope design is that for good resolution, and hence useful high magnification, the objective lens must subtend a large angle above the specimen. But this is exactly the condition that produces atrocious aberrations. What Foucault did was insert a silvered-glass mirror into the optical train, as shown in Fig. 12.25, to which he applied *retouches locales* to counteract the aberrations of the objective lens. The resulting images were sharper, more luminous and could support greater magnifications. Foucault presented his microscope to the Société Philomathique in early 1859, and to give an indication of its performance on a well-known specimen, Foucault showed the Society a hasty sketch of red blood corpuscles made by Bulard.[58] The microscope was hardly compact, however, because the mirror was located

A 40-cm telescope

40 cm from the other optical parts, and microscope design improved taking other routes.

While Foucault was working on his microscope, Le Verrier made a mistake. He kept back Foucault's pay sheet, cutting off his salary for a second time. The excuse was that Foucault's personal work was not benefiting the Observatory because there was talk that his 33-cm telescope might go to the Algiers Observatory.[60] The Minister lost his temper, and to emphasize his anger took up the pen personally to write back to Le Verrier. 'I remember quite clearly the *definitive* solution adopted with respect to M. Foucault after a host of parleys,' he wrote. 'The agreement must be honoured, and if you were to persist, I think you would attract accusations of trouble-making and persecution. My categorical advice therefore is payment of the salary owing.'[31]

A 40-cm telescope

Foucault and Le Verrier were both stubborn, but the time was ripe for compromise. The autocratic director normally got his way, but he realized he had gone too far; and Foucault was at a point where he needed to work at the Observatory. Keys and salary were handed over, no doubt, but more importantly, Foucault now had a budget, and could attack a 40-cm mirror. Progress was fast. The two men having declared peace in 1859 February, the telescope was under test by June and was presented to the Academy by Le Verrier in July.[61]

It is not recorded where Foucault sourced the glass for his smaller mirrors, but a 40-cm round is a substantial slab. However, crown-glass blanks of that size could be cast by the famous glass works at Saint Gobain, whose origins dated back to Colbert, and the breaking of the Venetian monopoly on mirror making. Saint Gobain routinely produced large pieces of glass for the Fresnel lenses in lighthouses; but they were unannealed, that is to say the molten glass was not cooled with any great care, and the glass was full of internal stresses. As glass was cut away, stresses were released, and the disc could buckle, albeit minutely, and although this mattered little for a lighthouse lens, it could ruin the shape of a precision optical surface. However, Foucault will have reasoned that very little strain was released in the optical polishing, and that so long as this was the last operation on the glass, there was little danger. He was right.[62]

Foucault had seen the lighthouse at the Exposition Universelle made by Louis Sautter (Fig. 12.26).[63] Like Secrétan, Sautter was Swiss born. He had bought the factory where Fresnel had had his lighthouse lenses made, and it was in this factory, near where the Eiffel Tower now stands, that the Saint-Gobain blank was trimmed to size. The rear of the mirror was made convex with the glass twice as thick at the centre as at the edge. For a given weight, this made the mirror stiffer. The convex rear was polished so that the silver deposition could be monitored through the thickness of the glass, and a groove was cut into the edge into which rope handles were tied for easier handling.

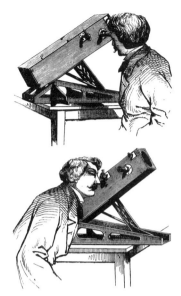

Fig. 12.24. Four-inch, silvered-glass telescopes in a mahogany frame were available from Secrétan's shop. According to a British source, they cost £10 (250 francs).[59] The length of eyepiece indicates that relay lenses were used. This produces upright images, which are desirable for terrestrial objects and would have increased saleability.

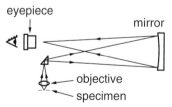

Fig. 12.25. Foucault's catadioptric microscope. *Retouches locales* were applied to the mirror to counteract the aberrations of the objective lens.

Fig. 12.26. Louis Sautter (1825–1912). For a native of land-locked Switzerland, his surprising business was the manufacture of lighthouses.

The grinding and initial polishing took place in Secrétan's workshop. Foucault outlined the procedures in his memoir on telescope making, written that autumn. It would be tedious to discuss more than a few highlights. A metal tool was used for the 40-cm disc, but more generally Foucault recommended using a cheaper, lighter glass tool for large mirrors. The ball was attached to a solid post set in the floor, and the mirror was placed on top. Most of the weight was taken by ropes and a spring attached to the ceiling. In this way less force was needed to move the mirror against the tool, and less heat was generated. This was important, because thermal expansion of the surfaces caused the ball to bulge and the mirror to flatten, resulting in excess removal of material where they remained in contact at the centre. If a particular radius of curvature was desired, progress was checked with a device called a spherometer, which can measure curvatures. Figure 12.27 shows a spherometer used by Foucault, made by Froment in his friend Deville's new metal aluminium. If the mirror curvature needed to be tighter, the length of the strokes was increased. More of the mirror weight was then taken by the ball edge and mirror centre, producing increased abrasion in these regions and greater curvature (Fig. 12.28a). To reduce the curvature, ball and mirror were swopped (Fig. 12.28b).

The final stage was the *retouches locales*, and these were done by Foucault himself at the Observatory. A space four or five times longer than the focal length was needed and the mirror was tested on its edge. He examined the mirror shape with only the first and last of his tests because they used the same test object. He described the preparation of his polishers in great detail,

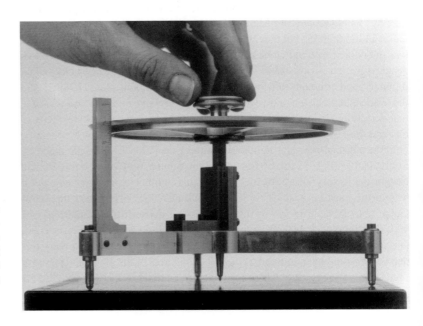

Fig. 12.27. Foucault's spherometer, made out of aluminium by Gustave Froment. The tricky part in spherometer use is to make sure that none of the feet lifts up when the central pin is lowered to touch the surface being measured.

(©*Observatoire de Paris*)

A 40-cm telescope

relating how they should be slightly more curved than the mirror for seamless *retouches,* and how in order to hold the rouge or Venetian tripoli satisfactorily, the paper glued on their surfaces should be brought to a bloom with a slurry of pumice in benzene. To establish a clear connection between cause and effect, he advocated systematic working, with strokes along diameters or chords for timed intervals. There is an infinity of ways that a sphere can be parabolized. The centre of the mirror can be deepened, its edges can be cut away, or these two extremes can be combined in any proportion. Deepening the centre is the easiest, and Foucault advocated this for mirrors less than 25 or 30 cm in diameter, but for larger mirrors he preferred to retouch both edge and centre, which minimized the amount of glass to be removed. The appearance of diffraction rings showed when the work was over.

Applying the *retouches locales,* Foucault must have felt at the height of his powers. It was precision work at the forefront of knowledge, requiring careful thought and a skilful hand and eye, but no mathematics. His friend Jules Lissajous commented:

> The workman gave him a surface with a sparkling shine, apparently finished, but in reality mediocre. In a few hours Foucault would bring it to the highest degree of perfection. He liked to compare himself to the sculptor who, with a few skilful touches, finishes the work of his apprentice and gives to the model of a medallion all its purity of line.[64]

Silvering was done in a copper basin rendered unreactive by an electroplated silver layer. Foucault stayed faithful to Drayton's process. It used costly ingredients such as oil of cloves and the strong-smelling gum galbanum, but the expense was minor for a telescope mirror. Solutions were made up in bulk, because the concentrations needed slight empirical adjustment for best results, and it was wise to use a mixture that had proved itself. The mirror needed to be cleaned scrupulously with nitric acid, *blanc d'Espagne* (an excipient which acted to absorb grime) and wipes of freshly carded cotton. After about five minutes in the silvering bath the thickening silver layer would suddenly change from transparent to reflective. A further period five or six times longer in the bath would produce a satisfactory deposit, which then needed buffing up with chamois leather for an hour or two. For Foucault, these procedures will have recalled daguerreotype-plate preparation.

Eichens mounted the 40 cm equatorially in wood (Fig. 12.29). The mirror rode on an air bag. The prism was kept small to minimize obscuration, but this meant that the focus was well inside the tube and out of reach of a simple eyepiece. Two relay lenses brought the focus outside the tube where it could be examined with interchangeable eyepieces for different magnifications. These lenses introduced spherical aberration, which Foucault eliminated with a final *retouche* to the mirror, which will have made it hyperboloidal. 'Contrary to normal usage, this method of compensation corrects the eyepiece with the

(a) Increase curvature

(b) Decrease curvature

Fig. 12.28. (*a*) During grinding, a longer stroke length made the mirror surface more curved because of the greater pressure and abrasion near the overhang (★). (*b*) To decrease the curvature, the ball and mirror were swopped over.

Fig. 12.29. Foucault's 40-cm telescope completed in 1859. The air bag has disappeared.

(©*Observatoire de Paris*)

objective, and has been applied with success,' Foucault informed the Academy. The telescope was tested on a ruled target 80 m away, and found to possess an optical power of 480 000. On the sky, it split the blue component of Gamma Andromedae.

The telescope was turned to the Moon. 'In the opinion of MM. Le Verrier and Chacornac,' reported Moigno, 'such a beautiful picture of the Moon has never before been enjoyed from under the Paris sky.'[65] The directors of the observatories at the Cape of Good Hope and in Armagh enjoyed the view too, along with George Airy. The British Association had been dismissive concerning Foucault's telescope two years earlier, but Airy reported back to the Royal Astronomical Society in London:

> in very high terms... The image was very brilliant, and many minute and delicate shadings, both on the volcanic and on the more level parts of the moon's surface, were brought out with a distinctness which he had never before witnessed.[66]

Le Verrier, Moigno reported, was pleased, and 'justifiably proud of the victory won for France'. Carried away with enthusiasm, Moigno continued:

> M. Le Verrier has already authorized M. Foucault – or rather, already ordered him – to give the Observatory a mirror with one metre diameter. This mirror will therefore soon be a wonderful reality. Already it will show much more than Lord Rosse's mirror, and then will come a two-metre mirror which promises marvels.[67]

We shall find that Foucault did not quite live up to these hopes. The biggest telescope he completed had a diameter of 80 cm, and no doubt he began work on it almost immediately, because it was finished by early 1862.[68] At the same time he made other mirrors. Bulard went off to Algiers with a 50-cm one, while Toulouse Observatory acquired a 33-cm mirror.[69] No doubt there were others, not to mention the telescopes being sold to individuals by Secrétan.

A Spanish eclipse

'These late eclipses in the sun and moon portend no good to us,' bewails Gloucester in *King Lear*, but for astronomers eclipses are full of interest, especially total solar ones, where the Moon blocks the bright light of the solar photosphere, exposing fainter details. Total eclipses are rare and often inaccessible, so with clear summer skies in prospect, astronomers from across Europe flocked to observe the solar eclipse that swept through Spain and Algeria during the afternoon of 1860 July 18.

Two French expeditions left for the eclipse track. One, from the École Polytechnique, went to Algeria, while the second, from the Observatory, went to Spain. Initial plans were made by Hervé Faye, but after protracted disagreements with Le Verrier, it was Le Verrier who took command.[70] With him went Foucault; Chacornac; another Observatory astronomer, Antoine Yvon

A Spanish eclipse

Villarceau; and some additional people from other institutions. No less than thirty-three crates of equipment were sent off at the end of June, much of it needed to determine latitude, longitude and time at the observing station. For observing the eclipse itself, Foucault's 20- and 40-cm telescopes were taken because with their short focal lengths they were less prone to buffeting by wind and relatively light and transportable; and indeed, they subsequently travelled to other eclipses.[71]

The eclipse path curved through northern Spain (Fig. 12.30). The selected observing site was near the *Santuario* or sanctuary at an altitude of 1400 m on the flanks of the Sierra del Moncayo mountains not far from Tarazona.[72] Yvon Villarceau arrived first to set up the instruments, followed by Foucault and Chacornac, and lastly Le Verrier, who was greeted by thunderstorms. The Observatory director was not one to await his fate passively. He sent telegrams to learn what weather was approaching from the Gulf of Gascony and commandeered what waggons he could from the surrounding villages. With Foucault and Chacornac he climbed the further thousand metres to the mountain summit and saw clouds all around. The *Santuario* bells were ringing to guide those lost in the fog.

The probability of clear skies looked slim to Le Verrier. Early on the morning of the eclipse, he and Foucault left camp in mist and fine rain, and with waggons and equipment started down into the valley where the skies had seemed clear the previous day. *En route* they met astonished locals climbing to view the eclipse with the astronomers; but Le Verrier and Foucault pressed downwards, finally stopping at about 11 a.m. on a little plateau near the Tarazona cemetery. The eclipse did not start until almost three hours later, but there was a lot to be done setting up clocks and aligning telescopes. In addition, Foucault needed to prepare a portable darkroom, because he was going to attempt to photograph totality.

In the event, the flight from the mountains was unnecessary, because the sky was clear everywhere at the 'supreme moment' of totality, as Foucault later put it in one of his last newspaper articles.[74] 'The savants, the amateurs and the curious from all countries who distributed themselves along the path of the shadow', he told his readers, 'have nothing to regret concerning either the costs incurred or the fatigues endured to be present at a spectacle which is so short, so grand and so rare.'

The various eclipse expeditions had four principal scientific goals. The first was to record the instants when the eclipse began and ended, since these timings could refine knowledge of the terrestrial and lunar orbits. Le Verrier attended to this. The second was to look for a new planet orbiting so close to the Sun that it was normally hidden by the Sun's blinding light. The previous year, Le Verrier had hypothesized that such a planet might account for unexplained details of Mercury's motion, but strangely, it was only one of the minor

Fig. 12.30. Path of the total solar eclipse on 1860 July 18 (shaded).[73] Le Verrier's party travelled through Bayonne and Pamplona to reach their observing site near Tarazona. A British expedition set up at Rivabellosa, near Vitoria. The Vatican's Father Secchi observed from near Castellón de la Plana. Another French expedition, from the École Polytechnique, observed near Batna in Algeria.

Fig. 12.31. The solar eclipse of 1842. Two pinky-red, mountain-like prominences were visible close to the limb. The much larger corona formed a complete ring with striations and two tufts.

Fig. 12.32. The one surviving Foucault photograph which is not a daguerreotype is this amphitype of the Couvent des Carmes. The light-sensitive salts were held in an egg albumen layer spread on a glass plate. After development the emulsion was bleached to yield a grey appearance. Displayed against a black background, there was some resemblance to the daguerreotype, with the highly exposed grey regions appearing white, while the unexposed transparent areas appeared black. It is difficult to date this amphitype, but Foucault was certainly experimenting with albumen-on-glass negative photography in 1852 when a distant cousin in the provinces wrote to enquire about progress.[75] As usually used, the term amphitype applies when the emulsion is albumen-based, whereas ambrotypes have emulsions held in collodion (cotton dissolved in ether).

(*Société française de photographie*)

members of the expedition up in the mountains who looked for new planets. (None was seen. Mercury's peculiar motion was finally explained by Einstein with General Relativity.)

The other scientific aims concerned the physical nature of the Sun. It had long been known that during the darkness of totality a great halo or corona of light appeared around the Sun and Moon, and during an eclipse in 1842, observers had been struck by pinky-red protuberances at the base of the corona (Fig. 12.31). We now know that the corona is an extremely hot, tenuous outer volume of the solar atmosphere, and that the protuberances or prominences are loops and columns of cooler solar material penetrating into the corona, but in 1860 it was unclear whether the corona and prominences were solar or associated with the Moon, produced perhaps by diffusion of sunlight in a supposed lunar atmosphere, or by diffraction. The decisive test was to measure whether they kept pace with the Sun's or the Moon's motion during the eclipse. Chacornac and Yvon Villarceau tackled this question for the prominences, which being sharp-edged, were easier to measure than the diaphanous corona. They used Foucault's telescopes with special eyepieces equipped with scales, and along with other astronomers, found that the prominences were solar. The question of the corona was resolved at the same eclipse by other expeditions, who showed that the corona was polarized and hence was not a diffraction phenomenon.

In the *Journal des Débats* Foucault described the shadow bands which ran across the ground in the seconds before totality, the pure-white corona, the slate-grey sky spangled by the brighter stars and planets, and the silver and copper colour of the light:

> At the sight of such an imposing spectacle one would like to control the emotion with which one is overcome, suspend the course of time, and concentrate all one's forces of attention on the grandiose scene which is unrolling with frightening rapidity.

Foucault's duty, however, was to photograph the corona and prominences, whose details could then be examined later and at leisure. Success was far from certain, because from previous reports it seemed the intensities would be too low, and indeed Foucault's friend Aimé Girard with the expedition in Algeria recorded nothing of totality. Nevertheless, the new, wet-collodion emulsions were some ten times more sensitive than the old daguerreotype plates, and it was of interest to try them out. Foucault adopted an ambrotype-like process, which was similar to one he had experimented with in the early 1850s (Fig. 12.32). He set up a *camera obscura* with a short focal-length lens and a guiding telescope and screw so that he could track the eclipse during the exposures.

The plates had to be prepared just before exposure, and used before they dried and lost sensitivity. During the 194 seconds of totality Foucault managed to take three plates with exposures of 10, 20 and 60 seconds. To his delight, the eclipse was recorded on all plates! It takes a cool head to work within a short and immutable time limit, and Foucault normally was cool, but overcome by emotion, he managed to jog the camera during the first exposure. Three extra images resulted, for which he estimated exposures of less than a quarter of a second. Each showed only the jagged lunar edge shadowed against the brightest, innermost part of the corona and prominences. The three images were identical, providing unexpected confirmation that the features were real, and not the result of emulsion blemishes or other external effects.

The longer exposures sampled deeper into the corona, recording an extent three times the solar diameter for the 60-second plate, as well as radial streaks (Fig. 12.33). A particularly bright streak emanated from a notch in the lunar edge, leading Foucault to conclude 'until fuller examination' that the corona was a diffraction phenomenon. He was perhaps influenced by his friend Hervé Faye, who held the same view, but he was wrong.

The 1860 eclipse was the first to be photographed successfully during totality, but Foucault was not the only one to do so. Among others were the Englishman Warren de la Rue, who managed to take two 60-second exposures, and Father Secchi, who took five 20-second exposures. De la Rue's photographs received extensive publicity, and it is a shame that Foucault's were neither preserved nor reproduced, because they were described as extending further into the corona; but jealous of his subordinates, perhaps, Le Verrier refused to publish their results in the Observatory's *Annales*.[77]

Fig. 12.33. A drawing of totality as it appeared from Tarazona.[76]

A 'really serious' telescope

His mirror-making trials, Foucault wrote, 'would only be of decisive importance from the day on which the mirrors thus produced attained dimensions superior to the biggest achromatic lenses.'[78] The biggest telescope lens in the world had a 52-cm diameter. It had been built by a certain Ignace Porro, an Italian living in Paris, and had been installed in an enormous, 15-m outdoor tube off the Boulevard d'Enfer, only a few steps from the Observatory.[79] Porro had offered this lens to Le Verrier in 1858 for the fabulous sum of 160 000 francs but there were doubts about its quality, Le Verrier was hostile, and then it seems the lens was broken. The 38-cm lenses in the refractors at Pulkova and Harvard were thus the biggest at professional observatories, and their size had been matched with the 40-cm reflector. Through prudence, perhaps, Foucault did not attempt Moigno's 1-m mirror, but set his ambition on 80-cm. Already, this was an increase by a factor of four in surface area and about eight in weight, but no particular difficulties were encountered in its construction and the telescope was a wonderful reality at the start of 1862 (Fig. 12.34). It was more-or-less a scaled-up version of the 40-cm telescope, though the focal length was decreased 10 per cent to 4.5 m in order to shorten the tube slightly.

Fig. 12.34. The 80-cm telescope (Foucault claimed a useful diameter of 78 cm). The wooden support has been modified from its initial form with horizontal and vertical axes into an equatorial mount adjusted for the latitude of Marseilles. Note how this telescope with its paraboloidal mirror is much stubbier than Lord Rosse's enormously long Leviathan (Fig. 12.8).

The glass had been cast by Saint Gobain in a mould specially prepared by Sautter and had been annealed, though not completely successfully. A full-sized tool would have required too much force for hand working, so the surface was ground spherical by Secrétan's workman using a 50-cm ball made of glass. The grinding took only a week. For the polishing, the most that a man could handle was a 22-cm paper-covered tool. The mirror was polished sytematically, in concentric circles, with frequent optical testing.[11] This operation also took no more than a week. Foucault applied the *retouches locales* at the Observatory. Eichens made a rapid mount out of pine with vertical and horizontal axes, and whenever the weather was favourable, the telescope was trundled out onto the Observatory terrace for observing. It was the first silvered-glass telescope that was 'really serious', Foucault told his readers in the *Journal des*

A 'really serious' telescope

Débats. The columnist in *Le Temps* newspaper optimistically hoped that the instrument would finally secure Foucault's election to the Academy.[80]

Chacornac was the usual observer with the telescope, though Foucault and others peeked too.[81] After several nights of turbid skies, the atmosphere calmed on March 20 and Chacornac was able to confirm the existence of a very faint companion to the star Sirius, which had been discovered two months previously by the American telescope-maker Alvan Clark. (The huge difference in brightness astounded astronomers, since it indicated that stars could have enormously different diameters. We now understand that Sirius's companion is a condensed object little bigger than the Earth, called a white dwarf.) Chacornac searched for a nebula (an extended patch of light) which it was claimed had disappeared. He observed a transit across Saturn's face of one of its moons, Titan. To greet the new telescope, nature sent another comet, which Chacornac drew (Fig. 12.35). He also drew the Ring and Whirlpool nebulae. Spiral structure had been discovered in the Whirlpool by Lord Rosse with the Leviathan, and it would be fascinating to see Chacornac's drawing, but unfortunately it is lost, though the description claims many more stars than Lord Rosse had seen.

Le Verrier was delighted with the telescope, and in reward arranged Foucault's promotion to *Officier* (Officer) in the Légion d'honneur (Fig. 12.36).[82] Nevertheless, the long waits for clear nights had made it very clear that Parisian skies were not the right location for France's most powerful telescope. The

Fig. 12.35. The nucleus of Comet Swift–Tuttle observed from Paris with Foucault's 80-cm telescope at 9 p.m. on 1862 August 23. A few years later the Italian astronomer Giovanni Schiaparelli showed that this comet followed the same orbit as the Perseid meteor shower, so confirming the suspected association between meteors and comets. Comet Swift–Tuttle returned in 1992.

Fig. 12.36. Foucault was made *Officier* in the Légion d'honneur for his work on the silvered-glass telescope.

(Family collection)

Fig. 12.37. Foucault's plan of the turret built to house the 80-cm telescope in Marseilles.

(Observatoire de Marseille)

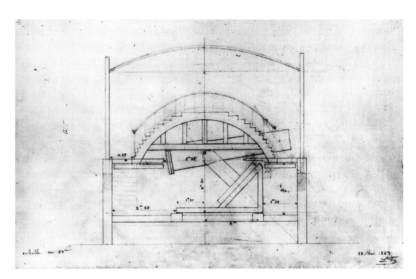

Fig. 12.38. Foucault's 80-cm telescope in the turret built to house it in Marseilles.

(Observatoire de Marseille)

Minister was approached and permission was rapidly given for the 80-cm to be moved to clearer skies in the south of France.

The towns of Montpellier, Marseilles and Toulon all made bids to be the permanent home for Foucault's masterpiece. Marseilles offered the most money, but there were worries that observations would be affected by the *mistral*, the strong wind that can blow down the Rhône valley. Luckily the *mistral* was blowing during an exploratory visit that Le Verrier and Foucault made to the Mediterranean coast. In Montpellier they were greeted at the railway station by Alfred Donné, who was now Rector of the university there. Donné was surprised and delighted to see his friend, whom he had not realized would accompany Le Verrier. Foucault and Le Verrier attended a meeting of the local learned society and were fêted with an outdoor dinner. However the *mistral* was just as strong in Montpellier and Toulon as in Marseilles. There was no further hesitation: Marseilles' offer was accepted. Figure 12.37 shows Foucault's plan for the dome with its unusual turret form and movable internal bridge to enable the observer to access the eyepiece.

The turret was completed during the summer of 1864, the telescope was shipped from Paris in September, and by December 28 observations had begun. Figure 12.38 shows the completed turret with telescope installed. The telescope served until 1965. Highlights of its career include the search for nebulae and double stars, as well as a first attempt in the 1870s to measure the angular size of stars. It was Fizeau's idea, but it was executed by the then director of the Marseilles Observatory, Édouard Stéphan.[83] Stéphan masked Foucault's mirror with two parallel, eye-shaped slits (Fig. 12.39). These acted as the slits in a Young's interference experiment (cf. Fig. 5.6) and with white starlight gave an interference pattern in the eyepiece something like that shown in Fig. 5.8. In most interference experiments the slits are close together in order to produce wide fringes, but with Stéphan's widely separated slits, the fringes were very small, having an angular extent of about $\frac{1}{6}$ arcsecond. If the angular size of the star causing the fringes approaches this value, the fringes will be smeared. Stéphan examined all the bright northern stars and saw no smearing from any of them, thus showing that none had angular diameters as big as $\frac{1}{6}$ arcsecond. Fifty years were to pass before a stellar angular diameter was successfully measured, by A. A. Michelson and F. G. Pease at the Mt Wilson Observatory in California.[84] Fringes from the giant star Betelgeuse disappeared when the slits were 2.5 m apart, indicating an angular diameter of about $\frac{1}{20}$ arcsecond. Stéphan had not been so far from success.

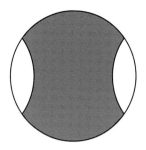

Fig. 12.39. Stéphan took advantage of the Marseilles telescope's large diameter in an interferometric search for stellar angular diameters. The mirror was masked to provide two widely separated slits which produced interference fringes with an angular separation of 0.16 arcsecond. Stéphan examined all the bright northern stars, but the fringes were never smeared, showing that in every case the star's angular diameter was less than the 0.16-arcsecond fringe spacing.

Modern telescope making

Some prescient remarks were made at the birth of the silvered-glass telescope. Having heard about the new invention, Donné wrote from Montpellier to congratulate Foucault. He first teased Foucault, answering the rhetorical question as to whether telescope building was occupying all his time with a comment that once again indicates Foucault's delight in the ladies: 'But no, because I know you still have some moments for *soirées* on the Rue de l'Odéon where it seems to me you have the air of having dates with little Marie.' Donné continued:

> ...you know, thinking again about your devil of a telescope, I thought that with your ingenious procedure there would be, so to speak, no limit to the size of the mirror, and that it could probably apply to other materials besides glass, to one of these glazes made today which are harder than marble and can take a perfect polish. Thus one could construct mirrors in masonry as big as the dome of the Panthéon or the Montmartre escarpment... one could even make mirrors of several pieces of glass...[85]

These predictions, most of which were echoed by Steinheil the following year, have proved essentially correct.[15] Foucault's idea of test-and-correct continues to underpin optical fabrication. With the advent of astronomical photography and the rise of astrophysics and extra-galactic astronomy, reflecting telescopes with fast focal ratios have become the workhorses of everyday astronomy. Foucault's ideas have permitted the construction of enormous mirrors, such as the 5-m mirror in the Hale telescope on Mt Palomar, the 6-m mirror at Zelenchuskaya in Georgia and the four 8-m mirrors comprising the European Very Large Telescope at Paranal in northern Chile. To be sure, some details have evolved. Glass has been replaced, as predicted by Donné, by ceramics, and by zero-expansion ones whose size does not change with temperature. This is advantageous both in the dome when the night-time temperature

Fig. 12.40. (*Upper*) A modern knife-edge tester. The illuminated pinhole can be seen shining just to the left of a small square hole formed by four knife edges. (*Lower*) Using the knife edge to test a 60-cm telescope mirror at the Mount John University Observatory, New Zealand.

Fig. 12.41. 'Foucault-gram' of a 30-cm mirror. (*Left*) Direct image of the mirror; its maker is faintly visible in a reflection from its flat *rear* surface. (*Right*) The knife-edge test indicates that the mirror is almost perfect, with only slight overpolishing at its centre, and a common, but unimportant, narrow turned-down edge.

(*Sarah Wheaton*)

changes and in the workshop, where the mirror can be tested immediately despite the heat generated by the *retouches locales*. Mirrors are now tested directly on the polishing machine, using laser interference techniques and zonal tests in which different regions are examined independently through the use of masks. Diamond cutters on computer-controlled lathes and erosion by plasmas can produce aspherical, non-rotationally symmetrical optical surfaces, while at a more prosaic level, silicon carbide has replaced emery as an abrasive for grinding, and mirrors are aluminized rather than silvered, though silver with a protective overcoat is returning to favour because of its greater reflectivity at short wavelengths.

Only now, a century and a half after Foucault's first telescope, are radically new ideas entering telescope construction. Foucault's air bag has transformed into computer-controlled activators which deform mirrors so as to counteract flexure and to correct for the smearing introduced by atmospheric turbulence. The idea that many mirrors could be ganged together to produce large collecting areas has been implemented, and in the next generation of telescopes, projects such as the European Southern Observatory's Overwhelmingly Large Telescope may achieve the size of the Panthéon dome and the *butte Montmartre*. The knife-edge test nevertheless remains an invaluable final test for focus quality, and if it had been applied to the mirror in the Hubble Space Telescope would have revealed its unplanned aberrations in seconds. Figure 12.40 shows a modern knife-edge tester in use. Because of its simplicity, the knife-edge test is extensively used by amateur telescope makers. Figure 12.41 shows the 'foucault-gram' of an amateur's excellent 30-cm mirror made from glass scavenged from a porthole.

Greater ambitions

The 80-cm mirror was the biggest one completed by Foucault, but it was not the end of his ambitions in either astronomical optics or telescope building, as we shall find in later chapters. But before we turn to these essentially technological developments, we must consider the final purely scientific experiment completed by Foucault. In a career characterized by precision and subtle effects, this experiment stands out as the single one in which its quality was expressed in an accurate numerical result. It is Foucault's measurement of the speed of light.

Chapter 13

The speed of light. II. The size of the solar system

The year was 1861 or so. Foucault was well on the way to completing a really serious telescope for the Observatory. Le Verrier was pleased. He decided to commit funds to a project which was as dear to his heart as it was to Foucault's. The project was the laboratory measurement of the speed of light.

We now know that the speed of light is a physical constant which is fundamental in electromagnetism and relativity, but in the early 1860s it seemed no more fundamental than the speed of sound. 'If the beautiful experiment by which M. Léon Foucault will soon attempt a direct measurement of the speed of light is crowned with full success,' said Le Verrier in 1861, 'this most critical question of modern astronomy will soon be answered.'[1] Why was the speed of light a question of astronomy rather than physics, and why was it so crucial?

The solar parallax, or distance to the Sun

The quantity that was really at issue was the size of the solar system. Kepler had drawn a scale map of the solar system, but actual distances were unknown. If one distance, such as the Sun–Earth separation, could be determined in absolute terms, all other distances would follow.

The Sun–Earth distance changes by a small percentage during the year because the Earth's orbit is slightly elliptical, so it is usual to refer to its mean value. This is the astronomical unit, and is expressed in metres or kilometres. In the nineteenth century, however, a related quantity called the *solar parallax* was used, defined as the angle subtended by the Earth's equatorial radius when viewed from the Sun (Fig. 13.1). It was from the value of the solar parallax that the speed of light was determined in two different ways discussed in Chapter 8. To recall, one method took the diameter of the Earth's orbit obtained from the solar parallax and divided it by the light-crossing time given by eclipses of Jupiter's moons (Fig. 8.1). The other derived from the value of stellar

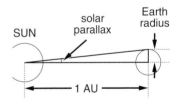

Fig. 13.1. The relation between the astronomical unit (AU), the radius of the Earth and the solar parallax. The solar parallax is just an alternative, though indirect, way of specifying the average distance to the Sun. A larger solar parallax implies a shorter astronomical unit, and vice versa. Not to scale. The solar parallax is actually about one four-hundredth of a degree.

aberration (Fig. 8.3), which in appropriate units equals the ratio of the Earth's orbital velocity to the speed of light. This yields the speed of light because the Earth's orbital velocity can be calculated by dividing the circumference of the Earth's orbit (derived from the solar parallax) by its accurately known orbital period of one year.

The speed of light was thus determined from astronomical measurements, not laboratory ones. The value differed by about 1 per cent depending on whether the stellar aberration or the light-crossing time was used, but this difference was minor compared to the much greater uncertainty arising in both methods from the poorly known solar parallax.

The determination of the solar parallax was characterized by Airy as 'the noblest problem in astronomy'.[2] In principle, the most straightforward method was direct triangulation of the Sun from two points on Earth with known separation. In practice, Mars and Venus were triangulated, because they lie closer to the Earth and can be measured more accurately when in *opposition* and *conjunction*, respectively (Fig. 13.2). (Opposition is when a planet and the Sun lie in opposite directions from Earth, and conjunction when they lie in the same direction.)

Venus at conjunction is much closer to the Earth than Mars at opposition and offers prospects of correspondingly better accuracy. However, planetary orbits are inclined to each other by a few degrees. Alignments are rarely exact, and the Earth-facing dark side of Venus is invisible against the dark of the sky. Very occasionally, however, the alignment is sufficiently accurate that Venus appears as as a black spot creeping across the face of the Sun. During this so-called transit, Venus can be triangulated.

Transits of Venus were thus the acknowledged best way to determine the solar parallax. The necessary alignments are rare; they occur once every century or so, and two transits then take place separated by 8 years. The last transits of Venus had occurred in 1761 and 1769, and many expeditions had observed the latter transit, including Captain James Cook's from Tahiti. The best estimate of the solar parallax was believed to be the 8.57-arcsecond value derived by Encke in Berlin from observations of these two events, but there was considerable unease concerning his result, which was heavily dependent on uncorroborated observations from Norway.

Another approach to the solar parallax came through Le Verrier's life work in astronomy, which was to analyse solar-system orbits using Newton's Laws of gravitation and motion. This heroic project took ideas which had led to the discovery of Neptune and extended them to the whole solar system.

To understand what Le Verrier was attempting, consider first the acceleration on a planet caused by the gravitational pull of the Sun. Its magnitude is given by Newton's gravitational constant, G (which characterizes the strength of gravitational forces), multiplied by the solar mass, M_{Sun}, and divided by the square of the distance separating the planet and the Sun. If this was the only

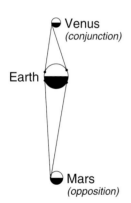

Fig. 13.2. Astronomers determined the solar parallax and distances in the solar system via triangulation of Venus when at its closest point, conjunction, 0.28 AU away. Mars at its opposition distance of 0.52 AU was also used. Triangulation of the more-distant Sun would have been less accurate. Not to scale. The small angle in the Earth–Venus triangulation is actually about one-sixtieth of a degree.

The solar parallax, or distance to the Sun

acceleration on the planet, its orbit would be an accurate ellipse, and from the time taken to complete one orbit it is possible to write an equation which relates the product GM_{Sun} to the Sun–planet distance, but does not set the numerical value of either. Only the mass of the Sun enters this equation, not that of the planet. The solar mass enters the corresponding equations for every planet, and through it the relative sizes of the planetary orbits can be determined. As Kepler had discovered, a scale map of the solar system can be drawn – with Jupiter orbiting five times further from the Sun than does the Earth, for example – but with no indication of actual distances or masses. Because the Sun is the major mass in the solar system, orbits are to a first approximation elliptical, as found by Kepler, and it was from analyses along these lines that Newton had justified Kepler's discoveries theoretically.

In reality, the gravitational attraction of the Sun is not the only force acting on the planets. Each planet exerts a gravitational pull on every other planet, and on the Sun; the biggest pull is from Jupiter, the most massive planet. In consequence the planetary orbits deviate slightly from ellipses, and coded within these deviations is information on the separations between the planets and their masses. Le Verrier's genius was to write down equations describing all these perturbations in such a way that the masses and separations could be deciphered from the observed irregularities of the orbits.

Le Verrier's analysis was complicated by additional constraints. Observations were made from a moving platform, Earth, and yielded planetary *directions* with only rudimentary information on distances. His equations involved almost 500 terms. The masses in these equations were always multiplied by Newton's gravitational constant, G, but G was poorly known: early nineteenth-century determinations had hardly improved on the 7 per cent uncertainty of the famous experiment conducted in 1797–98 by Henry Cavendish.[3] However, the *product* GM_{Earth} was well known, because it is set by two accurately measured quantitites, the rate at which a falling body accelerates and the radius of the Earth. In Le Verrier's equations, the known quantity GM_{Earth} was therefore the Rosetta Stone which revealed numerical values for the distances and the products GM_{Sun}, $GM_{Mercury}$, etc. These products could be divided to obtain the masses of the individual planets relative to the Sun because G cancels, leaving $M_{Mercury}/M_{Sun}$, M_{Venus}/M_{Sun}, etc. Since relative distances were given by Kepler's Laws, Le Verrier needed to write his equations in terms of only one absolute distance. For this he adopted the Sun–Earth distance, represented by the solar parallax.

After formulating these lengthy perturbation equations, Le Verrier reviewed and, as necessary, re-reduced planetary observations from the previous hundred years. He was then ready to tackle the Sun's motion in the sky, which of course is actually a reflection of the Earth's orbit. This had to be treated first, because if the orbit of the observing platform was evaluated wrongly, the locations of the planets would be in error too. Mathematical methods applied

Fig. 13.3. A confident Le Verrier. His life work in astronomy was to analyse solar-system motions in terms of Newtonian mechanics and gravitation. The project was conceived in 1847 and completed only weeks before his death thirty years later. When fully published, it occupied over 4000 pages in the *Annales de l'Observatoire de Paris*.

(©Observatoire de Paris)

to the perturbation equations yielded the solar parallax value and planetary mass ratios which predicted wobbles in the Sun's motion that best matched the observed ones.

A first report in 1853 was followed by a definitive memoir in 1858. Newtonian gravity, Le Verrier found, could explain the Sun's motion provided that the masses of the Earth and Mars relative to the Sun were respectively about one-tenth greater and smaller than accepted, and that the solar parallax was 8.95 arcsecond, more than 4 per cent bigger than Encke's value. His subsequent analyses of Venus and Mars in 1861 supported these conclusions.

Here then was the crux of the matter. The speed of light and the distance to the Sun are related quantities. Measure one and you have the other. After Fizeau's toothed-wheeled experiment between Suresnes and Montmartre, Arago had written:

> By repeating this experiment with mechanically more perfect apparatus, some day it will be possible, without leaving Paris and its environs, to measure this solar parallax which, around the middle of the last century, gave rise to voyages that were so long, so distant, so arduous and so expensive.[4]

Le Verrier's increased solar parallax implied that the Sun–Earth distance was 4 per cent less than was thought. In consequence the actual value of the speed of light should be smaller by the same fraction compared to the then-accepted value, which as we have seen in Chapter 8 was 308 300 km/s. Would a laboratory measurement made with mechanically more perfect apparatus agree with Le Verrier's prediction?

A laboratory measument

Foucault set about modifying his 1850 spinning-mirror experiment to provide the answer. This time the apparatus was set up in the Observatory. According to Moigno, the equipment was ready in late spring 1861,[6] but over a year was to pass before Foucault obtained a result because he was embroiled in other activities. The 80-cm mirror needed its final *retouches*, and he was busy with heliostats and regulators, as we shall discover in Chapter 15, as well as administrative tasks such as drafting the case for the award of the Légion d'honneur to Eichens.[7]

The apparatus was built by Froment and incorporated three innovations. First, Foucault dispensed with steam and condensation. The spinning mirror (Plate XXII) was driven with air supplied from a patented, table-sized bellows and regulator valve provided by the renowned organ builder, Aristide Cavaillé-Coll (Fig. 13.4).[8] Cavaillé-Coll was at that moment completing a new organ for St Sulpice, Foucault's parish church; but he was a friend and neighbour around the corner on the Rue de Vaugirard. The bellows and newly designed valve

Fig. 13.4. Aristide Cavaillé-Coll (1811–99). He built over 500 organs in France, including that in Notre Dame de Paris, and 100 abroad, including one in Manchester Town Hall. He also supplied organs and bellows for acoustical experiments. Foucault may have known Cavaillé-Coll since 1840, when his late friend Belfield-Lefèvre had ordered an organ from Cavaillé-Coll for despatch to India.[5]

(Association A. Cavaillé-Coll)

A laboratory measument

provided pressures stable to 1 part in 1500 and enabled Foucault to adjust and maintain the speed of the spinning mirror precisely. The second innovation was a 400-toothed wheel (Plate XXIII) which rotated accurately once per second and was used to set the speed of the spinning mirror, as will be explained below. The third innovation was a chain of five concave mirrors (Plate XXIV) which extended the light path without the significant loss of intensity that would have resulted with a single mirror of practicable size so far from the spinning mirror.

Figure 13.5 shows the optical arrangement of the apparatus.[9] It was much as Foucault had suggested twelve years earlier. Light from a heliostat illuminated a silvered-glass reticle which had been ruled 'with very great care' by Froment with 31 transparent lines, each one-tenth of a millimetre apart. Following Foucault's lead, we will first consider the apparatus with the spinning mirror stationary. The light passed from the reticle to this mirror, situated about 1 m away, and was imaged by a lens onto the first relay mirror a further 4 m distant. Four more mirrors extended the one-way light path to 20.2 m. The final relay mirror returned the light back along the path it had taken, and when everything was properly aligned, a final image of the reticle formed on top of the reticle itself, in just the same way as an image of the wire formed on top of itself in the 1850 experiment. However, a beam-splitting glass plate reflected the final image into a microscope where it could be measured with the crosshairs of a calibrated micrometer. The toothed timing wheel was positioned in the same plane as the reticle image so that both were simultaneously in focus through the microscope.

Figure 13.6 shows the view down the microscope. The diagonal cross-hairs of the micrometer are superimposed on the image of the reticle. The timing wheel is seen below; rotating once per second, its teeth will have appeared smeared. When Foucault opened the air valve and the spinning mirror began to turn, the scene in the eyepiece will have started to flash because the mirror is correctly angled for a return reflection only once per rotation. This flashing background acted as a stroboscope, and by adjusting the air flow, Foucault was able to make the teeth appear stationary. The spinning mirror was then rotating

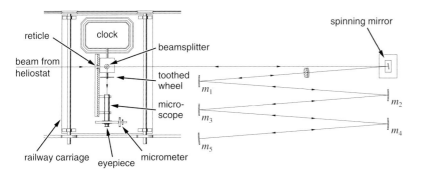

Fig. 13.5. Plan view of Foucault's 1862 laboratory measurement of the speed of light. Both a rotating mirror and a toothed wheel were involved. Not to scale. The one-way light path from the spinning mirror through the relay mirrors m_1 to m_5 was 20.2 m.

at the same rate that the teeth were passing, or 400 r.p.s. He found this speed could be maintained to within 1 part in 10 000 for several minutes.

Spinning the mirror also deflected the image of the reticle, because the mirror turned slightly during the time it took for the light to make the 40.4-m round trip along the mirror chain. The equipment was finally operational at the beginning of the summer of 1862. Foucault used the micrometer cross hairs to measure the image deflection, which yielded the speed of light when coupled with the mirror speed, the length of the mirror chain, and the distance beween the spinning mirror and the reticle image. Cloudy skies impeded work, but Foucault was surprised to find that his measures were more discordant than was reasonable considering the precision of the various components in the apparatus. It took quite some sleuthing to discover that the micrometer was less accurate than claimed. Foucault's surprising remedy was not to replace the micrometer with a better one, but to put the reticle, beamsplitter, toothed wheel, microscope and micrometer on a small railway carriage, whose distance from the spinning mirror was adjusted until the deviation of the image of Froment's trustworthy reticle was 0.7 mm, as determined from the reticle scale itself. This solution did offer the advantage that it was a macroscopic quantity that varied between one determination and another. 'The measurement then being of a length of about 1 metre,' Foucault noted, 'the smallest fractions still retain a directly visible size and no longer allow any room for error.'

It was now a matter of waiting for sunlight, but Le Verrier was pushing for a result. Mars was in opposition and a new determination of the solar parallax was imminent. He wanted Foucault's measurement to be published first.[10]

The skies cleared in mid-September (Fig. 13.7). Le Verrier read Foucault's brief note reporting his result to the Academy a few days later.[11] 'Derived from accepted numbers,' Foucault wrote, 'this speed should be 308 million metres per second, and the new experiment with the spinning mirror gives, in round numbers, 298 million.' He estimated that the uncertainty was no bigger than half a million metres per second. The solar parallax was increased from 8.57 to 8.86 arcsecond. Le Verrier's prediction was confirmed in its essentials. The solar system had shrunk by a thirtieth!

A triumph which no one will contest

Foucault's result earned immediate praise. The following week Babinet told the Academy that Foucault's measurement was 'a great scientific event', and 'a triumph which no one will contest'.[12] He went on to demonstrate that the opposition of Mars could at best provide an accuracy of about 1 part in 40, far inferior to Foucault's claimed 1 in 600. Foucault made plans to obtain even better accuracy by extending the light path to 'several hundred' metres with a longer train of transfer lenses (rather than mirrors), and even chose where to install the apparatus in the Observatory, but time overtook him and these

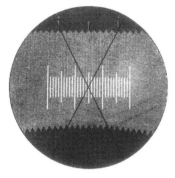

Fig. 13.6. Eyepiece view for the 1862 speed-of-light experiment. Unlike the 1850 experiment (Fig. 8.12), there is a scale, allowing quantitative measurements. The background will have appeared dim blue, because the scale was ruled in a silvered-glass plate, and silver layers transmit blue light faintly. The lower toothed edge is the curved rim of the timing wheel. It is unclear what purpose was served by the straight, upper toothed edge, which presumably was affixed against the reticle.

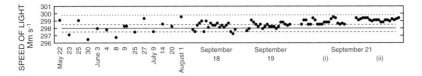

Fig. 13.7. Foucault's approach to the speed of light during 1862. Preliminary, less precise experiments in spring and summer gave way to definitive measurements in September. The full line and dashed outriders represent Foucault's final numerical value and its uncertainty, 298 000 ± 500 km/s; he appears to have discounted the measurements taken on Sunday, September 21. The astronomically derived value for the speed of light in the 1850s was 308 300 km/s, well outside the limits of this plot. The modern value, defined exactly since 1983, is shown by the dotted line.

intentions came to nothing.[13]

However, people *did* contest Foucault's triumph.[14] The 0.7-mm deviation was small. Fizeau's manuscript papers list other worries, including eccentricity of the surface of the rotating mirror, refraction caused by vortices of disturbed air in its vicinity, and deformation by centifugal forces. Most importantly, if the beamsplitter faces are not plane and parallel, the crucial scale of the image can be altered.[15]

There was also a human side to this distrust. Everyone knew that Le Verrier was overbearing and expected Foucault to find a smaller value. Had this pressure biased Foucault? Moigno addressed the point directly when Foucault was dying five years later:

> M. Le Verrier was pressing M. Foucault hard to determine this parallax by the direct measure of the speed of light. 'M. Foucault', [Le Verrier] added, 'would bring me his numbers each day, and I encouraged him to persevere until he arrived at the true figure.' Obviously M. Le Verrier was voluntarily or involuntarily comparing M. Foucault's determinations with his value of 8″.92, ... and if we did not know M. Foucault's character, we could well believe that [M. Le Verrier] did not leave him in peace until he had found the value 8.86, very little different from his own.[16]

Others, who did not know Foucault's character, may reasonably have entertained doubts concerning this most-difficult experiment; in addition, it was the first numerical result in Foucault's career, and he had no history of previous measurements to attest his competence as a quantitative experimenter. 'My last memoirs were very poorly received,' he wrote in a private letter. Making reference to a recent enlargement of the metropolis, he continued, 'One can push back the boundaries of Paris, but one does not change the distance to the sun with such impunity.'[17] Within ten years it seemed worthwhile to undertake new laboratory measurements of the speed of light. Like Foucault's, they were financed by the Observatory. The experimenter was Fizeau's protégé Alfred Cornu, from the École Polytechnique.

Cornu tried both Foucault's spinning mirror and Fizeau's toothed wheel, and preferred the latter. During experiments conducted in 1872 he obtained a value for the speed of light, estimated no more uncertain than 1 part in 300, of 298 500 km/s, extraordinarily close to Foucault's value. 'I shall not try to hide the fact that Foucault's number seemed suspect to me ... ' wrote Cornu. 'This coincidence, which was far from what I expected, I must admit, in great

Table 13.1. Definition of the metre adopted by the *Conférence Générale des Poids et Mesures* in 1983 October.

Authoritative text:

Le mètre est la longeur du trajet parcouru dans le vide par la lumière pendant une durée de 1/299 792 458 de seconde.

English translation:

The metre is the length of the path travelled by light in vacuum during a time interval of 1/299 792 458 of a second.

part removes the doubts that I had concerning the method of the turning mirror...'[18] In a later experiment over a 23-km baseline, he obtained 300 400 km/s, with a claimed accuracy of 1 in 1000.[19]

Cornu's measurements were made with an eye towards imminent transits of Venus in 1874 and 1882. Some eighty expeditions spread across the globe for the first transit, but a definitive solar parallax was not forthcoming; in particular, photographic measurements, from which much was expected, proved only slightly superior to visual observations. The speed of light was also gaining importance in physics. It had been shown to be close to a ratio of fundamental quantities in electricity and magnetism in 1856, and in 1868 the Scottish physicist James Clerk Maxwell had predicted that light was an electromagnetic wave. Subsequent speed-of-light determinations need not concern us, but by 1983 methods had become so accurate that it became sensible for the world's metrologists to redefine the metre in terms that made the speed of light an exact, defined quantity, 299 792 458 metres per second (Table 13.1). This modern value is very close to Foucault's 1862 result, though we can see that his estimated error was slightly optimistic. A spinning mirror experiment performed today – and with an electric motor, it is simple enough for the school laboratory – yields the length of the metre, not the speed of light.[20]

The astronomical unit is now one of astronomy's most accurately measured quantities. It is 149 597 870 660 m with an uncertainty of a few hundred metres. The corresponding solar parallax, rounded to only a few digits, is 8.79 arcsecond, vindicating both Le Verrier's prediction and Foucault's measurement. However, these values do not derive from laboratory experiments of great perfection, as expected by Arago, but from planetary-probe and planetary radar ranging data, as well as observations of the meridian passage of planets made since 1911 by the U.S. Naval Observatory. Arago's prediction was not completely unfounded, however, because knowledge of the speed of electromagnetic waves is crucial for the interpretation of ranging delays, but the spacecraft projected across the solar system have involved voyages far longer, far further from Paris, and of far greater expense than Arago, Le Verrier or Foucault could ever have imagined.

Chapter 14

Recognition

Foucault was both outsider and insider. He was outside the Academy, outside the faith no doubt, and until 1855 he had been outside employment. As a journalist he reported as an onlooker, insensitive to the effects of his criticisms. In contrast, he had his circle of friends, who met at his house every Thursday to discuss developments in science;[1] like gatherings were held by friends Henri Sainte-Claire Deville at the École Normale on Sunday mornings, and Joseph Bertrand at his home on the Rue de Rivoli.[2] Foucault was sufficient of an insider to be taken under Dumas' wing, to undertake commissions for the Ministry of Finance, and to sit on committees for the Exposition Universelle. His wardrobe included court dress (in blue embroidered with gold) reflecting formal visits to the Emperor. His worth was recognized, which is one of life's precious prizes. Finally, he had a bachelor's devotion to his society meetings: the Société Philomathique on Saturday nights, and later the Société Française de Photographie, or French Photographic Society, of which he was one of the ninety-three founding members in 1854.

The Société Française de Photographie

The Photographic Society was, according to its statutes, 'a purely artistic and scientific association' and its members entered into 'no engagement other than to work according to their ability for the development of photography'. Like the Société Philomathique, it was modelled on the Académie des Sciences, with a restricted membership, meetings where papers were presented, and published proceedings. Its founding members included artists such as Baron Gros, instrument makers such as Chevalier and Secrétan, numerous photographers and a slew of scientists including Victor Regnault (the Society's first president), Edmond Becquerel, Aimé Girard, Moigno, and of course Foucault himself. Foucault was soon elected to the Society's committee, with the remark that 'the Society should congratulate itself on the election of M. L. Foucault...whose new position at the Observatory...requires constant use of photography.'[3]

Constant use of photography may have been the intention, but in the event neither Foucault nor the Observatory devoted much effort to astronomical photography. This is surprising. Reflecting telescopes are particularly well-adapted to astronomical photography on account of their achromaticity and the wide field of view that results from their short focal lengths. These reasons underlie reflectors' ultimate popularity; but despite some snapshots of a lunar eclipse taken by colleagues with the 20-cm telescope, by 1866 Moigno was bemoaning the Observatory's dilatory involvement in celestial photography.[4]

Though minimally active in photography, Foucault was fully active in the Photographic Society.[5] 'He followed the work of the Society with an assiduity of which we were all proud,' Girard recalled.[6] He was repeatedly re-elected to the Committee,[7] he sat on commissions reporting on technical matters, and he belonged to every selection panel for the Society's approximately annual exhibitions from 1859 until his death. With his delight in well-executed engravings, he will have relished sifting through the submitted photographs, even though his judgments may have been primarily technical. He occasionally took the chair at the Society's monthly Friday-night meetings and made comments after papers that were soundly based on his personal experience, concerning, for example, photomicrographs or shutters that fall under gravity. He sat on various prize committees, including one that has remained famous. This was the competition sponsored by his antiquary neighbour, the Duc de Luynes.

According to one biographer, 'M. de Luynes took to photography with the vivacity of an amateur and the reasoned zeal of a chemist.'[8] Like Arago, de Luynes saw the immense potential of photography in natural history, ethnography and archaeology, but the difficulty of printing numerous, true and permanent copies remained unsolved. In 1856 he gave the French Photographic Society 10 000 francs for the purpose of 'stimulating the zeal of persons devoting themselves to this significant question...'

It was decided to hold two competitions. The first was for the system that gave the most permanent results. The second was for the best method of photoreproduction in commercial printing. The panel of judges included engravers, archivists and photographers; scientists besides Foucault were Victor Regnault, Balard, and the omnipresent Edmond Becquerel.[9] We need not detail the saga of the competition. The closing dates were put back several times, and promising candidates were required to demonstrate their process in front of a sub-committee including Foucault.[10] Both prizes were ultimately awarded for a carbon-based process devised by Alphonse Poitevin, a chemical engineer. The judges' unanimous recommendation for the second award was not announced until the spring of 1867, and was one of Foucault's last acts for the Society. The committee failed to appreciate the potential of one entry, from Charles de Berchtold in Vienna, which involved half-toning. 'This procedure has not seemed... of sufficient significance', reads their final report, but half-toning has become the principal method of photomechanical reproduction.[11]

Prizes unwon

Committee membership in societies is recognition of a sort, but it is partly self-imposed. External approval was indicated by Foucault's appointment to the Légion d'honneur and government committees such as the Comité Consultatif des Arts et Manufactures (Consultative Committee for Arts and Manufacturing), which approved patents, advised on tariffs and classified insalubrious industries.[12] He was decorated by the Emperor of Brazil, Dom Pedro II, who was fanatical about science and astronomy. The Royal Society awarded him its prestigious Copley Medal.

No money attached to these rewards. In the mid-nineteenth century, science was only beginning to be institutionalized in state and industrial laboratories. Researchers were mostly private individuals, whose wealth would limit their experiments, as happened for Foucault and Fizeau in the 1840s. It was for this reason that the Duc de Luynes and other patrons of intellectual life endowed prizes and 'encouragements'. The Academy awarded dozens. Why did Foucault never win any?

Some prizes required the submission of memoirs on a given topic. In the *Journal des Débats* Foucault criticized the topics set as either unrealistically hard ('*Derive the equations for general motion of the Earth's atmosphere*') or extremely tedious ('The question set is in no way difficult: it is just a question of a lot of patience; we will see if it attracts a physicist who is sufficiently devoid of inspiration and sufficiently keen to win the prize to devote himself to such barren research').[13] In addition, the Academy had become unwilling to believe anyone worthy of its prizes, except in medicine, as Table 14.1 illustrates. In 1859, a 'sad and disappointed' Moigno deplored that not one of the six big prizes for the exact sciences had been awarded, and even small prizes had been withheld. 'Therefore nothing, in the judgment of the Academy, has been invented or improved between 1st April 1857 and 1st April 1858 ...' He went on to give a long list of inventions culled from the year's *Comptes rendus*, including Foucault's silvered-glass mirror, and asked rhetorically whether it 'and so many other excellent things are not even worthy of a gold medal of FOUR HUNDRED AND FIFTY francs!'[15]

Foucault did however compete for one big prize. Louis-Napoléon's predilection among the sciences was for electricity (Fig. 1.9). Less than three months after his *coup d'état* he announced a competition open for five years with the enormous prize of 50 000 francs for the most useful application of electricity. As with de Luynes' prize, the judges were unwilling to make a decision when the competition closed. Preliminary prizes were awarded, and the competition was extended for a further five years, closing finally in 1863 May. By this time Foucault had refined his electric arc regulator (see next chapter). The judges, chaired by Dumas, spoke highly of this lamp, but awarded the prize to Ruhmkorff for his induction coil.[16] In this they showed greater fore-

Table 14.1. Prize monies awarded by the Académie des Sciences in 1852. Moigno commented that the Académie des Sciences should swop name and buildings with the Académie de Médecine.[14]

Mathematics	0 fr
Statistics	477 fr
Physics	0 fr
Mechanics	900 fr
Physiology	850 fr
Insalubrious industries	0 fr
Medicine & surgery	28 200 fr

Table 14.2. Learned societies that elected Foucault to foreign membership, and other honours.

Chevalier,	
Légion d'honneur, *Paris*	1850
Pontifical Academy, *Rome*	1853
British Association	1854
Copley Medal,	
Royal Society, *London*	1855
Société Royale des	
Sciences, *Liège*	1856
LL.D., Trinity College,	
Dublin	1857
Société de physique et	
d'histoire naturelle,	
Geneva	1859
Imperial Academy of Sciences,	
St Petersburg	1860
Académie des Sciences et	
Lettres, *Montpellier*	1862
Officier,	
Légion d'honneur, *Paris*	1862
Royal Astronomical Society,	
London	1863
Royal Society, *London*	1864
Royal Society, *Edinburgh*	1864
Prussian Royal Academy of	
Sciences, *Berlin*	1865

Fig. 14.1. Membership certificates.
(Family collection)

sight than the de Luynes committee that Foucault was sitting on. Already, over 500 miniature Ruhmkorff coils had been built for igniting the gas in the first internal combustion engine, invented a few years earlier by J. J. E. Lenoir.[17] Tens of millions of induction coils are now made annually for motor-car engines.

The Bureau des Longitudes

In 1862 Foucault was appointed to the Bureau des Longitudes.

The appointment appears surprising, because following the reorganization of 1854, which wrested the Observatory from the Bureau's control, the Bureau's principal function was to calculate and publish its annual almanac, the *Connaissance des Temps*, which tabulated celestial positions of use to astronomers and mariners.[18] Foucault had little to contribute to this task. However, the appointment followed a period in which Poinsot and other deceased members were not replaced because of the thorny issue of Le Verrier.

Foucault was by no means the only victim of Le Verrier's exactions as Observatory director – 'The results...of such an infernal character... are truly horrifying: some driven to suicide, others to madness, certain tortured with unequalled tenacity, a large number of broken careers...', was the opinion of one of his most outspoken critics, the fellow celestial mechanician Charles Delaunay (1816–72).

Le Verrier had much to contribute to the accuracy of the calculations underlying the *Connaissance des Temps*, and to foreign scientists his non-appointment would be inexplicable, but none of the Bureau's surviving members could stomach the idea. When finally forced into action, their recommendations to the Minister were Ernest Laugier (husband of Arago's niece Lucie), Delaunay and a military man called Peytier. The Academy too refused to recommend Le Verrier. A stroke of the Minister's pen ended this embarrassment. The number of members was increased from nine to thirteen, producing seven unfilled places. Everyone was appointed: Delaunay, Peytier, Laugier, Faye, Yvon Villarceau – and Le Verrier and Foucault. For Foucault the appointment was primarily a sinecure, with a salary of 3000 francs; Yvon Villarceau's oration at his funeral on behalf of the Bureau made no mention of any specific contribution, though soon before he fell ill, the Bureau did commission him to make a telescope lens.[19]

The Academy's doors must soon be opened

For an intellectual, the approbation of one's academic peers is an important reward. Over the years, learned societies in London, Geneva, Berlin, St Petersburg and elsewhere elected Foucault as one of their distinguished foreign members (Table 14.2, Fig. 14.1), but many years passed before he was elected to the principal academy in his native land, the Académie des Sciences.

We have seen that Foucault was less than respectful of this august company. As the following opinion from the late 1860s shows, he was not alone:

> ...when these savants, whose name is known across Europe, sit in the Institute's great meeting room, when one contemplates their wide brows expanded by deep thoughts, and also their rare hairs, one can easily imagine that one is in a sanctuary...

> But the reality of the meetings does not match this ideal. While the Permanent Secretary reads the correspondence with an air that is as much boring as bored, what a lot of chatter showing that little attention is being paid! And how much one regrets seeing these learned persons get much more heated over a personal quarrel than when science is in question![20]

Despite this, election to the Academy was highly prestigious, and during the nineteenth century it was slowly becoming a reward for what had already been achieved, rather than an indication of future promise. Track record was all-important. Candidates were expected to lose the first time, but gained seniority for subsequent elections.

At his first candidacy in 1851 Foucault had recently reported his pendulum experiment ('one of M. L. Foucault's most important credentials', according to one commentator[21]) and had polled surprisingly well (Chapter 9). Our commentator found this election 'very curious', but what followed was no less so.[22]

Academicians enjoyed robust health in the 1850s, and it was a long wait before death put an end to Cauchy's mathematical lucubrations in 1857, thereby creating a second vacancy to which Foucault could aspire. In early 1858 Foucault wrote to the Academy asking to be a candidate.[23] Cauchy's seat had been in the Mechanics Section. A few days later Moigno reported rumours that the Section would only consider theoreticians as candidates. 'We are happy to learn that this gossip is wrong,' he continued, and looked forward to a candidate list incorporating men such as Foucault, Froment and Breguet as well as professional engineers and theoreticians such as Adhémar Saint-Venant, chief engineer for bridges and carriageways, and the mining engineer Émile Clapeyron.[24] It was wishful thinking. When the Mechanics Section announced and ranked its candidates three weeks later, only professional engineers and theoreticians were on the list. This caused some commotion amongst the Academicians, but there was not enough support – or willingness to interfere in the affairs of another section – for Foucault's or other names to be added.[25] Nevertheless, when it came to voting, three Academicians wrote in Foucault's name.[26]

One problem for those seeking election to the Academy was that places were few and candidates were many. Cagniard-Latour was the next to die, in 1859. This prompted Moigno to list eight worthy successors (Table 14.3) and suggest that the Physics Section, with only six seats, was far too small considering the way that physics was flowering.[27]

Fig. 14.1 (cont.). Membership certificates.

(Family collection)

Table 14.3. Fortune favours those at the beginning of the alphabet! The Abbé Moigno's list of worthy candidates for election to the General Physics Section in 1859, and the date at which they were elected, if at all.

	Election date
Ed. Becquerel	1863
P. Desains	1873
Fizeau	1860
Foucault	1865
Jamin	1868
de la Provostaye	—
Verdet	—
Wertheim	—

Table 14.4. Votes for election to the Academy's Physics Section on 1860 January 2. Fizeau and Becquerel were clearly preferred.

	Votes
First round	
Fizeau	24
Ed. Becquerel	20
Foucault	14
de la Provostaye	1
Second round	
Fizeau	30
Ed. Becquerel	25
Foucault	3

The Physics Section placed Fizeau top of its list of candidates, with Foucault and Edmond Becquerel second equal, and added an additional name beyond seven of those guessed by Moigno. It took two meetings to discuss the candidates' work. When the time to vote arrived, Fizeau was elected on the second round (Table 14.4).

The next vaguely relevant vacancy occurred a year later in the Geography and Navigation Section. Astronomy provided the basis for surveying and navigation, and Foucault was working at the Observatory, but this was a weak claim to a seat in the Section, which unsurprisingly did not put his name on their candidate list.[28] (The gyroscope would not be used for navigation for another forty years.) Nevertheless, when it came to voting, a quarter of the Academicians wrote in Foucault's name, placing him a clear second (Table 14.5).[29] Moigno commented:

> The fact is that without having been put on the list, without even having been considered as a serious candidate, M. Léon Foucault obtained 16 votes – and what votes! The votes of MM. Boussingault, Dumas, de Senarmont etc. This seems to me to prove in the most striking way that a seat for this skilful physicist is marked out at the Academy, whose doors absolutely must soon be opened to him.[30]

Throwing off the yoke

Two years passed before the next vacancy. It was in the Physics Section, created by the death of the misanthropic Despretz in the spring of 1863. Foucault revised the booklet summarizing his research, which grew to thirty-four pages, but it is unlikely he made the weary round of visits to canvas votes – 'My scientific credentials... will thus be my sole recommendation...,' he wrote later.[31]

The ranking of candidates by the Section was a vital factor influencing Academicians' votes. With Fizeau elected, Foucault's serious opponent was now Edmond Becquerel. The Section's five members were Edmond's father Antoine, Pouillet, Fizeau, Babinet and J. M. C. Duhamel (1797–1872). The first two were in Becquerel's camp, while the last two were Foucault supporters. Probably no one was sure where the moody Fizeau stood, but since it was well-known that he had fallen out with Foucault, the presumption was that he might support Becquerel. Now Duhamel was temporarily absent from Paris. As a matter of fair play, the ranking and election should have been postponed until his return, but the Academy decided with unprecedented haste to proceed to the election within two months of Despretz' death. This was a ploy by the Becquerel supporters. With Duhamel absent, Becquerel would be first ranked, or at worst ranked first equal, if Fizeau turned out to support Foucault.

To Foucault, his chances must have looked slim. Even if the depleted Physics Section ranked him and Becquerel first equal, the Academy's decision for a vote-rigging early election had been made by thirty votes to eighteen, which hardly looked promising.

Donné had introduced Foucault 'into the intimacy of my family'. Foucault was fed up with the Academy and its relentless scheming. He expressed his feelings without reserve in a long letter to a female correspondent in Montpellier who appears to be Donné's wife, Marie Antoinette. The letter was enlivened by a jocular tone and distorted names such as Brequerel and Fozeau, but carried a serious message:

> Now I knew only too well that a long-established opposition had been spying on my acts and gestures and was wasting no opportunity to recruit amongst the simple folk who vote for sons, son-in-laws, nephews and distant relatives ... Opinions formed in advance, pride ruffled and mixed with family interests thus made the situation untenable, and I was ruminating in my corner on the way to get out, when the means was suddenly offered by the Academy itself, a superb, admirable and unrepeatable opportunity if I did not hasten to seize it as it flew by.[32]

Bringing forward the election was blatantly improper. The minutes of the secret part of the Academy's meeting on May 11 show that the Physics Section in fact ranked Foucault and Becquerel first equal. Despite their falling-out thirteen years earlier, Fizeau had supported Foucault, at least to the extent of not ranking him below Becquerel.[33] But Foucault could not know that this is what would happen. As the Academicians were about to hear the Section's reports on the various candidates, Pierre Flourens, the Permanent Secretary, rose to read out a letter which Foucault had just addressed to the Academy's President, asking him:

> to be my spokesman in front of the Academy, and to express my profound regret that I am unable in such circumstances to offer it the humble tribute of my long research.[34]

What a snub! A near-universal murmur of surprise arose from the public benches.[35] Nine Academicians nevertheless wrote Foucault's name on the ballot papers the following week.[36]

Having withdrawn from this fifth election, Foucault felt liberated. His letter to Montpellier continued (Fig. 14.2):

> My goal was fully achieved. My adversaries having scaled their efforts according to the expected resistance had their turpitude laid bare; close enemies had wasted their venom for nothing. The entire Academy having committed an error that was patent to everyone, I took advantage of it to throw off the yoke of being a candidate once and for all.
>
> As to the effect produced, it exceded all my hopes. Astonishment extended to stupefaction; an appearance of discussion was wanted: there was but a gloomy silence. It will certainly be remembered.

Table 14.5. Votes for election to the Academy's Geography and Navigation Section in 1861. Foucault was not a candidate, but nevertheless polled second.

	Votes
First round	
de Tessan	25
Foucault	16
Paris	9
d'Abbadie	7
Peytier	2
Second round	
de Tessan	38
Foucault	13
Paris	7
d'Abbadie	1

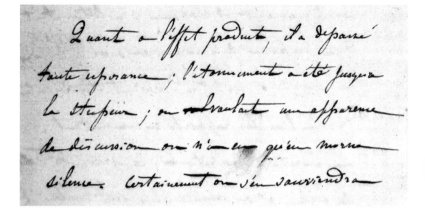

Fig. 14.2. 'As to the effect produced...,' Foucault wrote concerning his withdrawal from the Academy election in 1863, 'astonishment extended to stupefaction.'
(Family collection)

> Now, shall I tell you whether they are pleased with me? No, not completely; but neither did I have the intention of pleasing them. What was important was to emancipate myself through adopting a new attitude.

The Rector's wife replied in style:

> Thank you Monsieur Foco ... I can only congratulate you sincerely for having *burnt* your bridges in such a noble fashion! There was no need for so much plotting and so much agitation; but the shadow of your name is enough to *terrorize* people, and he who has a *running* nose *wipes* it, as the proverb so vulgarly says![37]

Persuaded by friends

'Foucault had no great ambition to belong to the Academy,' Donné later wrote; the Thursday-morning gatherings at his house were far more important to him.[1] Having burnt his bridges so spectacularly, he sat tight when the next suitable vacancy appeared the following January through the death of Clapeyron, who had been the Academy's choice in 1858 for the Mechanics Section.

The election progressed without haste. Perhaps there were no obvious candidates, but the grey-haired Mechanics Section seems to have been singularly slap-dash as it began to draw up its list of possible candidates. In March Saint-Venant had to write in asking to be considered, as though he was some unknown beginner, whereas he had polled second to Clapeyron in 1858. A man called Burdin who had moved to Paris and was therefore no longer entitled to remain as one of the Academy's correspondents also had to remind the Section of his existence, as did a minor celestial mechanician named Passot.[38]

In this semi-vacuum, others were weighing up their chances. The Academy was aware of the advantages of high-placed friends; Maréchal Vaillant, one of

Napoléon III's courtiers, had been elected in 1853 virtually unopposed as an *académicien libre*. Colonel Favé, besides being the Emperor's aide-de-camp, was professor of artillery at the École Polytechnique with a number of publications to his name and innovations which had proved useful in the Crimea. Artillery is based on mechanics, so Favé decided to print a *Notice* summarizing his achievements and sound out possible supporters.[39] He had been the conduit for the Emperor's generosity towards Pasteur as well as towards Foucault, so he approached Pasteur. A saddened Pasteur replied frankly:

> What I was far from expecting, Monsieur, was that if one day I was to receive the satisfaction of learning that you were a candidate for a vacant seat in the Academy, I should be amongst those who would not contribute to your success. It is however the dreadful sorrow that is my lot. I do not see any way of not voting for M. Foucault if he presents himself, concerning which I know nothing, because one never really knows what he has in his soul. If his election is not certain, I shall have the consolation of being able to support you in the second round.[40]

Though Foucault may have written off the Academy, friends believed his place was there. It was necessary to persuade the Academicians more effectively than in 1858 that he was an appropriate candidate for the Mechanics Section. Joseph Bertrand began the task. As Moigno reported, 'He has loyally and vigorously pitted his great academic reputation against the blind and heart-breaking unpopularity of M. Léon Foucault.' Bertrand wrote a 'glorious panegyric' of 'twenty eloquent pages' preaching Foucault's case in the *Revue des Deux Mondes*, an influential monthly magazine with a large readership.[41]

Since the vacant seat was in the Mechanics Section, Bertrand's title was *Des progrès de la mécanique* (On Progress in Mechanics). His article opened:

> The Gospel says: 'He who believes will not be condemned.'[42] Savants should take inspiration from this maxim, and say: 'He who discovers will not be condemned.' Any method that leads to discoveries should be gratefully welcomed, and whoever brings a new truth should always be warmly received.[43]

Bertrand elaborated. Geometry and mechanics had been revolutionized by Descartes and Lagrange respectively when they introduced mathematical formalisms which permitted analyses in a general, algebraical way. These powerful techniques had acquired 'an authority... almost a tyranny' which had led to the idea that Cartesian geometry and Lagrangian mechanics were the only acceptable theoretical justifications. But it was no less valid, and much closer to the phenomena of nature, to think of geometry in terms of lines, triangles, etc., and of mechanics in terms of motions and Newton's laws. 'In mechanics, nothing allows us to dispense with considering things as they are,' Bertrand wrote. He went on to draw an analogy between Foucault and the deceased Poinsot. With his polhodes and herpolhodes Poinsot had given a physical idea of the motion of arbitrary three-dimensional rotating bodies. His results were of course contained within Lagrangian mechanics, but it was not from Lagrangian mechanics that they had been derived.

So too for Foucault. There was no new physics in the pendulum and gyroscope experiments, but in the centuries since Galileo and Newton, no one had realized that the Earth's rotation could be revealed so easily. Poinsot and Foucault both had 'the same deep feeling for reality'. Credibility is weakened by excessive praise, so Bertrand also outlined some faults shared by the two men:

> ...they give the well-trodden paths the same wide berth, sometimes unjustly so. Both of them express a disdain, which I do not share, for estimable work... which has the feel of an exercise done by an industrious schoolboy, and to add a final trait which is common to them both, both discount everything in science which they do not find clear.

Bertrand was not so unsubtle as to mention that Poinsot had been a long-standing member of the Academy, in the Geometry Section; but the conclusion that Foucault therefore deserved a place in the Mechanics Section will have been obvious to every reader.

Foucault was flattered. 'This article... is the most beautiful day of my life,' he wrote to de la Rive in Geneva, but he was not persuaded to stand again. 'It is already sometime ago that I dropped that from my sights.'[44]

The Mechanics Section continued to procrastinate over its choice of candidates. Then, on December 26, Saint-Venant withdrew. In *Comité secret* the Section finally presented a list of two candidates, Édouard Phillips and Eugène Rolland, who were very much a mirror of the Section, except in age. However, there were calls from outside the Section to add Favé to the list. Favé had rendered service to the Bureau des Longitudes and had defended the *Muséum* (the natural history museum) against charges of maladministration.[45] Academicians from these institutions wished to show their gratitude and believed he could be elected.[46]

The debate continued the following week and was agitated. The Foucault faction argued that if Favé were added to the list, Foucault should be too.[47]

Figuier tells us that Foucault made decisions quickly, and having thrown off the yoke of candidacies, it would have been uncharacteristic for him to change his mind. But no doubt friends had been working on him since the spring, and Bertrand's article in the *Revue des Deux Mondes* may have been aimed as much at Foucault as at the Academy. Foucault agreed to tempt his chance once more. He had no time to produce a new *Notice*, so he added a few extra sheets to the one from 1863. He distributed this on the very day of the Academy's next meeting, prefaced by a letter to the President:

> The empty seat in the Mechanics Section... seems in the opinion of the Section to be open only to a savant who has achieved distinction in analytical mechanics...

Nevertheless, discussion has taken a turn which tends to widen the field; I therefore beg permission to accede to the counsel of friends in the Academy who consider my recent work in applied mechanics as being of a nature that will attract their votes.

Stimulated by the controversy, Academicians turned up in force. The discussion must have been agitated yet again, because after two rounds of voting, both Favé's and Foucault's names were added to the two proposed by the Section, by 30 and 35 votes respectively. 'The odds of the election are at the moment with M. Foucault,' noted Moigno, reverting to the terminology of horse racing.[48] The Academicians discussed the merits of the four candidates, and after an hour and three-quarters of *Comité secret* disbanded exhausted.

Sixty-one of the Academy's 64 members were present the following week. As they worked through the scientific business, tension mounted. Finally it was time to vote. An absolute majority was required, and in the first round Favé was one vote short. Rolland, the lowest scorer, was eliminated. In the next round Phillips' support mostly switched to Foucault, though one of Favé's supporters appears to have switched as well; but there was still no absolute majority (Table 14.6). Where would Phillips' two votes go in the next round?

The answer was unexpected. Favé attracted one vote, while the other became an abstention. Favé and Foucault had each polled 30 votes. A tie! The President (a professor at the *Muséum*) consulted the regulations and decided further voting should be postponed until the following week. 'The emotion was very great during the vote and each of its twists,' reported Moigno; 'the wishes of the onlookers were visibly for M. Léon Foucault.'[49]

We can imagine the agitation during the ensuing week. The balloting will have reminded Foucault of the interminable rounds of voting during his first election fourteen years earlier (Table 9.2), and he can only have thought how stupid he had been to submit to the pains of a further candidacy. Favé quite likely had similar thoughts. He was linked to the Emperor, of course, and votes for or against him were being interpreted by some as political comments on the Empire, which was under increasing criticism. 'M. Favé, a modest and excellent man, has found himself dragged into an intrigue which is not of his making', remarked one observer.[50]

Who had abstained in the previous round of voting, resulting in the tie? This question was answered when the final votes were cast on January 23. The public benches were crammed. As the voting was about to begin, Pouillet announced that 'in the interests of science', he would not vote for either candidate, since in his opinion, neither was standing for the appropriate section. Our observer attributed less honourable motives to Pouillet's absence: dislike of Foucault, and anger with the Second Empire.

The onlookers were hushed as the Academicians voted. The result remained close, but Foucault obtained 31 votes, beating Favé's 28. A tumult of approval burst from the onlookers; there was even clapping, rapidly repressed

Table 14.6. Success at last! Foucault's hair's-breadth election to the Academy in 1865.

	Votes
1865 JANUARY 16	
First round	
Favé	30
Foucault	20
Phillips	10
Rolland	1
Second round	
Foucault	30
Favé	29
Phillips	2
Third Round	
Favé	30
Foucault	30
1865 JANUARY 23	
Foucault	31
Favé	28
blank ballot	1

by the President. Several Academicians rushed outside to announce the news, followed by most of the public. The election, one journalist remarked, was 'unique in the annals of the Institute' and still provoked comment four decades later when Bertrand died.[51]

The Academy's recommendation was approved by Napoléon III within two days. 'There are few instances where the approval has been so prompt,' wrote Moigno, 'and this speed overabundantly proves that this election has been welcomed... in high places with great feeling.'[52] No doubt the Emperor had some consoling words for his defeated aide-de-camp, but it was not until 1876 that Favé was elected to the Academy.

One of the joys of his life

Foucault described membership of the Academy as 'one of the joys of his life', but a fortnight after the election, his elation was tempered by the death of Gustave Froment, with whom he had worked so well.[53] True to form, Foucault was a conscientious Academician, missing only eight of the 127 meetings before he fell ill, presenting the work of non-members, and acting as a go-between.[54] He was elected to committees awarding Academy prizes. Ironically, in one case the committee decided there was no meritworthy new work; while in another, the set subject was the measurement of refractive indices, requiring patience but no inspiration.[55]

Why was Foucault's route to election so tortuous? Partly it was just bad luck, because chance affects all appointments, but 'his airs of a pasha with three pigtails', as our observer put it, were not to everyone's taste. Nor did everyone consider him to be a true physicist because he had no formal training beyond the *baccalauréat* and little competence in higher mathematics. 'We are amateurs,' he used to joke with Donné. 'We did not attend the École Polytechnique, we did not attend the École Normale, we are not the pupils of this or that master.' As such he was suspect in realms dominated by the *grandes écoles*, such as the Mechanics Section, which from 1847 to 1885 was composed exclusively of *polytechniciens*, with the single exception of Foucault. Most importantly, all commentators blame antagonisms engendered by his newspaper articles. Joseph Bertrand later wrote that 'important people' in science had tried to catch the young journalist's attention:

> seeking perhaps less his opinion than his praise. Coldly polite, attentive only to the truth, Foucault judged discriminately, after study and reflexion, without giving any thought or promise as to whether he would please. This young man... dared to provoke impatience through his calm assurance, irritation by his bold frankness, and exasperation sometimes with his subtle irony, in those who thought they were his masters, but whose fate was often to be forgotten.[56]

Chapter 15

Control: the quest for fortune

In 1855 Foucault devoted a *feuilleton* to a new turbine which had been designed by his hydraulic engineer friend L.-D. Girard to provide power for the Menier chocolate factory at Noisiel-sur-Marne, east of Paris. Novelties included the absence of fixed blades, and a rotor – which was 5 m in diameter and weighed 6 tonnes – that had helical channels that widened sideways to accommodate the decelerating water. But what most appealed to Foucault were the resources and speed pertaining in industry:

> The helical rotor had no childhood; the author was not subjected to the humiliation of a model at one-tenth scale; from the idea he went straight to full-scale implementation ... in less time than it takes for an academic commission to emit a simple opinion.[1]

Attracted by this example, and perhaps encouraged by royalties from Secrétan's commercialization of the telescope, Foucault turned his energies from pure to applied science, to what is now called control engineering. He patented all his inventions in France, as well as in England and Belgium for key developments.[2] The motivation was not pecuniary *per se*, as his friend Jules Lissajous later explained:

> Foucault looked forward to a brilliant fortune; not that he was at all avid for money, but money for him meant independence. This complete and absolute independence which he longed for so ardently was to be the means of realizing his great ideas with quality instruments in a model laboratory. After so many years of painful exertion, it would have meant working in comfort.[3]

Controlling light

We have already met the instrument-maker Jules Duboscq, one of whose products was the stereoscope (Fig. 15.1). In the early 1850s Foucault had been commissioned by his provincial cousin to buy some stereographs from Duboscq ('I want nothing that can shock,' the cousin warned), and at about the

Fig. 15.1. Stereoscopic daguerreotype of the Parisian optician and instrument-maker, Jules Duboscq (1817–86). The stereoscope was invented by Wheatstone and perfected by Brewster, but Duboscq made it a popular success after it had charmed Queen Victoria at the Great Exhibition in 1851. A scientific triumph of stereoscopy was to show that the Moon was palpably spherical.

(George Eastman House)

same time Foucault was immortalized in three dimensions by Duboscq (Fig. 15.2). However, Foucault knew Duboscq for another reason. His father-in-law J. B. F. Soleil had supplied parts for Foucault's electric arc in 1849, and as successor to Soleil, he was manufacturing an electric arc that was much more compact and commercial than Foucault's mammoth prototype (Fig. 7.12 vs. 7.8). It made sense for the two men to work together on the electric light, but the project that they completed first concerned the control of natural light.[4]

Fig. 15.2. Stereoscopic daguerreotype of Foucault taken by Duboscq. The images are reversed. Lissajous commented that Foucault's strabismus and unequal eyes gave 'something strange and very characteristic' to his look.[3] Although printed at reduced size, this and the previous figure will fuse to 3D in a stereoscope that has narrow eye-piece separation.

(National Museum of Photography, Film & Television/SSPL)

The heliostat

Let us recall that the function of a heliostat is to reflect a beam of sunlight into a fixed direction despite the unstoppable westwards march of the Sun. In his younger years Foucault had made full use of Silbermann's heliostat (Fig. 4.5), which was manufactured by Soleil, and then Duboscq; but it was a dainty mechanism capable of moving only a small, light mirror and the reflected beam was little more than 5 cm wide. The development of negative-on-glass photography had spurred a need for photographic enlargments, in which Duboscq was a pioneer, but the light in a 5 cm beam was quite insufficient.

In early 1861 Duboscq produced a heliostat for photographic work with a massive, 80-cm diameter mirror, which was enough for 27-cm negatives, but the solar tracking was done by a human operator.[5] This was the same constraint as with the first photo-electric microscope, as Foucault must have realized as he pondered how to make a clock drive turn a massive heliostat mirror automatically.

In Foucault's rugged and simplified design, a solid pillar and fork took the weight of the mirror, which was 30 cm long by 15 cm wide. Operation is simple to understand from Fig. 15.3. The law of reflection requires the mirror to be oriented symmetrically with respect to the continuously changing direction of the incoming solar rays and the desired fixed reflection direction; or expressed another way, the perpendicular to the mirror must bisect the angle between the

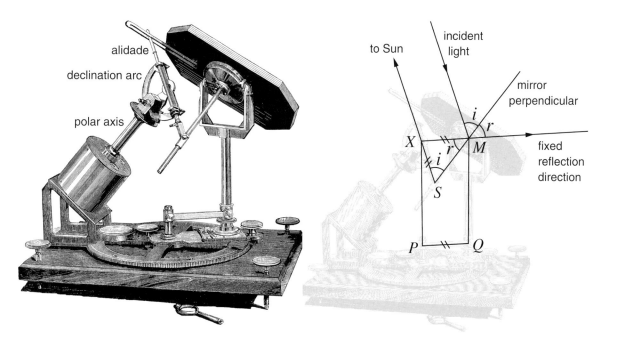

Fig. 15.3. Small heliostat designed by Foucault and built by Duboscq in early 1862. A clock in the cylindrical cannister drove the polar axis at the solar rate and the alidade then tracked the Sun. The direction XM of the reflected beam was set by adjusting the parallel arm PQ. The mechanical construction made SXM an isosceles triangle with $i = r$, as required for a fixed reflection direction.

Fig. 15.4. The Foucault–Duboscq heliostat for photographic enlargment produced in the autumn of 1862. The design was simplified by removing the elevation adjustment on the reflected beam. The mirror measured 80×40 cm and was mounted on rollers to reduce friction. A spring in the pillar devised by Duboscq helped drive the mirror through difficult positions where a large force was required.[8]

incident and reflected rays. Foucault derived motion from a clockwork equatorial drive with polar axis and declination arc, which, once set, tracked the Sun with a pointer, or alidade. Silbermann's heliostat could operate anywhere, but in a first simplification, Foucault angled the polar axis for a definite latitude.

The moving alidade positioned the mirror via sliding joints. One joint connected to a rod attached perpendicularly to the mirror, and maintained the reflection condition. Another connection oriented the long axis of the mirror so that the reflected beam was not flattened by the oblique angle of the incident rays.

Foucault's heliostat was not completely novel; it was a modification of a design devised by the Dutchman Willem 's Gravesande in the eighteenth century.[6] However, it was much easier to set and the reflected beam was more stable, unflattened, and could be directed to northerly azimuths. Foucault patented the design in 1862 March, and on the very same day Duboscq showed the instrument to the Academy.[7] Besides photography, the new heliostat was claimed useful for lecture demonstrations and astronomy. We shall meet this last application again in the next chapter.

Seven months later Foucault and Duboscq produced an even bigger model designed specifically for photographic enlargments, with an 80×40-cm mirror capable of producing a 30- or 40-cm beam (Fig. 15.4). This giant heliostat was shown to the Academy in its waiting room, where according to Moigno it excited universal admiration.[9] Nevertheless, Foucault's heliostats do not seem to have been a commercial success, and the patent was abandoned within a few years.[4] Off-beat uses such as lighting shady rooms are unlikely to have produced sales,[10] and with photographic plates becoming smaller and more sensitive, mammoth beams were no longer required. The limitation of a particular latitude may have been irksome, and as we shall now see, the electric arc was finally beginning to offer an easy and reliable light source for both laboratory and lecture room.

Back to the arc

Though occupied with the heliostat, Foucault did not forget the electric arc. There was a market to be tapped. Over the previous decade, Duboscq had sold 280 of his regulators,[11] but the moment was especially timely because a new and practical source of electricity had become available to replace the inconvenient Bunsen cells. At last it seemed that the arc light would move from specialized laboratory applications into widespread practical use. The electrical source was the generator (Fig. 15.5) manufactured in Paris by the Alliance Company under the direction of a M. Auguste Berlioz.[12] The machine derived from one made by F. Nollet, a professor of physics at the military school in Brussels; a similar machine was made in London by F. H. Holmes, who had

Fig. 15.5. In the 1860s the Alliance Company's magneto provided a less troublesome source of electricity than Bunsen cells. A 3 or 4 horsepower steam engine provided enough power to run an arc light.

worked with the Alliance Company in Paris. The Alliance generator was a magneto; that is, it included permanent magnets, arranged in a fixed crown, which induced alternating currents in a rotating winding. For use in electroplating, a commutator was required to rectify these alternating currents into direct ones and to transfer them to the outside, but simple slip rings sufficed for uses where alternating currents were acceptable, such as powering arc lights. The magneto was robust and easy to maintain, and could be driven by a steam engine, which was an established technology.

During the 1850s many people had devised electromagnetic regulators, but the problem with most of them was initiating the arc. For this, the carbons had to be touched together and then separated, but Foucault's mechanism, like all the others, could only move the electrodes closer, so the light had to be started manually, which was especially inconvenient if it was high above wharves or atop a ship's mast.

The idea that the regulator needed to be able to separate the electrodes as well as approach them was obvious, and in Paris a man called Victor Serrin had produced a self-starting regulator in 1860 (Fig. 15.6). The weight of the upper electrode provided the motive force to bring the electrodes closer. To keep the luminous region stationary, gearing ensured that the electrodes' relative motion was in the 2:1 ratio of their erosion. In addition, the lower electrode was mounted on an upwardly sprung-loaded arm which was normally attracted downwards by an electromagnet wired in series with the arc. If a gust of wind blew out the arc, or the current was accidentally interrupted, the lower electrode jumped up to restrike the arc.

Cosmos reported that in 1861 June, Foucault attended a demonstration at the Invalides of a magneto and Serrin arc lamp powered by Lenoir's internal

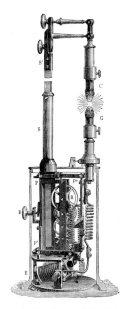

Fig. 15.6. One form of Serrin's self-starting arc lamp from the 1860s.

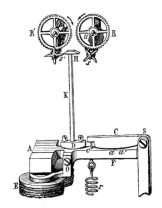

Fig. 15.7. Robert-Houdin's *répartiteur*. As the end, **A**, of the armature, **AF**, pivots towards the electromagnet, **E**, the magnetic attraction increases sharply, but is opposed by an increased torque from the sprung-loaded finger, **C**, because the point of contact *a* moves rightwards to *a′*. The position of the armature and the pallets, **H**, therefore alters uniformly as the current changes despite the strong gradient in the magnetic attraction.

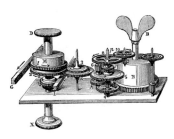

Fig. 15.8. Simplified sketch of the wheelwork in Foucault's improved arc-light regulator. The spindle, **pq**, does not provide any connection between the three sets of gears mounted on it. The planetary gears are marked **6** and **k** and permit the spring boxes **M** and **N** to drive the electrode-supporting beams **G** and **I** in either direction.

combustion engine.[13] The equipment had been assembled 'for a rich amateur in Brazil', which is surely a reference to the Brazilian emperor, Dom Pedro II. This august client, along with the demonstration of how easy it was becoming to produce electricity, were perhaps triggers that caused Foucault to begin reflecting again on regulation of the arc light and its commercial possibilities. He will also have been aware that the expiry of Staite and Petrie's French patent was imminent.

In his improved regulator, Foucault adopted a different approach for moving the electrodes. True to his predilections, he used clockwork. He provided *two* clock trains acting in opposite senses, one to separate the electrodes and the other to bring them together. To control the motion, pallets could block the escapement of one or other of the clock trains, or both. The pallets, in turn, were moved by an electromagnet wired in series with the arc current. In his 1849 regulator, the armature attracted by the electromagnet had needed only two positions – advance and stop – and simple snap action, as in a relay, had sufficed. Now three positions were needed – advance, stop and separate. The armature therefore needed to turn by an amount that changed uniformly with the current, unaffected by the sharp increase of force that occurs close to an electromagnet's poles. Foucault adopted the solution invented by a talented clockmaker called Jean-Eugène Robert-Houdin (1805–71). Robert-Houdin had branched out into illusionism and conjuring, and had made sufficient money from his *soirées fantastiques*, or magic shows, that he had been able to retire and devote himself to science and technology, though as a former entertainer, he was not always taken seriously.[14] Foucault, however, took Robert-Houdin sufficiently seriously to use his *répartiteur*, or apportioner, in his arc light. Figure 15.7 shows one form of this device, which had been presented to the Academy in 1855 and shown at the Exposition Universelle.[15]

Two clock drives involved a lot of gearing. So that the carbons could be moved either way, it was necessary to connect the drives together through what is called planetary gearing. In this, a gear on a first spindle carries secondary gears on an off-axis spindle, and engages through them with other gears on the first spindle (Fig. 15.8).[16]

Foucault first tried out these modifications on one of his old arc lights from 1849.[17] Something more compact and rugged was needed for a commercial product. This is where Duboscq came in, and the device was patented in the autumn of 1863 (Fig. 15.9).[18] The innovations claimed were the use of planetary gearing, the three-position control of the electrode motion, and lamination of the electromagnet cores and armature. This last improvement was needed for use with the Alliance magneto because its alternating currents produced large induction currents in solid soft iron and excessive heating. The design allowed the vertical position of the arc to be adjusted, which was important when aligning it with other optical apparatus, while the fact that the electrodes were moved by clockwork, and not gravity, meant that the regulator could be used in any orientation or on unstable platforms, such as ships. Presumably the clockwork needed to be wound only when the carbons were changed and so resulted in no additional difficulty when the regulator was installed in inaccessible places. With alternating current the electrodes are consumed equally and the electrodes must move at equal rates to keep the arc stationary. No doubt the regulator was available with the electrodes geared 1:1, but the obvious addition of a gearbox to permit switching between a.c. use and 2:1 gearing for a battery was only made some years later, on the eve of Foucault's death.

Lighthouse trials at La Hève

The safety of modern travel makes us forget the dangers that attended sea voyages until well into the twentieth century. On his way home from the Liverpool meeting of the British Association, Foucault wrote to Jules Regnauld describing something that had particularly impressed him:

> I haven't yet spoken, old friend, of my visit on board a magnificent steamship, the *Arctic*, destined for passenger service between Liverpool and New York. It was nevertheless one of the highlights of my trip... A thousand horsepower engine, it was all that one can imagine that is most majestuous, and especially most reassuring; I envied the lot of one of my travelling companions who should have continued his journey on this floating continent; I even envied the sailors who had to swab its august decks. Well, my dear friend, the *Arctic* is no more; taken broadsides by a French steamer somewhere off Newfoundland, it sank into the abyss in one piece with three-quarters of the passengers. Yet again, the sea is a traitor, and once you embark, the possibility of death must be faced.[19]

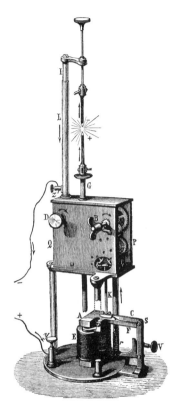

Fig. 15.9. The Foucault–Duboscq arc light from the 1860s was the *ne plus ultra* of arc lights but because of its complexity was poorly adapted to industrial use.

The electric arc could improve maritime safety through brighter lighthouses and better illumination of ships at night.

The French lighthouse board had experimented with electric arcs in the late 1840s, but had abandoned the project because batteries were too complicated to use in remote locations. The magneto returned thoughts to the electric arc on both sides of the Channel, with the first fully operational electric light shining forth at the South Foreland lighthouse, near Dover, on 1858 December 8. Duboscq's regulator was used, powered by a Holmes machine. A few years later, two Alliance magnetos were bought for tests in France at the twin lighthouses at the Cape of La Hève, near Le Havre, where the traffic-laden River Seine joins the English Channel (Fig. 15.10).

Through his association with Sautter, Foucault will have been well aware of what the lighthouse authorities were doing, and vice versa, and of course Duboscq's arc was in use at South Foreland, so it is no surprise that the Foucault–Duboscq regulator, along with Serrin's, was tested in the trials, which began just after Christmas in 1863.[20] For the greatest light, and security against

Fig. 15.10. The twin lighthouses at La Hève near where the dangerous Seine estuary enters the English Channel. In 1863, trials of Foucault's arc lamp were made in the southern lighthouse (right), but Serrin's regulator was adopted when the lights were converted to electricity in 1865. The lighthouses were destroyed by allied artillery in 1944 during the Liberation of France.

breakdowns, two of each regulator were installed on the cast-iron shelving inside the lantern of the southern lighthouse, one above the other.

Electric light was an immediate success. A single arc emitted the light of some 3500 Carcel lamps, over five times more than produced by the lamp burning rape-seed oil in the northern lighthouse. The electric light was more penetrating (though less so in fog), and only slightly more expensive. 'There is as much difference between the lighthouses as between a candle and a gas light,' declared one of the local pilots. However, Serrin's regulator was preferred. There were two problems with Foucault's regulator. First, additional insulation was needed with cast-iron shelving because its body was the conductor which took current to the lower electrode. More seriously, the regulator broke, twice. The mechanism separated the carbons too much, though why this should have caused damage is unclear. Duboscq later forestalled this breakdown with a stop, but it was too late. Serrin's regulator was adopted for the electrification of La Hève in 1865 and later for other French lighthouses.

A miracle in Neptune's Empire

The sinking of the *Arctic* was no isolated occurence. Ships were ramming each other all the time, especially at night. Towards the end of the 1860s, the Abbé Moigno ran a campaign to equip ships with electric searchlights, citing numerous French and foreign ships involved in night-time collisions: the *Warrior*, the *Royal Oak*, the *Calcutta*, the *Montebello*, *La Flandre*, the *Latouche-Tréville*, the *Prince-Pierre*, the *Général Abbatucci*, the *Marquis of Abercorn*, the *Lord Gough* ... over 2000 collisions were recorded in 1867.[21]

A breakthrough came via the Emperor's cousin, Prince Jérôme-Napoléon. In 1866 he acquired a new yacht, which, like its predecessor, he immodestly christened the *Jérôme-Napoléon*.[22] It was a converted aviso. (Avisos were mixed sail-and-steam ships of about 50 metres length and a few hundred tonnes displacement, with typically 150 horsepower engines.) The Prince wanted all the latest accessories for his new status symbol and this included the electric light. An Alliance generator was installed in 1867 March. Moigno reported enthusiastically that the lamp would be Foucault's in a projector made by Sautter (Fig. 15.11).[23] A fortnight later Moigno received telegrams from the yacht's captain, M. Georgette Du Buisson, reporting the light was a complete success, finishing with the optimistic assessment that collisions were henceforth 'impossible'.[24]

A year later, by which time Foucault was in his grave, electric light made 'its first miracle in Neptune's Empire', as Moigno poetically described it.[25] An aviso, *Le Renard*, had been despatched for England, presumably with some ill-considered message for the British government, because almost immediately it was decided to attempt to stop delivery. The *Jérôme-Napoléon* was sent in chase. Captain Georgette Du Buisson rapidly found the aviso with his searchlight (Fig. 15.12):

> On its side, the poor *Renard*, suddenly drowned in this flood of light whose source and nature were entirely unknown, did not know what fate awaited him, nor to which saint to pray. The ship that was lighting him from such a great distance could just as well shoot over a cannon ball... He heaved to and resigned himself to waiting.

Luckily, the *Jérôme-Napoléon* was a friendly ship.

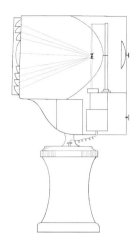

Fig. 15.11. The projector for Foucault's arc light embarked aboard Prince Jérôme-Napoléon's eponymous new yacht in 1867.

Fig. 15.12. The Prince Imperial's yacht identifies the aviso *Le Renard* by the light of Foucault's electric arc lamp.

The electric arc eclipsed

During the 1860s and 70s it became frequent for building sites, factories, squares, boulevards and other places to be lit with electric arc lights. Numerous new regulators were devised, but the arc light enjoyed only a brief ascendancy. Its principal problems remained. It was far too luminous for most uses, and it was too harshly coloured ('hard and silvery', was Foucault's characterization[26]). The development of incandescent filament lamps by Swan in the late 1870s solved both problems and eclipsed widespread use of the electric arc, though it did find a rôle in specialized tasks such as high-magnification microscopy. Nor did the electric arc penetrate to every lighthouse because of the inconvenience of running generators at isolated sites. The luminous sources in lighthouses are now mostly metal-halogen lamps, similar to those used in car headlights.

Regulating motion

Construction of a motorized drive so that the 80-cm telescope would track the stars smoothly and accurately was one of the problems that turned Foucault's thoughts back to mechanics and the production of uniform motion. The standard way of regulating a motor was to attach Watt's centrifugal governor (Fig. 15.13). This governor is in fact just a conical pendulum (Fig. 7.2), except that it has two bobs for dynamic balance. If the motor speeds up, for example, the bobs turn faster and their separation increases. Through a sliding sleeve this actions a lever, which in turn applies a brake or reduces power to the motor, so reducing its speed towards the desired value. In the steady state, the bobs open to an angle which depends on how fast they are turning: the higher the speed, the greater the angle.[27] This is not a condition for precise control, where the governor should spin at almost the same speed whatever the spread of the bobs.

Various attempts had been made to improve the isochronism of the conical pendulum. In the seventeenth century Huygens had trialled a suspension which flexed against a curved surface. More recently an engineer called Pecqueur had put springs in the suspension rods so that they lengthened at higher speeds; but springs are unreliable because they can weaken. Another engineer, Farcot, had produced an approximate correction by hanging the weights from off-axis pivots. Foucault produced another solution described by Moigno as 'very simple and very unexpected'.[28] The Watt governor finds its equilibrium angle when the inward component of the tension in the rods provides the required centripetal force to keep the bobs in a circular orbit. Foucault's analysis was to say that the lack of isochronism arose because the bobs were too weighty when they were turning fast. He devised a system of weights and rods which pushed upwards on the sliding sleeve with exactly the right corrective force. 'This condition had never been realized before me,' Foucault wrote 'and

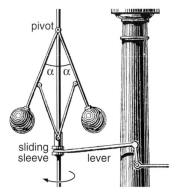

Fig. 15.13. The governor invented by the Scottish engineer James Watt (1736–1819). Rods suspend heavy weights from a pivot which is set into a vertical shaft rotating with the engine. As the shaft spins, the balls fly out, raising the sleeve and moving the lever. If the speed is too great, the lever applies a brake or reduces the power to the engine. If the speed is too low, the brake is loosened or the power increased.

yields a regulator which is completely and rigorously isochronous; or in other words, a completely new industrial product.'

Secrétan built the prototype apparatus, which was tested at the Observatory. Foucault described the underlying principle to the Academy on 1862 July 21 in a very vague and general letter, because he had not yet patented his invention.[29] (The patent followed a month later, written in Foucault's own hand.) Figure 15.14 illustrates the device, which was described as suitable for astronomy and 'chronography', i.e. driving recording cylinders and the like. The regulator actioned an air brake to control the running speed. In the patent, Foucault claimed that the speed stayed constant to 1 part in 5000, or somewhat better than one second per hour.

In the following months Foucault completed a regulated drive for the Marseilles telescope. It is to be seen in Fig. 12.34; its principle appears similar. Later Foucault made another, smaller regulator of this type with a 60-kg drop weight for a 12-inch (33-cm) refracting telescope that Secrétan had recently built for the Observatory.[30]

Le Verrier had torn down Arago's lecture theatre, but with time he had realized the importance of publicity. Foucault showed off the Marseilles regulator (and his speed-of-light experiment) as one of many exhibits at a *soirée* organized by Le Verrier at the Observatory in the spring of 1863. The ever-upbeat Moigno reported that the regulator was 'a truly marvellous mechanism, which gained the admiration of ... all the savants present'.[31] The Minister of Public Instruction was among the spectators, as was Maréchal Vaillant, head of the Emperor's household and the Bureau des Longitudes. Both men must have been glad to see Foucault addressing the practical problems of astronomy, as expected of the Observatory physicist. Le Verrier was happy too: 'I now highly value M. Léon Foucault,' he wrote to the Minister a few months later.[32] But the waters would roughen again, as we shall find.

Sautter too was interested in uniform motion for turning the shutter and filter cages that produced identifying flashes and colour changes in the biggest lighthouses. He also wanted to regulate the steam engines that powered lathes and other machine tools in his factory. Foucault now entered what was to be the final phase of his life, in which he worked frenziedly on the problem of regulation. 'He thus worked long hours,' Henri Sainte-Claire Deville recalled; 'but, occupied with work, he wrote little.'[33] However, it was only his reports to the Academy that became rare;[34] his progress with governors was recorded in over fifteen patent documents, where, uninhibited by the brevity required by the *Comptes rendus*, his discoveries were set out at length. It would be wearisome to describe their every detail, but it is instructive to study their general thrust.

Science often progresses from the specific to the general, and Foucault soon realized that his arrangement of rods and weights was just one of many ways to provide the upthrust required for isochronism. He devised more compact arrangements with inverted counterweights. He rewrote the isochronism

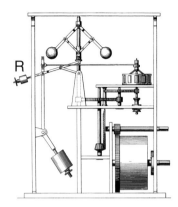

Fig. 15.14. Foucault's first governor was clockwork driven. Only rods and weights were involved in setting the working speed, which could be adjusted by moving weight R. Excess power was dissipated by an air brake consisting of a fan spinning at 30–35 r.p.s. in a cylindrical box with movable shutters. To overcome any stickiness of operation, Foucault kept all control parts in vibration. This oscillation was maintained by a slight ripple machined into the inner faces of the collar on the sliding sleeve.

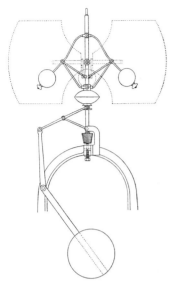

Fig. 15.15. Foucault's governor for turning the filters and shutters in Sautter's biggest lighthouses (manual control sufficed for lesser lights.)[36] Excess power was dissipated by two fan blades (dotted) whose inclination was controlled by the flyballs via link rods. The flying-saucer shaped weight above the sliding sleeve increased the operating speed, which could not be adjusted. The large spherical counterweight and connecting rods provided the variable upthrust required for isochronism.

(Bibliothèque nationale de France)

condition in a more general form. He also improved steam valves and their linkages to reduce friction and provide a more gradual cut-off.

In early 1864 Sautter reported progress to the Society of Civil Engineers.[35] A weight-driven clockwork motor for lighthouse use was controlled by a fan-type governor like that shown in Fig. 15.15, and it had been possible to halve the mass of the motor to 100 kg. The filter cage could be clutched and unclutched without any appreciable change in the motor speed, which kept time as well as an ordinary clock.

Improved isochronism made the regulating action more sensitive. A smaller change in speed would cause a much bigger change in the flyball separation and the position of the control lever. The same control could therefore be obtained with a smaller governor. At Sautter's factory the 45-kg Watt governor on a 16-horsepower steam engine had been replaced by a Foucault governor weighing only 5 kg. A clock dial was attached to the steam engine, which managed to keep time to within a second in half an hour, despite a distant steam valve and abrupt load changes.[37] Another Foucault governor was giving 'very good results' at the foundries at Fourchambault on the River Loire.

Governors for the navy

Foucault's innovations now came thick and fast, and we can see some of them in four governors which he contracted to build for the French navy in the summer of 1865. One governor was for the sawmill in the naval dockyards in Brest, while the others were for three avisos: the *Renaudin*, which was being refitted at Brest, and the sister ships *Bouvet* and *Guichen*, which were under construction at the Rochefort shipyard on the northern margin of the Bay of Biscay.[38]

In an obvious attempt to experiment, every governor was different (Fig. 15.16). Let us look at the *Guichen*'s governor first. Here Foucault provided additional suspension rods and weights above the attachment pivot to make a pair of symmetrical arms which, like perfectly balanced see-saws, would stay wherever they were put, at least when the shaft was not rotating. In this way, the regulator was removed from the influence of gravity and could be used at sea; but governor action required an inward centripetal force to keep the flyweights in a circular orbit. This was provided by springs. However, Foucault realized that the Watt governor could be made independent of gravity by the simpler expedient of hanging the bobs from the sleeve rather than the shaft. He called these *fixed-plane* regulators because with appropriate connecting rods, the weights moved in a fixed horizontal plane and had no tendency to settle in any particular position in the absence of rotation or a spring. This arrangement is *almost* the one illustrated for the sawmill and the *Bouvet*, where the weights in fact move in shallow arcs for reasons that will be explained below.

A spring pulls with a force that is proportional to its elongation from some unstretched length. Foucault had realized that this law would result in isochronism in a fixed-plane governor provided the springs were attached to the flyweights' centres and arranged so that they provided zero pull when the flyweights were aligned along the shaft (though this was not a position the weights could actually take up). This was easily achieved when a hook provided an offset attachment point, as in the sawmill governor, but any unequal pull between the two springs could twist the sleeve against the shaft, increasing friction and wear. It was simpler and more symmetrical to attach the springs between the weights. Foucault gave extensive instructions on how to roll helical springs out of brass, but the fatter ones would fall off when the flyweights were at their closest. Foucault therefore moored the springs at the *edges* of the flyweights, as in the *Bouvet* and *Guichen* regulators. The springs were then too taut for good isochronism, but Foucault made an approximate correction in the *Bouvet* governor by lengthening the suspension rods so that the weights moved in a shallow arc slightly below the level of the pivot.

Springs had another advantage. Since their mass was small, they reacted faster than when a counterweight had to drop under gravity. Governor response was therefore prompter. If this was desirable for the abruptly changing load in the sawmill, it was crucial for the avisos. Heavy seas could raise a ship's propeller out of the water, causing the engine to race; but worse, when the racing propeller fell back into the water, the shock could shear the propeller shaft. In many shipping lines, steamships were breaking their shafts on average once per year.[39]

Avisos were hardly the pride of the French navy. Unstable, slow, and in need of frequent refuelling, they were ill-adapted to carrying urgent messages. Nevertheless, the contract must have pleased Foucault, because Napoléon III was expanding and modernizing his navy, and if the regulators performed well, there would be further sales. However, he must have held doubts about success, because two days after signing the contract he declined an invitation to the country, writing:

> the fact is that I am obliged by a properly legal contract to deliver to the Navy Ministry in September some regulators of a new design for which the first parts are not yet forged. If I were imprudent enough as to come away at such a time, I would be over confident; errors would be made, and the job would certainly not be finished on time.[40]

His caution was to prove justified.

The governors were quickly made. The first, for the *Renaudin*, was delivered barely two months after signature of the contract, while the last and most complicated, for the *Guichen*, was delivered the following January. The pieces were cast at Sautter's factory, which was equipped for work on this scale. The contract price was 1250 francs for each device.

The governor in the sawmill presumably gave satisfaction, because early in 1867 Foucault delivered a similar, smaller governor to the Brest shipyard for use with a mobile, 20-horsepower steam engine in the foundry.[41] Sea trials showed that his governors did not perform so well afloat.

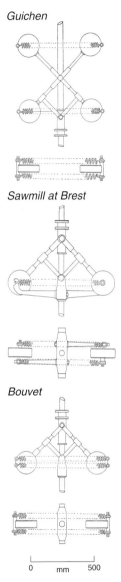

Fig. 15.16. Foucault governors for the French navy. The design speed was 100 r.p.m. for the sawmill regulator and 150 r.p.m. for the others. The *Renaudin* governor was like the Brest one, except that each flyweight had two springs.

The *Renaudin* governor was tried out during a voyage to Gabon in West Africa. 'The Foucault moderator... was of no use to us,' the captain reported dismissively. The cause was ascribed to sticking steam valves rather than the governor, but by the autumn the governor had nevertheless been abandoned. It seems the *Bouvet* governor performed better, and it was still in place when the ship sank off the coast of Haiti (as San Domingo had become) in 1871.

Greater efforts were made with the *Guichen*'s governor. First sea trials were held between Rochefort, Lorient and Belle Île, but parts of the governor unscrewed. Repaired, it performed no better: the engine speed oscillated. 'The governor was thus completely ineffective,' the examining commission reported, asserting boldly, 'Such a result, in complete disaccord with the remarkable behaviour observed for devices on land, can only be explained by incorrectly manufactured springs.' Further trials were curtailed by the *Guichen*'s departure on tour, but the performance of Foucault's governors had been sufficiently disappointing that the navy's engineering inspectors decided Foucault should be asked how to improve them. It was noted that since delivering his governors:

> M. L. Foucault has modified his original design, in which he has recognized the faults, and he has devised a new arrangement for which experiments, he says, have fully established the practical value.

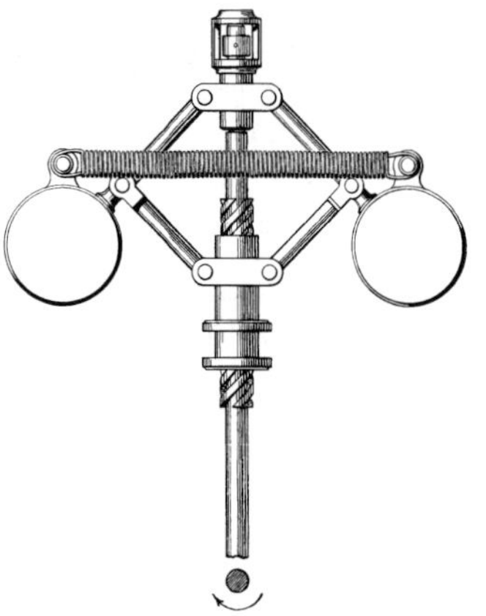

The *Guichen* governor was called back to France for modifications, but by the time it arrived, Foucault was dead.

Oscillation

Foucault reported ten steam-engine governors in service at the beginning of 1865,[42] but it is clear his naval governors were not the only ones that oscillated.

Some engineers blamed the perfect isochronism, which they felt made the governor too sensitive, since a very small change in speed would slam the flyweights from one end of their throw to the other. 'M. Foucault's regulator,' said one, 'has the air of a chemical balance made to tip on a milligramme which has been put in the hands of a coal merchant.'[43] Foucault must have recognized the justness of this rebuke, because some of his later regulators had their isochronism degraded, for example, by the slight lowering of the flyweights in the Brest sawmill governor. Foucault identified steam valves and loose rods and transmission belts as other causes. Ultimately, he decided that the principal problem was the large flywheel attached to all steam engines for the purpose of smoothing out the jerkiness resulting from the small number of cylinders. His centrifugal governor was controlled by speed, but with a heavy machine or flywheel, the speed changed only slowly, so the governor applied its correction late. 'It is almost inevitable that the machine will exhibit a periodic variation in speed,' he concluded.

Fig. 15.17. A Foucault governor responsive to acceleration as well as speed due to the variable coupling provided by the screw threads on its shaft and sleeve. (The threads had to be cut with the handedness appropriate to the rotation direction.) Another innovation is that the rods pivoted about separate points, eliminating the expensive, interleaved forks needed for two rods to pivot about the same point. Foucault patented this governor in 1865 December, but it was to be his nemesis at the Exposition Universelle in 1867 (see next chapter).

Foucault's solution to this problem is shown in Fig. 15.17. A cunning expedient made this governor respond to acceleration as well as speed. He put the suspension pivot on a bearing so that the weights and rods could turn independently of the drive shaft. At constant speed, friction nevertheless kept the flyweights rotating at the shaft speed. To make them respond to the shaft's acceleration as well, he applied a coarse screw and thread to the shaft and the inside of the sliding sleeve. When the shaft was turning at constant speed, no force was transferred, but when it accelerated or decelerated the screw applied a twist to the sleeve and flyweights. The weights' separation changed and the steam valve was actioned. Foucault's hope was that like some of his earlier governors, this new regulator could simply substitute for old Watt governors on a wide variety of existing steam engines.

The engineer in charge of these modifications for the navy after Foucault's death, a certain M. Madamet, advised that periodic oscillations were reduced with the new design, but not entirely eliminated. Extra springs, empirically applied, could improve performance, he said, and indeed he had already applied such springs (with Foucault's approval) to the sawmill and foundry governors. The *Guichen*'s modified governor was installed aboard a transport ship, the *Marne*, and trialled off Brest in 1869. The sea was calm and did not pitch the propeller out of the water, so performance was compared with the *Marne* steaming upwind with sails furled and then downwind with sails hoisted. The governor held the engine steady at 58 r.p.m. despite a 40 per cent difference in the speed of the ship. The examining commission reported that 'the governor works well', but advised that the acid test would be trials in rough seas, which were never forthcoming.

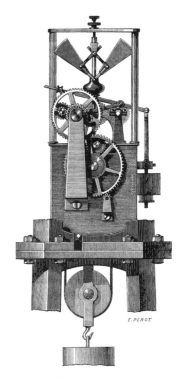

Fig. 15.18. A Foucault governor sold by Secrétan for telescope tracking. Eichens later produced a very similar governor but with an erect counterweight (Fig. 16.4).

Knowledge of true principles

In one of his patents Foucault remarked that previous attempts to improve governors had been *ad hoc* and had enjoyed limited success 'through lack of essential information which can only come from knowledge of true principles'. Thrilled that he had discovered a general way of making the Watt governor isochronous, Foucault supposed his solution would be universally applicable. Lissajous described how these hopes were dashed:

> But once Foucault had embarked on the industrial path, he understood too late that the task was harder than he had thought; it was not just a question of replacing an old device in each machine with a new one, it was necessary to alter the construction according to the arrangement of each engine, and according to the requirements of each industry. He had thought the problem limited, the field circumscribed, but each day he found himself confronted with new solutions and unforeseen circumstances.[3]

Fig. 15.19. Small Foucault governor with a centrifugal fan manufactured by Secrétan. It was sold adjusted so that the fan blades turned at 8 r.p.s. and the pulley at the front made 1 r.p.m. The operating speed could be refined with the fan-blade weights.

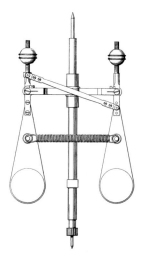

Fig. 15.20. 'This excellent regulator, so poorly known', was the opinion of one astronomer concerning Foucault's final, highly simplified fan governor.[49]

The proper theory for control engineering is mathematical, but as Bertrand commented, '[Foucault] is little acquainted with the language of algebra; he understands it, but he does not speak it.'[44] Cured of his earlier disdain, Foucault must have made great efforts with the phrase book, because his patents contain far more mathematics than his previous publications; but it was algebra and the analysis was limited to the equilibrium case. (As in his thesis, the derivations were marred by mistakes, resulting in numerous errata.[45]) The performance of a governor cannot in general be separated from that of the machine it is controlling and the delays involved before both react. The correct analytical approach is not the algebra of equilibrium but the differential calculus of change, and the secret of smooth control is not only whether the feedback signal is prompt but also whether it is damped. Airy had made some tentative explorations of the dynamical behaviour of governors in 1840, but it was not until the year after Foucault's death that the Scottish physicist James Clerk Maxwell derived some clear stability criteria for governors and set a firmer basis for the development of control theory.[46]

Governors with fans

If Foucault's governors failed to perform satisfactorily when they were mere appendages to a massive machine, they were more successful where the load was an appendage to the governor and excess power was dissipated with a fan. Foucault's first isochronous governor was of this form, as was the one he built for Sautter's lighthouses. Figures 15.18 and 15.19 show two later models commercialized by Secrétan in 1866.[47] Both are fixed-plane regulators: though the blades separate, the height of their centre of gravity stays constant. The first was intended for equatorial tracking and was supposedly very stable and reliable, comprising only weights and levers, though a later report suggested poor performance.[48] The second incorporated springs and so was potentially less stable. The inversion of the fan blades made the apparatus 'more agreeable to the eye', which may have been a reasonable marketing decision but was hardly a sound scientific justification! It was also available attached to a recording cylinder for timing experiments in physics and physiology.

Figure 15.20 shows a final fan governor which was patented during Foucault's terminal illness. Its principal features were its mechanical simplicity and reduced number of joints. The moving sleeve had gone, along with the associated friction. The cross rod (made, like the fans, of aluminium) ensured that the blades splayed symmetrically. 'I attach great importance [to these improvements],' Foucault wrote; and he was right, for with carefully made springs, speeds stable to parts in tens of thousands were achievable, and the device was still available for sale at the beginning of the twentieth century.[49]

Chapter 16

Unfinished projects

Reflectors or refractors?

Although an achromatic doublet produces a much better refracting telescope than a simple lens, the stellar images are still surrounded by a purplish halo because the dispersive powers of crown and flint glass do not cancel exactly at all wavelengths. Foucault imagined that his reflecting telescopes with their perfect achromatism would always be superior to refractors, but in 1860 his opinion changed.

He had just finished his 33-cm telescope mirror, while at the same time Secrétan had completed a similarly sized lens for an equatorial refractor at the Observatory. Foucault examined this lens with his tests and found some defects in its figure. 'But, contrary to my expectations, I had to admit that, in certain aspects, the refractor was manifestly superior to the reflector,' wrote a chastened Foucault.[1]

The reflector was always better as far as angular resolution was concerned: for example, when it was a question of resolving a close double star into its separate components. But for extended objects, such as planets, it was sometimes a matter of distinguishing subtle brightness differences which in the refractor became discernible because they translated into changes in tint through their purplish halos. 'Seen in even a mediocre refractor, Jupiter's bands appear more strongly accentuated than in the best reflector,' sighed Foucault.

Le Verrier seized on Foucault's changed opinion concerning refractors, and in the spring of 1861 ordered him to work afresh on the Chance discs.[2] But Foucault was busy with the speed-of-light experiment, and at the same time it was becoming clear that he would complete the 80-cm mirror successfully. Le Verrier decided to change track, and early in 1862 wrote to the Minister asking for approval to set Foucault loose on a 1.2-m mirror. In addition, there was the problem of Arago's dome. The 15-inch (38-cm) Lerebours lens that had been installed there only a few years previously was devitrifying. Foucault suggested that a replacement 18-inch lens could be cut out of the Chance discs, but Le Verrier was unwilling to sacrifice such uniquely big pieces of glass; and in a wonderfully ambiguous phrase, proposed to 'postpone definitively' work on the Chance discs.[3]

The Saint Gobain glassworks saw commercial advantages in being associated with the world's largest telescope and were willing to cast and anneal a 1.2-m blank at cost price. The sum was small – less than 6000 francs – and Le Verrier could proceed without a special allocation from the Minister. A first blank was too small, but within two years the Observatory was in possession of a beautiful disc that had been cut down in Sautter's factory to 1.23-m diameter and had had its back polished smooth.

Lacking money to continue, Le Verrier's next tactic was coercion. 'Must we now put it [the blank] in store?' he asked the Minister. 'We shall be ridiculed if we do not polish it.'[4]

Foucault held reservations. ('M. Foucault has embarked on his big mirror... we pushed him to do so,' Le Verrier admitted.) The 80-cm mirror had been polished by hand, but there is a limit to a man's strength. During his visits to the British Isles, Foucault will have inspected the machines on which Lassell and Lord Rosse had polished their metal mirrors. He suspected the 1.2-m disc would require the same, at a cost of some 30 000 francs; but ever prudent, he first wanted to polish a 40- or 50-cm mirror mechanically.[5]

Government permission

Foucault grew impatient waiting for funding. He wrote to de la Rive:

> What I need now is a rich supply of ducats, not smoke. With ducats one does what one wants, one pursues science for science's sake, one makes telescopes and one's own observations without needing the government's permission.
>
> And since the regulator is promising to become pretty productive, I am mining that vein without worrying about what the Burgraves [decision makers] say.
>
> So I shall probably make my dream come true: I shall have a handsome laboratory, instruments, apparatus; and who knows, perhaps also pupils raised in this new school of liberty.[6]

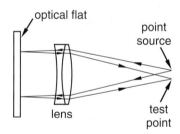

Fig. 16.1. Foucault's autocollimation set-up for testing a lens. The glass plate acting as a mirror must be accurately flat so that only lens defects aberrate the image at the test point. (The two-way light path doubles the aberrations.) Foucault applied his three optical tests to this image (cf. Figs 12.11, 12.13 and 12.15).

Le Verrier was not one to campaign on a single front. Besides twisting the Minister's arm, he enlisted many of his senatorial colleagues in an Association for the Advancement of Astronomy and Meteorology which immediately promised 50 000 francs towards polishing the Chance discs.[7] This will have returned Foucault's thoughts to lens-making. Exactly how he had applied his tests to Secrétan's 33-cm lens is a moot point, because an aberration-free telescope objective must act like a parabolic mirror and ideally requires testing with a parallel beam or an infinitely distant source. Perhaps Foucault used a pinhole source at a finite distance and compared the resulting differential solid with that produced by his 33-cm mirror. However, what is clear is that at some time, probably during 1864, he put into practice an idea which he had had from the inception of his work in optical fabrication, and that was to apply the method of *retouches locales* to the construction of an optical flat, or rigorously flat mirror, a construction which both Gambey and Arago had declared impossible. A flat mirror is lens making's holy grail, because with it a lens can be examined with Foucault's three tests just like a mirror, in what Foucault christened the *method of autocollimation* (Fig. 16.1).

Foucault made a 35-cm diameter optical flat. Lord Rosse had tested flat mirrors by examining the reflection of a watch dial in them, but Foucault used an adaptation of his first test. The flat was fine-ground with emery in Sautter's factory to produce a surface with a velvety sheen. With symmetrically applied strokes of the tool, the mirror could only take on a rotationally symmetric shape. It was then illuminated obliquely by the light from a pinhole placed a few metres away and the reflected image was examined with a small telescope. The presence of concentrations of light, or *caustics*, in the reflected beam indicated whether the mirror was slightly concave or convex, and appropriate *retouches locales* were applied. Finally, the flat was polished, using a slightly convex polisher, methodical strokes and frequent testing. Unlike the telescope mirrors, for which the final polishing took only hours, polishing the flat took three days, and must have been tedious.

A beautiful optics institute

With his optical flat, Foucault knew that he would be able to use *retouches locales* to make achromatic doublets for refracting telescopes. Le Verrier, perhaps, conceived the idea of 'a beautiful optics institute' to exploit Foucault's discoveries, and pressed him to associate formally with Secrétan. Negotiations were difficult, it seems; but on 1865 March 22 a contract was signed, and two months later Le Verrier ordered a 9-inch (24-cm) objective, well within the test capacity of Foucault's 35-cm flat.

Buoyed by the existence of this flat, Le Verrier believed that the way was finally open for polishing the Chance discs, but Foucault was typically prudent and wanted 30 000 francs for preliminary studies.[8] Nevertheless, Le Verrier's campaign with the Burgraves bore fruit in 1865. The National Assembly voted almost 400 000 francs for completion of the 1.2-m reflector as well as a giant refractor fashioned from the Chance discs.

It was probably then, with a 9-inch lens to complete and the Chance discs in view, that Foucault decided it was time to train up a pupil, even if the school was not one of complete liberty. The man he selected was Adolphe Martin (Fig. 16.2), with whom he had recently sat on a Photographic Society committee. Martin was a chemistry teacher at the venerable Collège Sainte Barbe (adjacent to the Panthéon) with a long-standing interest in photography. He

Fig. 16.2. Adolphe Martin (1824–96). In 1852 he published a book on photography under Charles Chevalier's imprint and invented the ambrotype process, which was essentially the one used by Foucault in Spain.[9] In the 1860s he became Foucault's confidant concerning optical fabrication. Following Foucault's death, he published a series of papers on these techniques in the *Comptes rendus* and filled Foucault's place on the Committee of the Société Française de Photographie.

(Bibliothèque nationale de France)

was writing a doctoral thesis on optical instruments. The thesis was most un-Foucault-like, being theoretical and mathematical, but Martin will have recognised the advantage of being associated with Foucault's practical approach, and vice versa. When the thesis was finally complete, Martin dedicated it to Foucault, with 'respectful homage from his devoted pupil'.[10]

No doubt Foucault and Martin completed some smaller lenses first, but the 9-inch objective was delivered to the Observatory within a year. Performance was not satisfactory, and a few further months were required before renewed delivery in 1866 September (Plate XX). Sainte-Claire Deville later disclosed that the perfectionist Foucault had not been completely satisfied even at this stage: the lens still exhibited slight negative spherical aberration which he had not wished to eliminate before tests on the sky, and the degree of achromatism needed checking.[11] But once the objective had been installed in its telescope, everyone else was satisfied and Le Verrier had refused all entreaties concerning finishing touches. 'If I let him have his way,' Le Verrier said, laughing, 'I should no longer have a telescope, for M. Foucault is never satisfied with his work.'[12] Foucault then hatched a plot. He would take the lens away, but to hide the abstraction would substitute another, unfinished one. The first symptoms of his illness put paid to this little crime.

By mid-1867 Foucault had completed another substantial-sized lens, a 19-cm (7-inch) objective destined for the newly founded Lima Observatory. Tested on Gamma Andromedae and other objects in the Parisian sky, the unanimous opinion was that the new objective was superior to a similarly sized and more expensive lens made by the respected Munich opticians Merz & Mahler.[13] However, revolutions in Peru delayed completion of the Lima Observatory, and a dozen years later the lens was still languishing in its packing crate.[14]

Although Foucault completed his 19-cm lens at about the time he fell ill, he will have formed the moral certainty that he was capable of figuring a 75-cm lens well before then. The Chance discs were lathed to approximate shape and given a first polish. He was then able to see that they would not do; they were too thin. This meant an achromatic doublet made from them would have a very long focal length, which would require a mechanically challenging and expensive tube length and dome size. Foucault suggested softening the discs with heat to reduce their diameter and increase their thickness; additionally, the streaks in the flint would disappear. Le Verrier could not agree. He had obtained funding for a 75-cm lens, and that is what he had to deliver. In the face of this stalemate, work stopped.

The siderostat

The developments of the previous years all pointed in a different direction, a direction which had been taken by Fizeau and Foucault when they had photographed the Sun in 1845, and by the École Polytechnique expedition to the 1860 eclipse. In both cases the Sun had been photographed immobilized in the mirror of a Silbermann heliostat. Foucault had even checked the flatness of the Polytechnique mirror for his chemist friend Aimé Girard, who had taken the photographs in Algeria.[15] Foucault's idea was to build a permanent instrument along similar lines.

The advantages of the *siderostat*, as Foucault called it, were considerable. There would be no flexure in the stationary parts and heavy instruments like spectroscopes could be attached easily. It was true that the flat mirror would sag as it turned, but this could be minimized by making the glass very thick.

There were, however, challenges to be met. The heliostat mirror for a photographic enlarger did not need to be especially flat or accurately driven, but for astronomy both had to be of high quality. These were no longer problems. Foucault knew how to make an optical flat. He knew how to mount a heavy mirror from his heliostat, and could drive it accurately with a fan governor. Nor was focal length a critical issue in a stationary telescope. Foucault proposed that Saint Gobain's 1.2-m blank should be made into a siderostat mirror with the Chance discs fashioned into the associated refracting telescope.

Long discussions ensued with Le Verrier, who found the idea perilous. It involved too many novel features – figuring the Chance discs, and making, mounting and driving a giant optical flat. He asked for a written proposal. Foucault demurred, and instead proposed a smaller siderostat, for which Eichens made a wooden model; but Le Verrier's not unreasonable opinion was:

> But when our colleague was already responsible for the construction of a large refractor, worked upon, abandoned, and then worked upon again, and on which almost nothing had been done; when he was already responsible for a reflector on which nothing at all had been done; when these first two projects were established by legislation, and the Minister had rightly written to us to spare no effort to ensure their success: we ask, would it have been reasonable to consent to their abandon in favour of a third project?

Going alone

Foucault's inventions and commercial arrangements were not great money spinners. When he died, he was owed profits of 2700 francs by Secrétan (though almost half of this had to be paid on to Martin), while he actually *owed* Duboscq some 1800 francs. Thus Foucault did not amass a brilliant fortune; but his salaries totalled over 9000 francs per year from the Bureau des Longitudes, the Institut and especially the Observatory, where Le Verrier had given him a pay rise in 1865. His personal expenditure was modest; at his death his estate was worth some 80 000 francs, including 8200 francs in cash.[16] His finances were thus healthy enough. He was able to rent an appartment for use as a laboratory at 42, Rue de Fleurus a few steps from the Rue d'Assas, and Donné later recalled having seen him pay craftsmen between 6 and 10 francs per day to install his instruments. Foucault decided to build his siderostat privately. Returning to scientific goals, he planned a protracted observing campaign on the Sun, whose physical constitution was still far from elucidated in the 1860s, as earlier chapters have indicated, let alone the mechanisms of energy production and its conduction through the solar substance. In the autumn of 1866 Foucault suggested a new, safer method for solar observation. Instead of a filter at the eyepiece, which, as William Herschel had found, could shatter in the heat, he suggested silvering the front surface of the objective, since he knew from long experience that silvering reflected most of the heat and light, transmitting only a safe dim-blue solar image.[17] To obtain a horizon above surrounding buildings, he decided to install his siderostat on the first floor of the house on the Rue d'Assas. Eichens made a second preparatory model in wood, since in the Rue d'Assas the telescope could not lie to the south (cf. Fig. 5.15). Meanwhile Foucault had changes made to his house, later described by Donné in a passage which also reveals Foucault's desire for domestic order and comfort:

> No apartment was more comfortable, more agreeable, or even more elegant than that of this illustrious physicist. Just recently he had refurbished his rooms and arranged them appropriately for the observations of the Sun that he was proposing to make, and as part of this, his laboratory was arranged with savant care, resulting in what between us we called his 'house of precision'. Everything worked perfectly in his flat: the doors, the windows, the locks and the fireplaces – all opened and closed easily, precisely, and noiselessly. To protect himself from the din of the Rue de Vaugirard, he had double frames in the windows... small curtains of purple-brown silk adorned each window, and could be drawn over the panes along a smooth rail. Even art was not neglected in this delightful retreat, at least as far as the harmony of colours was concerned... The door panels were framed in ebony, the tone of which married superbly with the lilac colour of the wallpaper and the muted shades of the plasterwork. It was from beside the fire, and seated in a good armchair, that the happy inhabitant of this pretty room should have conducted a series of new observations of the disc of the Sun, sheltered from the cold in winter and from the heat in summer. One might find these preparations excessive for a serious scientist, but Foucault was not only considering his comfort in setting himself up thus. He said, quite rightly, that it was the best way to pursue exact observations, which needed to be made over many hours in order to arrive at a convincing result.[18]

We can speculate that the exact and long observations that Foucault hoped to make were measurements of velocity shifts in spectral lines indicative of rotation and even convection. Sadly, no sooner were the changes to the house complete than Foucault fell ill.

The Exposition Universelle, 1867

Louis-Napoléon's second Exposition Universelle was even more of a showcase for his reign than the first. The site was four times bigger. Frédéric Le Play conceived a new classification for the 52 000 exhibitors which was reflected in a building of concentric ovals (Table 16.1, Fig. 16.3). Royalty attended in such numbers that the writer Prosper Mérimée quipped they would have to be put 'two to a bed'.[19] Less-elevated foreigners also came: Thomas Cook transported over 12 000 English tourists who each paid £1 16s for a four-day trip. Royal or commoner, visitors were awed by fountains of eau-de-cologne, a hydraulically operated passenger lift, a machine that converted handfuls of rabbit hair into felt hats, and by a 47-tonne Prussian cannon, which, had they known it, was a harbinger of Bismarck's plans for Paris a few years later. In all, the exhibition attracted well over ten million visitors, more than twice the number in 1855, and a profit was made.

Many of Foucault's inventions were on display. Duboscq's stand showed the arc light and an enormous heliostat with a mirror 1.5 m long and 70 cm wide. Dumoulin-Froment (Froment's successor) exhibited a replica of the gyroscope (Plate VIII). Ruhmkorff's coils sparked across long gaps, thanks to Foucault's mercury cups. An equatorially mounted telescope with a 16-cm silvered-glass mirror was to be seen on the Secrétan company's stand, along with silvered-glass mirrors in other apparatus, and 9-, 11- and 13-cm objective lenses constructed by Foucault.[20] Foucault regulators were also displayed, but Secrétan was not the only exhibitor to do so. The previous year, Marc Secrétan had passed control of the firm to his son Auguste, who had decided to dispense with Eichens' services. This was foolhardy since Eichens was a lynch-pin in the business. ('Although... built by the business known under Secrétan's name,' Foucault had written a few years earlier concerning Observatory instruments, 'in reality all these items of precision engineering are the work of M. Eichens who for many years has been head of the workshop, and, let it be said, of the business.'[21]) Eichens set up on his own, and obviously enjoyed Foucault's support because he exhibited a Foucault fan regulator at the Exposition which was only slightly different from one of the ones he had built at Secrétan's (Fig. 16.4; cf. Fig. 15.18).

Once again Foucault was a judge, one of the five for Class 12, devoted to precision instruments and equipment for teaching sciences. The exhibition opened on April 1 and the judges had little time before the prizegiving ceremony on July 1. In Class 12 there were entries to assess from over one hundred French exhibitors as well as foreign makers. 'The judges were up to the job, but they were hurried, overworked, bewildered; they were not given the time to make soundly based decisions,' lamented Moigno. When the prizes were announced, gold medals had been awarded to well-known names – Duboscq, Ruhmkorff, Dumoulin-Froment, Secrétan, Steinheil, Chance Brothers, Deleuil,

Table 16.1. Exhibition categories devised by Le Play for the 1867 Exposition were divided into ten groups (Roman numerals) and subdivided into 95 classes. The classes within Group II are shown (Arabic numbers). Foucault sat on the prize jury for Class 12.

I. Works of art
II. Liberal arts
   6. *Printing & publishing*
   7. *Stationery, bookbinding & artists' supplies*
   8. *Graphic arts*
   9. *Photographic supplies*
  10. *Musical instruments*
  11. *Medical supplies & instruments*
  12. *Precision instruments & teaching supplies*
  13. *Maps & related apparatus*
III. Furniture & objects for the home
IV. Clothing & jewelry
V. Extractive industries
VI. Industrial equipment
VII. Food processing
VIII. Agriculture
IX. Horticulture
X. Physical and moral improvement of the populace

Fig. 16.3. The 1867 Exposition Universelle on the Champ de Mars, adjacent to where the Eiffel Tower now stands. Le Play's classification of exhibits was reflected in galleries of concentric ovals, with each class assigned to a particular oval.

Sautter; while Foucault's regulator earned Eichens the Exposition's most prestigious award, the *Grand prix*, of which only sixty-six were given.[22] But others of merit had been overlooked. 'Never have judgments been greeted with so little favour,' wrote Moigno protesting about the decisions taken by Class 12. 'Ill-informed, [the judges] gave in to the pressure of past renown, personal sympathy, blind friendship, current rivalries...'[23] Harsh words, deriving from unreasonable conditions which the perfectionist Foucault will have detested.

The Exposition Universelle brought additional agonies. Large industrial machinery was displayed in the building's tall, outer oval. Individual machines were powered from common drive shafts via belts and pulleys. To power their shaft, the United States exhibitors had brought along a top-notch steam engine made by the Corliss company in Providence, Rhode Island. Unfortunately, the commissioner in charge of the American section was a long-time expatriate with little interest in his countrymen and despotic in his treatment of them to the extent that it was said that 'every one of them would be glad to pull on a rope to hang him'.[25] Although the commissioner knew that the Corliss engine was coming, he asked the French organizers to provide a steam engine, which they did, made by the Flaud company in Paris. This engine was equipped with Foucault's latest governor, which despite displaced springs, a threaded sleeve and a local flywheel, broke into oscillations, provoking unflattering comparisons from professional engineers.[26] In truth, the governor's task was not easy because the engine had simultaneously to power such diverse equipment as woodworking and nail-making machines, with abruptly varying loads, and a delicate loom, where uniform speed was critical. Foucault was present at the Exposition's opening at 6 a.m. every morning to coax better performance out of his governor before proceeding to the tribulations of judging Class 12. Paul Worms de Romilly, an engineer, later analysed Foucault's governor theoretically and showed that as designed, oscillations were inevitable, but he believed that had Foucault lived, he would have eliminated them by appropriate shaping of the vanes in the steam valve.[27] In this he was perhaps right, because the Corliss engine, which was renowned for its good speed control, had steam valves that snapped shut on a spring, and its governor controlled the trip point, rather than the degree of opening. The American exhibitors were understandably exasperated with Foucault, Flaud and their commissioner as the Corliss engine – which won a gold medal – ran idle behind the ill-functioning French machine.[28]

Most sinister rumours

The calvary of the Exposition Universelle was supplanted by another. Foucault had been feeling tired and overworked. On 1867 July 10 numbness of the hand prevented him from signing his name. This became public knowledge at the Academy's meeting three weeks later. 'During and after the meeting,'

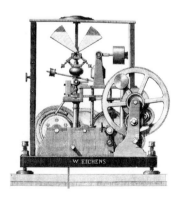

Fig. 16.4. The Foucault centrifugal regulator exhibited by Eichens at the 1867 Exposition Universelle. This regulator was subsequently applied to several large equatorial telescopes.[24]

wrote Moigno, 'most sinister rumours were circulating concerning the health of M. Léon Foucault.'[29] An arm was paralysed, it was said, and speech was difficult. Alarmed, Moigno sought out Madame Foucault, and then Dr Guérard of the Academy of Medicine, Foucault's physician. Moigno renounced seeing the patient from fear of tiring him, but did learn that he went out for a walk most days, and that two days earlier had dined with Jules Regnauld. Moigno's view was that the rumours were exaggerated: Guérard and the later-famous neurologist Jean-Martin Charcot had visited Foucault that very morning and noted a significant improvement. 'Everything leads to the hope of a complete and early recovery,' the Abbé concluded. But soon vision was affected too.

Donné attributed the symptoms to fatigue from 'your devil of an Exposition' and invited Foucault to Noyon to rest. 'Rest of mind and body is torture for one such as you,' Donné acknowledged, but he told Foucault that he had done enough to assure his own glory, which was 'eternal', and that with hospitality and distraction, cure was assured. The Rector went on to cite an officer friend whose identical symptoms had disappeared under this benevolent therapy.[30]

Foucault however held no illusions: his medical studies had been too complete. All the witnesses testify to the horror of his gradual deterioration. A former colleague from the *Journal des Débats* wrote:

> This destruction of the mind was sufficiently slow and gradual that he could not have failed to feel its bitterness. One could even say that he grieved for himself when, via a gesture or a word arduously spoken, he expressed his despair as he felt more and more enveloped by darkness. Which of his final friends will forget the refrain, 'Still sick!' which he repeated in such a harrowing fashion as he let his exhausted forehead drop onto his feeble hand?[31]

As the end approached, even the Empress was concerned (Fig. 16.5).

These sad descriptions suggest a diagnosis of multiple sclerosis (actually first described by Charcot in 1866) which, unlike a cerebral tumor, can involve periods of remission, and occasionally progresses very rapidly.

A depressing little ceremony took place at the Rue d'Assas on Monday, November 4. Foucault had called together his notary and a handful of friends to make his will.[33] It is a sad and touching document because although Foucault was able to dictate his wishes, no doubt slowly and painfully, he was too incapacitated even to scratch a mark on the paper. His friends had to certify the document as his true intention. In France, testatory freedom is limited. The majority of his estate devolved to his mother and sister; his will concerned his scientific apparatus. He bequeathed his gyroscope 'with its two boxes' to the Collège de France. The 1862 speed-of-light apparatus went to the Observatoire de Paris. Since the experiment had been funded by the Observatory, this was, as Foucault noted, 'really a restitution'. (The authorities nevertheless levied 90 francs inheritance tax.[34]) The pendulum and drive from the 1855 Exposition

Fig. 16.5. The Empress Eugénie enquired after Foucault's health with 'a touching kindness' during a visit to Sainte-Claire Deville's laboratory.[32]

Universelle were given to the Conservatoire Impérial des Arts et Métiers. All other instruments and apparatus were bequeathed to Jules Regnauld. Finally, the patent for the electric light was made over to Duboscq on condition that 5 per cent of sales income was paid to Foucault's sister and mother.

The approach of death returned Foucault to the Catholic religion – or perhaps brought him to it for the first time, because Moigno noted that the approach was 'slow, and several times resumed'.[35] The Abbé continued:

> His mind, encumbered but healthy, must have applied itself successively to the problems of creation, meditation, redemption, forgiveness and absolution. His conversion was voluntary; he wholeheartedly welcomed the ministry of the pious priest who surrounded him with friendship.

Extreme unction was received.

A death-bed conversion seems surprising for one of Foucault's scepticism and iron will, but his mind had been eroded by a long illness and family pressure was being brought to bear (the priest was one 'in whom [Foucault's] mother had full confidence').[36] We must certainly treat cautiously claims that his aphasia was lessened during talk about God and Christ.

The end came at 11.30 p.m. on 1868 February 11, a Tuesday.[37] His dying word epitomized the physical and mental torments of his final months: 'Malheur!' (Misfortune!)

The funeral service was held on the Friday. Advertisements in the newspapers asked mourners to gather at the house on the Rue d'Assas; they then moved on for an 11 a.m. service in the nearby church of Saint Sulpice where his parents had married over fifty years before.[38] Thirty-three Academicians attended, triple the number for a run-of-the-mill Academy funeral.[39] Foucault's final resting place was the Montmartre Cemetery in northern Paris (Fig. 16.6).

The pall bearers were all members of the Academy. One was Charles Combes, a fellow member of the Mechanics Section, while the other three – Le Verrier, Yvon Villarceau and Delaunay – were from the Observatory and the Bureau des Longitudes.

There were three graveside orations. Another member of the Mechanics Section, General Arthur Morin, spoke on behalf of the Academy. Yvon Villarceau spoke for the Bureau des Longitudes. Finally, Joseph Bertrand traced out 'in affectionate terms' the scientific work of his friend.[40] It was noted unfavourably that Le Verrier said nothing.[41]

Fig. 16.6. Foucault's tomb in the Montmartre Cemetery where his family had purchased a plot in 1859. The pillar behind his bust lists what at the time were perceived as his greatest discoveries: photo-electric microscope, pendulum, gyroscope, telescope, regulators.

Fighting

The fighting began less than two weeks later when Henri Sainte-Claire Deville read a note to the Academy about his friend's last work at the Observatory.[42]

Le Verrier's directorship was under increasing criticism for its dictatorial style and the personnel were in revolt. The Minister of Public Instruction had admitted to the Emperor: 'For the last four years I have not dared to look at the Observatory,' and a commission of enquiry had been put in place. Late in 1867, while Foucault was ill, Le Verrier had responded to the pressure he was under by writing a widely distributed brochure entitled *Work of the last Thirteen years: Present situation; ...Deplorable inaction of the higher administration...* As the title suggests, this was a polemical work, full of self-justification. 'A whole paragraph concerns the work done for the Observatory by M. Léon Foucault,' Sainte-Claire Deville announced in a strained voice to his hushed audience, 'work which does not seem to be recognized at its proper value.'[43] He went on to say that he had had extensive discussions with his 'unlucky friend' and was making his presentation 'for the honour of his memory and to ensure his rights to a great and still unpublished discovery.'

In his pamphlet Le Verrier had outlined how he had obtained 400 000 francs to build telescopes from the Chance and Saint Gobain discs. He had then ascribed blame for the lack of progress:

> When one undertakes work of this magnitude, the functionary in charge must be left completely free to concentrate on the task, otherwise failure will always follow. Now the able physicist on whom we counted was removed from the job and diverted to work for the Exposition Universelle. So, to our great sorrow, the work was suspended.

Le Verrier had continued:

> Moreover, the project is unimpaired, in the sense that since no work has been done, the money is untouched.

This angered Sainte-Claire Deville. Perhaps nothing had been spent of the 400 000 francs, but, he objected, Foucault 'had worked, had worked at length and conscientiously, on the question of large lenses; I can even say that he had resolved the problem in the most remarkable and most complete manner.' Sainte-Claire Deville hoped his intervention would clearly establish Foucault's authorship of methods for shaping lenses. He concluded by depositing a *paquet cacheté* by Adolphe Martin outlining Foucault's procedures.

Le Verrier was not expecting this attack and was absent from the Academy that day. The following week he was present right enough, ready to answer all charges in front of numerous spectators who were seeking entertainment on a grey March day.

As the candles flickered, Le Verrier began by claiming that he had resolved to publish a complete account of Foucault's work. He was particularly qualified for this task, he said, because of the numerous papers and manuscripts written by Foucault in his possession. He would show that Foucault's works were 'greater, linked together better, and went back to an earlier date' than

Sainte-Claire Deville had indicated; and so much so that his paper 'which said nothing new, was pointless'.[44]

Le Verrier went on to describe Foucault's early career and the long history of the Chance discs. Time was passing, and in a move which indicated what the Academy really considered important, the chairman interrupted him in order to conduct an election.[45] The new Academician selected, Le Verrier returned to the fray. The Chance discs were purchased, followed by successive periods during which Foucault thought or not to work them. Deferring further remarks to the following week, Le Verrier concluded, 'You will see... how much I was busy helping M. Léon Foucault... despite the embarrassment that these changes produced... with the Ministry.'

It was then Sainte-Claire Deville's turn. Rising from his heavy Academician's chair, he confined himself to a prepared text. He described Foucault's plans for the siderostat and his collaboration over it with Charles Wolf, one of the Observatory astronomers.

Le Verrier interrupted, furious. Another pointless presentation! *He* would tell the full story of the siderostat the following week. 'Ah! M. Deville, Léon Foucault would be very badly off if he only had advocates like you,' he sneered. 'Because you are so well informed, tell me the date of the invention of the siderostat!'[46]

Sainte-Claire Deville was confused; he did not know exactly, and withered under Le Verrier's offensive. Ashamed of the spectacle, some Academicians called for the Academy to pass into secret session; but smelling blood, the chairman, Charles Delaunay, chose to widen the attack to another issue. Why would Le Verrier not name the observers who were discovering new minor planets at the Marseilles Observatory? Le Verrier retorted that they were mere mechanicals carrying out his orders, who were happy with their anonymity. He added, 'For each [planet] they receive a rise of 250 fr and a gold medal.' Uproar ensued. Delaunay cleared the public benches and the squabbling continued in private. The meeting 'will never be forgotten by those who were present,' commented one newspaper.[47] 'Three hours of vain and sterile talk,' was the opinion of another.[48] A third seized the opportunity to repeat calls for Le Verrier's dismissal.[49]

Respecting the scientific legacy

Crowds flocked again to the Academy on the following Monday, hoping the spat would continue, but they were disappointed.[50] Le Verrier sent a note saying he was taking some holiday, for which he had 'the greatest need'.[51] The only excitement was the news that the Emperor had decided to honour Foucault's memory.[52] Prompted by Sainte-Claire Deville, Louis-Napoléon established a committee to complete the work that Foucault had had in hand and to print his papers in a collected volume. Besides honouring his protégé,

Napoléon III was following a tradition of government publication of the works of prominent savants. 'I want to pay the expenses of this publication,' wrote the Emperor, and he allocated 10 000 francs per year from his privy purse.[53] The committee was chosen solely from Foucault's close friends and scientific confidants: Eugène Rolland, whom he had beaten in the 1865 Academy election, Charles Wolf from the Observatory, Jules Lissajous, Jules Regnauld and Adolphe Martin. The absence of any Academician raised hackles at the Academy, but Sainte-Claire Deville growled back that 'if the committee has been composed thus, it was to accord with the wishes of our late colleague; there was a scientific legacy to respect there, and it has been respected.'[54]

While dying, Foucault had mournfully announced that he had ideas that would have kept him busy for twenty years, including renewed measurements of the speed of light and publication of a book on regulators.[55] It was later speculated that Foucault destroyed all jottings concerning these future projects.[56] Be that as it may, there was little new on paper, and the committee soon realized that they could only complete the siderostat. Eichens built the mechanical parts and Martin the plane mirror (Fig. 16.7).[57] The siderostat was installed in the Observatory garden, but received only desultory use for solar photography and photographic enlargement.[58] A copy was ordered by the Scottish aristocrat Lord Lindsay for his observatory at Dun Echt, but the siderostat did not prove to be the world-beating instrument that Foucault had imagined. A giant siderostat made for the Paris Exposition Universelle in 1900 had a catchy advertising slogan ('The Moon at a metre'), tracked well, and was briefly the world's largest-aperture telescope. However, siderostats have been of little in-

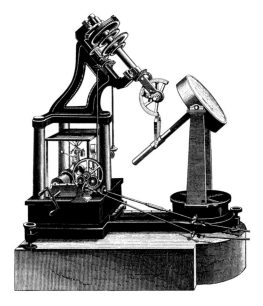

Fig. 16.7. The siderostat, completed at the end of 1869, was installed in the Observatory garden with a 20-cm diameter horizontal telescope, but it was not used extensively. The observer could move the field of view around the sky by turning the control rods that depart to the right. The planetary gearing mounted on the polar axle relayed orientation adjustments even when the siderostat was tracking. The regulator was later replaced by the simpler one illustrated in Fig. 15.20.

terest to professional astronomers, in part because their field of view rotates slowly as they track.[59]

The Saint Gobain blank and Chance discs fared no better. The structure for the 1.2-m telescope was commissioned from Eichens and completed in 1875. The optics were entrusted to Martin, but he was never able to figure the mirror successfully, and at one stage tried to hide its poor quality with a stopped-down eyepiece. By the late 1870s he had clearly lost interest in the project. (He had also been paid.) Part of the problem may have been that the mirror was thin and insufficiently rigid. In the 1940s it was reworked but was broken soon afterwards during silvering.[60] Martin and Eichens were also contracted to build the refractor from the Chance discs. The striated flint was replaced in 1879, and again in 1880, but the Observatory had difficulties financing a site for the dome and the project was abandoned.[61]

An inconsolable mother

The fall of the Second Empire at the battle of Sedan on 1870 September 4 put an end to the imperial privy purse. Publication of Foucault's papers was taken up by his mother Aimée. She appointed Lissajous editor, but ultimately the volume was brought to press by Charles-Marie Gariel (Fig. 16.8), professor of physics at the medical school, and husband to her grand-daughter Marguerite, Foucault's niece. The *Recueil* appeared in 1878 (Fig. 16.9), a couple of years before Aimée's death. '[I] had the satisfaction of finishing this publication during my life,' she wrote, but most of the copies were destroyed in a fire.[62]

Aimée had additional projects to commemorate her beloved son. She endowed a Mass for the repose of his soul.[63] She distributed memorabilia, giving a porcelain teapot of which her son had been particularly fond to a woman friend,[64] and a copy of the *Atlas* of micrographs to Jules Regnauld, inscribing it 'A token of sincere affection for my dear Léon's most ardent friend. His inconsolable mother...'[65] She felt that the floor occupied by her son could not be converted to any domestic use without commiting 'a sort of profanity'.[66] She sought the consolation of a shrine. The priest of Saint Sulpice, the Abbé Riche, came to her rescue. Inspired by Le Play's political ideas, he organized Wednesday meetings of young men in the hallowed rooms to listen to lectures on questions of social development and economics, and to hear and discuss essays on philosophy, history, science and literature. The meetings began about 1872 but presumably did not continue past the death of Foucault's mother on Christmas Day 1880. To prolong the memory of her son further, she had instructed in her will that the 'house of precision' should not be demolished until at least twenty-five years after her son's death. The block of flats now on the site carries a commemorative inscription and stone pendulum (Fig. 16.10).

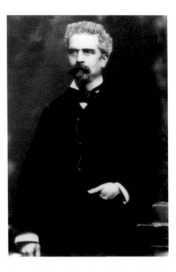

Fig. 16.8. Charles-Marie Gariel (1841–1924) who edited the *Recueil* of Foucault's scientific work.

(Family collection)

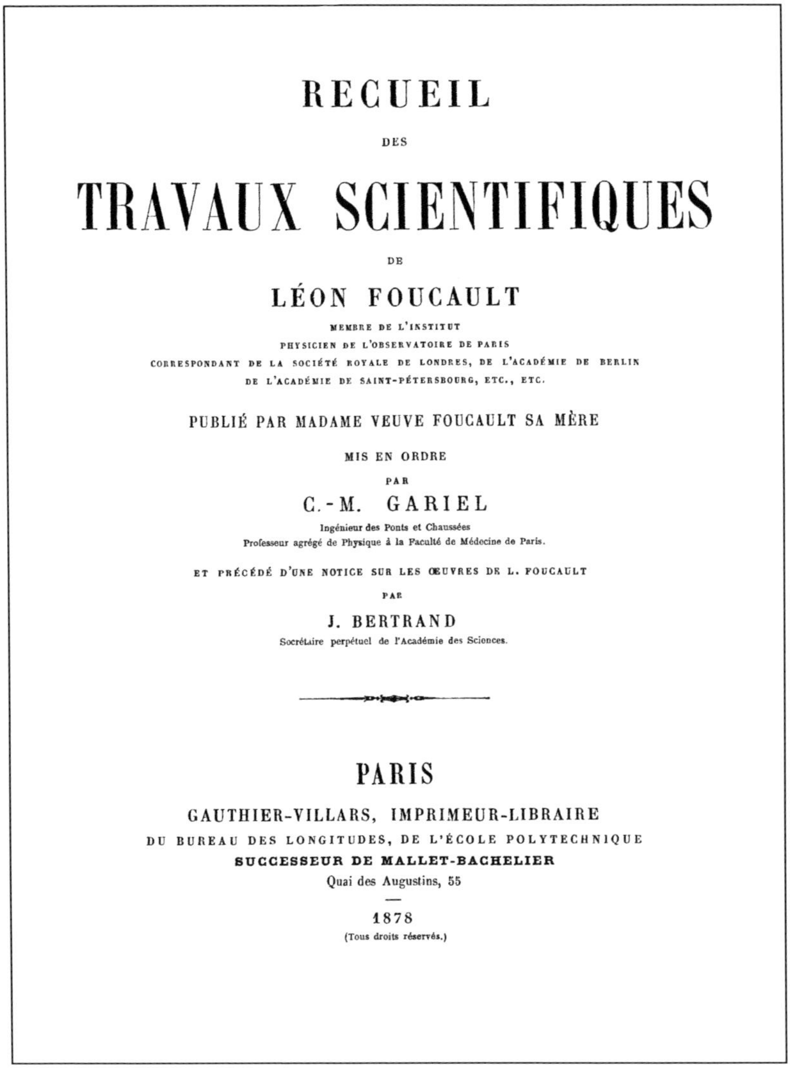

RECUEIL

DES

TRAVAUX SCIENTIFIQUES

DE

LÉON FOUCAULT

MEMBRE DE L'INSTITUT
PHYSICIEN DE L'OBSERVATOIRE DE PARIS
CORRESPONDANT DE LA SOCIÉTÉ ROYALE DE LONDRES, DE L'ACADÉMIE DE BERLIN
DE L'ACADÉMIE DE SAINT-PÉTERSBOURG, ETC., ETC.

PUBLIÉ PAR MADAME VEUVE FOUCAULT SA MÈRE

MIS EN ORDRE

PAR

C.-M. GARIEL

Ingénieur des Ponts et Chaussées
Professeur agrégé de Physique à la Faculté de Médecine de Paris.

ET PRÉCÉDÉ D'UNE NOTICE SUR LES ŒUVRES DE L. FOUCAULT

PAR

J. BERTRAND

Secrétaire perpétuel de l'Académie des Sciences.

PARIS

GAUTHIER-VILLARS, IMPRIMEUR-LIBRAIRE
DU BUREAU DES LONGITUDES, DE L'ÉCOLE POLYTECHNIQUE
SUCCESSEUR DE MALLET-BACHELIER
Quai des Augustins, 55

1878
(Tous droits réservés.)

Fig. 16.9. The *Recueil* or *Collected Scientific Works* was finally published in 1878, ten years after Foucault's death, and at his mother's expense. Accompanied by an *Atlas* of nineteen copperplate engravings, it cost 30 francs and remained listed in the publisher's catalogue until the mid-1890s.

Other memorials

Foucault was honoured with other stonework. Through Sainte-Claire Deville's influence, no doubt, a bust in Italian marble was commissioned for the courtyard of the École Normale Supérieure from the sculptor Gustave Garnier. Foucault's sister saw the original clay model and was 'satisfied by it', while Sainte-Claire Deville found it 'marvellously true to life'.[67] Among the copies

Fig. 16.10. Stonework commemorating the first pendulum experiment, corner of the Rues de Vaugirard and d'Assas, Paris.

made, one was for his mother, which was later donated to the Paris Observatory (Fig. 16.11). Following the burning of the Hôtel de Ville by the Paris Commune in 1871, Garnier went on to carve a full-length statue of Foucault, complete with electric arc and Panthéon pendulum; it adorns the Rue de Rivoli facade of the new Hôtel de Ville.[68]

There were other memorials, continuing long after his death. Foucault's is one of the seventy-two names of post-Revolutionary savants cast by Gustave Eiffel into the metalwork of his tower. The navy named a submarine after him, which in 1916 achieved the dubious distinction of being the first to be sunk by an aircraft. There is a Rue Foucault in Paris, a Léon Foucault Gymnasium at Hoyerswerda in Germany, and even a Foucault crater on the Moon.

However, Foucault's greatest monument is his pendulum, which continues to fascinate and inspire. It has been chosen as an instrument of death in detective fiction and in a post-modernist novel.[69] On a less macabre note, hundreds of Foucault pendulums swing all over the planet and draw a continuous stream of curious admirers. Known to everyone, it is the single insight of the pendulum experiment that has, in Donné's word, made Foucault's name eternal.

Fig. 16.11. 'Marvellously true-to-life' bust by the sculptor Gustave Garnier (1834–92). This copy was made for Foucault's mother. Others were made for his tomb, the École Normale Supérieure, the Institute and the Faculty of Sciences in Toulouse.[70]

Chapter 17

Commentary

Previous chapters have laid out Foucault's life with its triumphs and numerous tribulations. The story is interesting in itself and illuminates a bygone age, but what conclusions can we draw concerning Foucault and his achievement?

Uncertainties

I impress upon my students that a physical measurement cannot be interpreted without an estimate of its uncertainty, but the principle is generally applicable. How trustworthy are the details reported in this biography?

I hold no doubt that my description of Foucault's work is correct in its broad outline, since it is mostly based on his own publications. However, these are summary documents and often do not indicate how his ideas arose or developed. Written with the aim of making a scientific case in the *Comptes rendus*, or a didactic point in the *Journal des Débats*, they may even be completely misleading concerning temporal details. For example, Foucault claimed that is was only after he had completed his silvered-glass collimator mirror that he realized it could be used as a telescope in its own right. This beggars belief. Fragments of evidence suggest episodes which seem highly probable, but for which the evidence is not explicit: for example, that he suffered a breakdown in 1849, that he conceived his polarizing prism a decade before he made it, and that his catadioptric microscope was a time-filler. Inevitably, a few of these deductions will be half truths, or even plain wrong.

Some of Foucault's work is lost. We have seen indications that he made far more experiments with electricity, while in the *Recueil* Gariel printed drawings of an apparatus which converts the rotation of an axle into backwards-and-forwards oscillations of a shaft (Fig. 17.1). He could find no indication of why Foucault built this device, and neither can I.

Chance has played its role in determining what few letters and other original items survive. Foucault's interest in plants and music is indicated directly in only a single letter. There were surely other joys and vicissitudes about

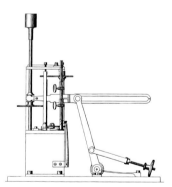

Fig. 17.1. What is it? Why was it made? Is it even the right way up? (Holes in the baseplate suggest it might have been mounted against a wall or ceiling.)

which we know nothing. Further, meaning may be unclear after 150 years. Plate XXI reproduces a painting by the engraver Adolphe Rouargue, who was Foucault's brother-in-law. There can be little doubt that it shows Foucault, his sister and her three children.[1] The girl with *marguerites* (daisies) around her hat will be the eldest child, Marguerite. But the picture is full of other symbolism besides daisies. Why has Foucault got an ear of wheat in his hat? Why is he examining the steamboat with his spyglass? (His father had bought some shares in a Loire steamboat company. Is he checking up on the family investment? Or does the steamer contain one of his governors?) What is the significance of the hollyhock and the unsteady chair? It is unlikely that these questions will ever be answered.

Worse than an object whose meaning is lost is a document that misleads. One inventive journalist told his readers that:

> Thrown out of the Observatory by M. Le Verrier, M. Foucault had to ask the Carmelites for hospitality in order to continue with his experiments, and he died from a chill contracted in the hall of the monastery![2]

We have seen that Louis Figuier is another great source of misinformation.

History is seen through national and linguistic eyes. In this book I, an English-speaker, have written the biography of a Frenchman, and discovered how the history of science is often perceived very differently in the two worlds. Is it Boyle's or Mariotte's Law that links the pressure and volume of a gas? Did Charles Babbage or Blaise Pascal build the forerunner of the computer? Was the steam engine invented by Denis Papin or Thomas Newcomen? Foucault currents are unknown outside lands of French influence, and as we have seen, this is not surprising, since they have very little to do with Foucault.

Biographers rarely discuss the limitations of their craft, but they work perforce from random and fragmentary evidence. Further, a biography is such a lengthy and engrossing project that the biographer inevitably develops an emotional relationship with the subject. Tears filled my eyes when I first saw Foucault's will and realized he had been too incapacitated to sign it. In ascribing motives and feelings I have frequently wondered whether I was writing about Foucault or myself. Biography is like a crime-scene report. At its best, it presents sufficient evidence to draw some valid conclusions, but it can never be the whole truth or completely unaffected by the outlook and attitudes of its author.

Foucault the man

'We do not like the man,' wrote one journalist, 'but the savant can pass for one of the glories of France.'[3] 'Dry and hardly affable at first sight,' said another.[4] Even if not as abrupt as Magendie, Foucault never hesitated to say what he meant quite directly. When seeking the chair of experimental physics at the

Sorbonne in 1853, he did not beat about the bush when he wrote to Napoléon III: 'I therefore ask Your Majesty to approve my application and designate me as the choice to the Minister of Public Instruction.'[5] His judgments were considered, and he expressed them unambiguously. Lissajous commented:

> Incapable of dissimulation, he spoke the truth to everyone in a form that was sometimes cutting, though never malicious, but with so much sharpness and shrewdness that the shot always found the weak point in the armour. May it please God that all whom he wounded thus have had the generosity to pardon him.[6]

Foucault's path to the Academy was a rocky one because the injured were not always forgiving.

Foucault's outlook and concerns are completely modern with his attachment to empirical evidence and reason, and comments against ecological destruction and the tastelessness of processed food. What is striking is the unwavering strength of his attachment to rationality, since most of us suffer from fits of unreason. Perhaps this was a reaction to his father's madness and his own uncertain mental equilibrium.

Though Foucault's public persona was austere and prickly, his loyalty to friends was absolute and reciprocated. Dry humour pervades his newspaper articles, but the outright jocularity of his few surviving letters show how impish he was in private. He ended a note fixing a lunch date with Jules Regnauld, 'All [wishes] to you from heart and stomach.'[7] He even laughed at his sufferings under Le Verrier, as an obituary writer recounts:

> ...but, I hasten to add, Foucault was never embittered by his problems at the Observatory, and his friends know the humorous and sometimes biting zest with which he recounted his misadventures...[8]

Foucault liked the indoor life, whether in laboratory, Academy or house. He described himself to Donné's wife as 'the big incorrigible child who takes his tea on the floor by the fire'.[9] Visits to Noyon for repair of his fragile nerves may have involved occasional excursions to the seaside, but Donné reported that more generally:

> he would not leave the house, or at most walk a few times in the garden. He led, one might say, the life of one of those cats in good houses who install themselves in the best place in the bedroom or living room. When some visitor called who did not interest him, he would retreat to a corner with his notebook and omnipresent pencil...[10]

Virtually nothing remains of his relations with his family, but we can glimpse that they were important to him. In his slightly homesick letter to Regnauld from London in 1854 he added:

> I think it would also please my dear brother-in-law if you were to pass by and tell him that I am still among the living... Probably he will pass

Table 17.1. Staff who had left the Paris Observatory since 1854. List prepared for the Commission examining Le Verrier's directorship in 1867.[13] Le Verrier was fired three years later.

---

6 astronomers

Faye	Desains
Liais	Simon
Puiseux	Chacornac

11 assistant astronomers

Besse Bergier	Barbier
Charrault	Lépissier
Leyssenne	Voigt
Serret	Lechartier
Vezy	Gernez
Babinet	

47 astronomer aides, calculators, assistants

Butillon	Vinches
Bouchet	Garrit
Reboul	Gélin
Thuvien	Thirion
Delaire	Flammarion
Saffray	Harlant
Bourdette	Loiseau
Lartigur (C.)	Hœltzen
Lartigur (H.)	Dussolin
Durand	Barbelet
Boillot	Monin
Bulard	Tixier
Lafon	Lucas
Dumerthé	Leroy
Lecocq	Descroix
Boblin	Parault
Dien	Perrin
Hermitte	Thillay
Noël	Combres
Delépine (twice)	Dubois
Rambosson	Massenot
Hirtsch	Caniard
Maerleyn	Neuville (Me)

---

on solid details concerning the trip of the mother, daughter and litter of children. You will have a magnificent theme to embroider to give me the news of the whole household.[11]

A time of transition

Science changed from an amateur to a professional activity during the nineteenth century. In 1851 Charles Babbage wrote, 'Science in England is not a profession: its cultivators are scarcely recognised even as a class,' and it was hardly better in France. Foucault started out as a happy amateur with science as a Sunday recreation; for ten years he earned a few francs from journalism, and he ended up in government employ for a final dozen years. He did not make this transition completely easily, because it compromised his independence. He continued to work at home even when a laboratory became available for him at the Observatory, only moving there when scientific requirements exceeded resources on the Rue d'Assas. There was perpetual tension with Le Verrier, who wanted managed research and plans of action. Le Verrier had sold more than he could deliver when he obtained 400 000 francs from the National Assembly for giant-telescope building. Foucault had founded a new approach to optical fabrication but his ideas were evolving and the way forward was uncertain. Conflict was inevitable, and it was hardly Foucault's fault. Nevertheless, he was one of the very few at the Observatory who managed to survive Le Verrier's exactions (Table 17.1). Primarily this was because Foucault was tenacious when crossed, as he had showed with Libri and Magendie many years earlier, and partly it was because he was protected by the Emperor and by his position as a civil servant. 'Sir,' he said to the Minister during the investigation of the Observatory in 1867, 'we remaining established civil servants can protect ourselves if necessary. I recommend the weak to you.'[12]

Though Foucault had no desire to lead a team, he was not misanthropic. His early career was moulded by collaboration with other scientists, and later he worked with instrument makers. He also undertook what is now an inextricable part of the way that science is done – he trained up a student. The trend was already emerging, with Dumas and Sainte-Claire Deville supervising students in their laboratories.[14]

The full rôle played by Louis-Napoléon in Foucault's career is unclear. When members of the Bureau des Longitudes had an audience with the Emperor, Moigno noted that Foucault 'alone occupied the Emperor's attention, that His Majesty took him by the arm and spoke for some instants with him', while Donné stated that Foucault was always careful to report his most interesting discoveries to his protector.[15] Was Foucault ever invited, like Pasteur, to the Emperor and Empress's famous *soirées* at their château in Compiègne? But now that science has been institutionalized, it is the rare head of state who would emulate Louis-Napoléon's personal interest in the work of individual scientists.

Creative processes

Foucault remained mute concerning certain intriguing aspects of his work.

When Newton was asked how he discovered the law of gravitation, he replied testily, 'By thinking of it.' Foucault was equally reserved concerning his creative moments. 'Foucault', said General Morin at his graveside, 'rarely took the trouble to explain the principles which had guided him.'[16]

Many contemporaries remarked on the originality of Foucault's approach. We can see three contributing factors. First, he had no training in physics beyond the *baccalauréat*. 'Often at the beginning of a new research topic,' wrote Bertrand, 'he had recourse to the erudition of his friends, and without any embarrassment had them teach him the basics of some classical theory.'[17] Lacking formal instruction, his thoughts were less fettered along conventional lines. Second, his temperament was one of independence and had no inclination to accept earlier ways as the one road to discovery. Third, as a commentator noted, 'He spent more time thinking than reading.'[18] Much scientific advance comes through diligent and plodding application – which were not to Foucault's taste – but thought and imagination are essential for novel discoveries (cf. quotation beneath frontispiece).

Foucault recorded equally little concerning the constructional challenges of his apparatus. As a forward-looking scientist, he used metric units in his memoirs, but my examinination of some of his surviving instruments show that they were engineered in pre-Revolutionary inches. He did describe how he achieved the dynamic balance of his spinning mirror, but if a lot is known concerning the problems of constructing a Foucault pendulum, it is not thanks to Foucault, but because the experiment has been replicated so many times. He described the broad principles of his optical test methods, and his early memoir on silvered-glass telescope construction in the Observatory's *Annales* is unusually explicit, but it is only Martin's posthumous descriptions that give any inkling of how Foucault developed these ideas to optical flats and lenses.

Foucault's science

Foucault's science was not a succession of triumphs. His first arc light failed because of a rival patent, and his second one faced stiff competition from Serrin's design. No one was very interested in his heliostat. His like of ponderable, mechanical ideas was reflected in his delight in horology and precision engineering. He disdained the abstraction of mathematics, though he did relent when he realized that analysis was essential for progress with governors. Despite a valid deduction about the mechanism of the daguerreotype, he was out of his depth when he approached questions that would have led towards the etherealism of modern physics, such as the reversal of the D lines, and the nature of liquid conduction. However, this makes him much more representative as a scientist, few of whom proceed from one success directly to another, or forge new paradigms.

His science spans both the fundamental and the applied and is rooted firmly in the nineteenth century. No one snaps daguerreotypes or makes photometric measurements with starched screens any more, sunlight had been banished from the laboratory, and control is now effected by modern electronics. Amateur telescope makers still use the knife-edge test unaltered, but commercial optical fabrication has moved on with new materials and improved test methods. The gyroscope gave rise after some decades to the gyrocompass, but that too has been superseded by optical gyroscopes and the Global Positioning System.

Turning to fundamental problems, the nature of light is understood to be more than just a choice between waves and particles and the solar distance is measured with different techniques. Further, the interference experiments with Fizeau and the speed-of-light measurements were made within a scientific context elaborated by others. Foucault never produced any fundamentally new understanding; he left no 'Foucault's Law'. In the *Revue des Deux Mondes*, Bertrand wrote:

> he shows more wisdom than depth: he does not produce works of synthesis, but inventions; he brings no new theories, but decisive and unexpected facts which clarify and confirm understanding.

He added, 'the savant simplicity of his methods would have been understandable two hundred years ago,' and they are equally accessible to us, nearly two hundred years later. We can see clear themes that develop in his work, some of which ultimately combined together in the siderostat. The pendulum, which put Galileo on the track of dynamics, provided Foucault's most abiding insight. Our understanding of the nature of space and the interpretation of the pendulum and gyroscope have been refined, but their fundamental characteristic endures: they reveal the Earth's rotation.

The two words that I would choose to summarize Foucault are *independence* and *precision*. Independence of personality, and independence of approach; precision of observation, precision of thought and word, and precision of construction. These are not bad goals for any of us. Foucault's talents were extraordinary, but they are not so far above our own base competencies that we are overawed by them. Faced with adversity, we can be comforted by the knowledge that Foucault encountered misfortunes too. His genius seems achievable. Posterity rightly remembers this remarkable son of the City of Light.

Colour plates

Plate I (*left*). A dandified Léon Foucault, c.1850. The daguerreotype image is reversed.

(©*Musée des arts et métiers-CNAM, Paris/Photo Studio CNAM*)

Plate II (*right*). Foucault's dexterity and tidy handiwork are evident in these adjustable mirrors in black glass (the left one is missing) probably made for the demonstration of interference over long path differences with Fizeau in 1845.

(©*Observatoire de Paris*)

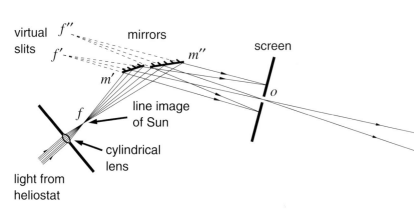
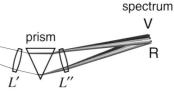

Plate III. Layout of the apparatus built by Fizeau and Foucault in 1845 to demonstrate interference at large path differences (see p. 62). Distorted scale: the distances between the mirrors and the screen, the screen and the prism, and the prism and the spectrum were all about 2 metres.

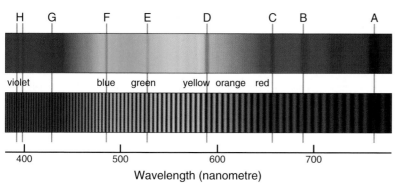

Plate IV. *(Upper)* The solar spectrum is crossed by dark bands lettered A–H by Fraunhofer. *(Lower)* These bands provided reference points for the superimposed interference fringes with large path differences produced by Foucault and Fizeau in 1845.

Plate V. Five solar spectra with different exposures recorded on a pre-exposed daguerreotype plate as part of Fizeau and Foucault's investigation of the daguerreotype response *c.*1844–45. In the three most-exposed spectra (uppermost) the pair of Fraunhofer H lines appears bright rather than dark because of solarization. The negative action of the reddest wavelengths is apparent where these three spectra are darker than the intermediate tint of the intervening regions of the plate, where the only exposure has been to uniformly coloured light either from an oil lamp or from the apparatus sketched in Fig. 5.21. It is also apparent that the neutral point between positive and negative action moves to redder wavelengths as the exposure increases. (To find the neutral points accurately, cut a narrow slit in a sheet of paper and slide it along each spectrum until there is no perceptible difference with the background.)

(Société française de photographie)

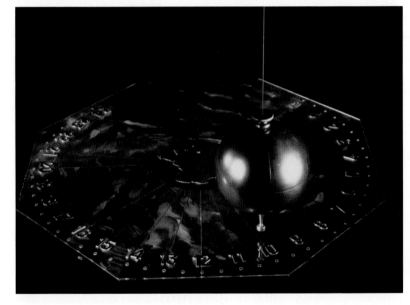

Plate VI. The lead-filled brass bob from Foucault's pendulum experiment in the Panthéon in 1851. Its mass and diameter are 28.2 kg and $6\frac{1}{3}$ French inches (172 mm). The baseboard and stylus are not original.

(©Musée des arts et métiers-CNAM, Paris/Photo J.C. Wetzel)

Colour plates

Plate VII. To avoid elliptical paths Foucault burnt a thread to launch his bobs from rest with respect to the Earth. The same method is being used here by the then Minister of Public Instruction to launch a replica pendulum set up in the Panthéon in 1902 by the astronomy popularizer Camille Flammarion (bearded) and the physicist Alphonse Berget (bespectacled).

Plate VIII (*right*). Foucault's gyroscope, devised in 1852. This copy was made for the Exposition Universelle in 1867 by Dumoulin-Froment, successor to Froment. A reviewer's comment is apposite: 'Many of these very erudite instruments are there [at the Exposition], slack and idle; they that would cause such astonishment...if they were working...; motionless within their shell of iron and brass, it is only their bizarre form which surprises...'[1]   However every detail has its purpose.

(©Musée des arts et métiers-CNAM, Paris/S. Pelly)

Plate IX (*below*). A full gyroscope set included a second torus and wooden stand. The tendency of rotations to align was demonstrated by turning the stand about a vertical guide peg. The spinning torus swivelled its axle into the vertical. The stand could also be orientated east–west. The rotor axis then moved along the meridian to align with the celestial pole.

(Science Museum/SSPL)

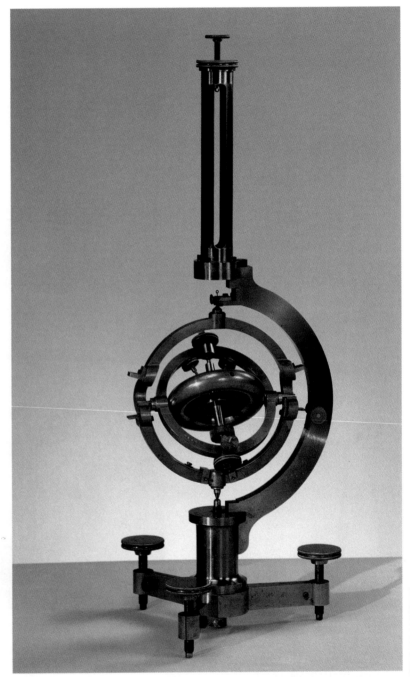

Plate X (*left*). Prototype of the electromagnetic drive devised by Foucault to maintain the oscillations of his pendulum at the 1855 Exposition Universelle. The magnet attracted the pendulum's iron bob during the inward part of each swing.

(©*Observatoire de Paris*)

Plate XI (*right*). A Foucault photometer, devised *c.*1855 for testing the lighting power of peat gas.

(*Museo
per la Storia dell'Università di Pavia*)

Plate XII. Foucault's 1855 lecture demonstration that mechanical energy can be converted into heat. Cranking the handle spins the copper disc between the poles of the electromagnet. This induces currents in the disc, causing it to heat up.

(*Museo
per la Storia dell'Università di Pavia*)

Plate XIII. A Ruhmkorff induction coil with Foucault mercury-cup contact breaker and a Fizeau capacitor in the hollow baseboard, c.1856.

(Physics Department & Museum of Scientific Instruments, Urbino University)

Plate XIV (*left*). Is this Foucault's first telescope from 1857? The mirror is $4\frac{1}{2}$ French inches in diameter; but with the circular stop inserted, the aperture is reduced to 88 mm. The mirror is no longer silvered and the eyepiece is missing.

(©Observatoire de Paris)

Plate XV (*right*). Equatorially mounted 10-cm telescope sold by Secrétan.

(Smithsonian Institution)

Colour plates

Plate XVI (*left*). The small telescopes sold by Secrétan were signed to guarantee authenticity.

(Smithsonian Institution)

Plate XVII (*upper right*). Wooden models of the *differential solids* used by Foucault when explaining his knife-edge test.

(©Observatoire de Paris)

Plate XVIII. The Marseilles 80-cm telescope with its wooden equatorial mount made by Eichens. To track the stars, the telescope was equipped with a Foucault governor (now lost).

Plate XIX (*left*). The polished back and grooved edge of the 80-cm mirror.[2]

Plate XX (*left*). Foucault's 24-cm (9 French-inch) achromatic doublet made using *retouches locales*.

(©Observatoire de Paris)

Plate XXI (*right*). Mysterious painting of Foucault, his sister and her children, *c.*1866. Why is Foucault examining the distant steamer with a spyglass? What symbolism underlies the ear of wheat in his hat?

(Family collection)

Plate XXII. The air-powered spinning mirror which Foucault used in 1862 for the first accurate laboratory measurement of the speed of light and distance to the Sun. The diameter of the mirror in its black barrel is 14 mm. To save weight, the rotor-balancing screws are set into a drum made of Saint-Claire Deville's new metal, aluminium (above the mirror barrel).

(©*Observatoire de Paris*)

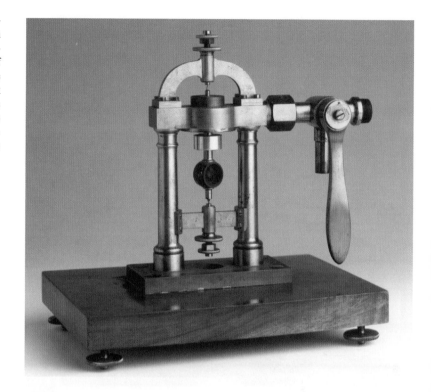

Plate XXIII (*left*). This clock-driven timing wheel was used to ascertain the speed of the spinning mirror stroboscopically. Fine teeth cut in the wheel rim were lit with flashes from the sweeping beam reflected by the spinning mirror. When the teeth appeared stationary, the mirror speed was exactly 400 r.p.s. (cf. Fig. 13.6).

(©*Observatoire de Paris*)

Plate XXIV (*right*). These five relay mirrors provided the 20.2-m path over which Foucault measured the speed of light. The two mirrors to the right retain remnants of metallization.

(©*Observatoire de Paris*)

Appendix A

Maps and chronology

France *c.*1852

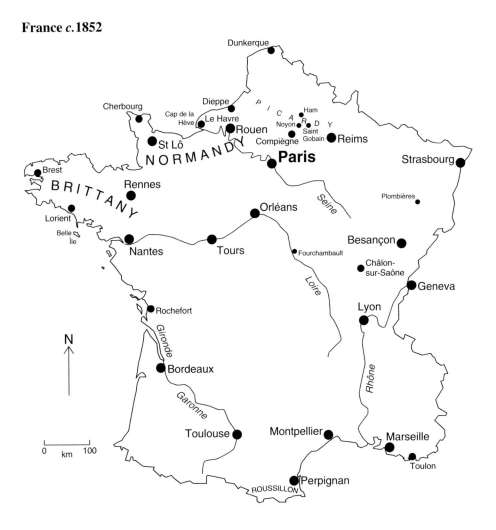

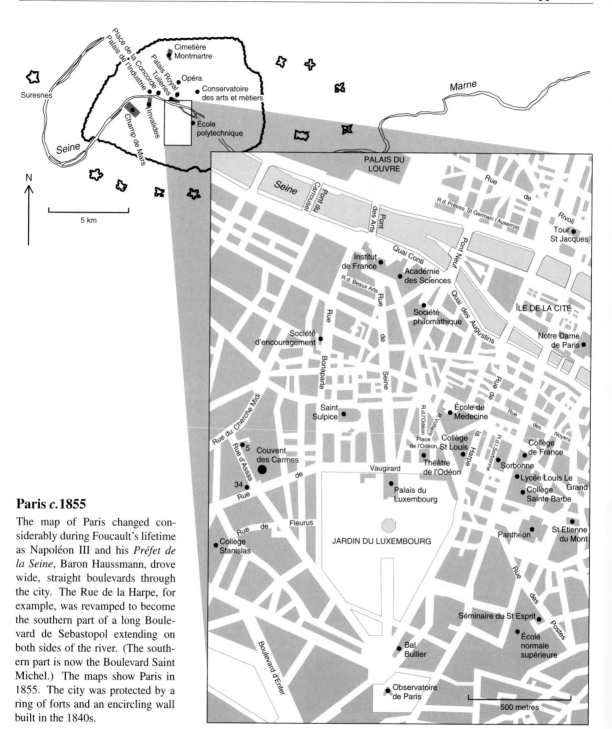

Paris *c.*1855

The map of Paris changed considerably during Foucault's lifetime as Napoléon III and his *Préfet de la Seine*, Baron Haussmann, drove wide, straight boulevards through the city. The Rue de la Harpe, for example, was revamped to become the southern part of a long Boulevard de Sebastopol extending on both sides of the river. (The southern part is now the Boulevard Saint Michel.) The maps show Paris in 1855. The city was protected by a ring of forts and an encircling wall built in the 1840s.

Summary chronology

1819 Léon Foucault born in Paris
1834 Father's *interdiction*
1837 *Bachelier ès lettres*
1839 *Bachelier ès sciences physiques* — Father dies — Daguerreotype process made public — Begins medical studies
1841 Has begun daguerreotyping — Uniform bromination of plates — Article in *Journal des Artistes*
1843 Daguerreotype improvements with Belfield-Lefèvre — Working on daguerreomicrographs with Donné — Measuring intensities with Fizeau
1844 Abandons medicine?
1845 Daguerreotype of the Sun — Starts reporting for *Journal des Débats* — Donné and Foucault *Atlas* published — Photo-electric microscope — Interference at large path differences and work on daguerreotype response with Fizeau
1846 Chromatic polarization with Fizeau — Improved dynamic range of daguerreotype with Belfield-Lefèvre
1847 Daguerreotype response with Fizeau — Criticism of Magendie — Electric arc self-regulates for first time — Conical pendulum — Transmission of time — Interference of calorific rays with Fizeau
1848 Second Republic — Babinet's report — Civil war in June — *Instruction pour le peuple* — Louis-Napoléon elected president
1849 Presents automatic arc to Academy; reversal of D lines — Dichoptic colour mixture with Regnauld — Arc light at Opéra — Rearranged Bunsen cell — Breakdown? — Elected to Société Philomathique
1850 Air–water light-speed experiment — *Chevalier de la Légion d'honneur*
1851 Pendulum experiment — First candidature for Academy — Vibrating rod — President (chairman) of the Société Philomathique — *Coup d'état*
1852 Gift of 10 000 fr. from Louis-Napoléon — Gyroscope — Second Empire
1853 *Docteur ès sciences physiques* — Journalists' access to Academy papers ends — Conductivity of liquids
1854 Crimean War begins — Liverpool meeting of British Association — Illuminating power of peat gas — Founding member of Société Française de Photographie
1855 Appointed Observatory physicist — Exposition Universelle — Demonstration of equivalence of mechanical energy and heat ('Foucault currents') — Napoléon III will pay for all future experiments — Copley Medal
1856 Improvements to Ruhmkorff coil — Donné worries about Foucault's mental state — Report on Chance discs
1857 Silvered-glass telescope — New polarizing prism — Improvements to mercury breaker — Dublin meeting of the British Association — Le Verrier advises that Foucault should be considered to have resigned
1858 33-cm telescope — Ellipsoidal and paraboloidal mirrors — Optical tests
1859 Catadioptric microscope — 40-cm parabolic telescope
1860 Work on 80-cm mirror — Photography of Spanish eclipse
1862 Heliostats — 80-cm mirror completed — *Officier de la Légion d'honneur* — Appointed to Bureau des Longitudes — Last *Journal des Débats* article — First isochronous regulator — Speed-of-light measurement
1863 Further work on governors — New self-regulating arc light
1864 Work on 1.2-m disc — Work on regulators — 80-cm telescope installed in Marseilles
1865 Elected to the Academy — Commercial association with Secrétan — Work on objective lenses — Governors for the navy
1866 Silvered objective for solar observations — Work on optics and governors
1867 Electric light applied at sea — Problems with naval governors — Exposition Universelle — Falls ill
1868 Léon Foucault dies in Paris
1869 Siderostat completed, financed by Napoléon III
1870 Le Verrier dismissed — Second Empire falls
1878 *Recueil* published

Appendix B

Extracts from the *Journal des Débats*

Most of Foucault's articles treated several different topics, each of which began with an introduction, continued with a description of the new work recently presented at the Academy, and ended with some of Foucault's own thoughts. The topics in each article frequently totalled some 4000 words. Without electronic entertainment, and with domestic chores attended to by servants, no doubt that was what the well-off and leisured subscriber to the Journal des Débats *wanted; but the length and turn of phrase may seem over-long to the modern reader. The extracts below have been selected to be short, and also for their charm or the insight they give concerning Foucault's personality and opinions. The first item, however, is a long one, and gives a good example of Foucault's careful and leisurely crafting of the clear expression of a physical result – and his blunt criticism.*

On electric currents

When one of two bodies contains more electricity than the other and when they are put in contact by a metal wire, equilibrium is established more or less rapidly and one says that current flows in the wire from the more electrified body to the less electrified one. In letting this current flow, the wire acquires new properties: it acts on the compass needle and gives off a certain quantity of heat.

The action on the compass needle betrays both the presence of a current and its direction. If, for example, the wire and needle are placed parallel, one above the other, the flow of the current will be revealed by a deflection of the needle to one side or the other according to whether the current entered at one end of the wire or the other. As for the liberation of heat, it is the same in both cases; and if the physicist had no other way of recognizing the presence of a current, he would not be able to determine its direction. The magnetic and calorific effects thus do not have, physically speaking, the same phenomenological relevance, and one can understand why experimentalists have generally been more interested in magnetic effects.

Nevertheless there are circumstances where the magnetic action is completely paralysed, and in consequence the physicist has no option but to look at the calorific action in order to know whether a current is flowing. For example, I am thinking of a conducting wire covered with an insulating sheath and folded back on itself and twisted like a rope with two strands, because thus arranged it can transmit a strong current. What does the compass needle do in the presence of this metallic braid, of which the two strands, carrying opposite currents, exercise equal and opposite effects?

It must and indeed it does remain stationary. The compass needle, normally so sensitive and precise, is in this case completely unable to say whether a current is flowing or not. It is then that it becomes appropriate to consider the heat liberated, because the calorific action of one wire in no way affects the calorific action of the other wire. On the contrary, the two effects add, and if the current set flowing is of sufficient intensity, the metal wires will heat to the point of glowing red, and even melt, even when they have been made of platinum or some other highly refractory metal. If the magnetic action sometimes fails, one can at least always count on the calorific effects.

Having established these facts, we are now going to use them to throw light, if possible, on one of the most controversial points in the theory of electro-dynamical phenomena.

When one takes two batteries that differ in every way both with respect to their electromotive forces and the resistances that are implicit in their construction, one can join them in two different manners: either by connecting the poles of the same name, or by connecting the poles of different name. The strengths of the currents that flow in the system have, in the two cases, values that are very different because the union of the poles of different name makes the two electromotive forces act in the same sense, while the connection of like poles puts the two batteries in opposition. When one has adopted this latter arrangement, that is to say, when one connects the two poles of the same name, or, what comes to the same thing, when one directs the two currents one against the other, one observes a resulting current whose intensity can be calculated from, and, as one says, as a function of, the electromotive forces and the resistances of the two batteries.

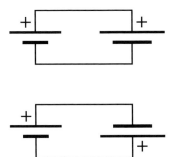

Figure B.1. Unequal batteries with like and unlike poles connected.

The calculation can be done in two ways.

The simplest method consists in comparing the two electromotive forces, in subtracting the smaller from the bigger, and in applying the force that results to the sum of the resistances of the two batteries. Then there is the other method which consists in supposing that the current from the weak battery passes through the strong battery, which rigorously is permitted, provided that one supposes at the same time that the current from the strong battery passes through the weak one under the impulsion of its full electromotive force. One thus obtains two currents in opposite directions, unequal in intensity and for which the difference gives the observed current.

Algebraically, the two methods differ only in the order in which the arithmetic is done, the subtraction applying either to the electromotive forces, or to the currents which result from them. Physically the difference is more important in that one cannot dispute the possible existence of two electromotive forces acting in opposite directions, whereas the co-existence of two opposing currents in a single body can certainly be questioned.

Nevertheless, the second method, although more complex, prevails over the first sometimes, notably in the situation where the opposition of the two batteries is complicated by the addition of an extra conductor which connects the two positive poles and the two negative poles in the fashion of the cross-bar of an H; the supposition of two opposed currents flowing simultaneously in a single circuit then proves a useful artifice in the calculation of the strengths of the currents which appear in different parts of the circuit. Now there remains the question of whether the agreement between the calculated currents and the values furnished by experiment justifies the physical reality of the existence of the two opposite currents which were introduced in the algebra as supplementary quantities. To throw some light on this important question we

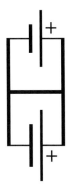

Figure B.2. Batteries with an H connection.

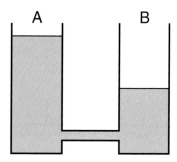

Figure B.3. Connected reservoirs.

are going to have recourse to a comparison based on the flow of liquids between two connected vessels.

We will give ourselves two reservoirs of water A and B in which the water levels are different, and let us make them connect by a horizontal tube. At the moment when connection is established, a current starts in the tube and transports liquid from reservoir A, where we suppose the level to be higher, to reservoir B, where the liquid level is lower. The difference of levels determines the speed of flow which is not proportional to the simple difference but to its square root.[1] When one wants to evaluate the dynamic force during this flow of water, one relates the height of the two fluid surfaces to a common horizontal plane, for example that which contains the horizontal tube, and one takes the difference of these heights. Nevertheless nothing prevents one proceeding differently. One could just as well consider successively what would be the flow in opposite directions in the absence of any counter pressure, solely as a consequence of the water in A first, then solely as a consequence of the water in B; one would thus have two hypothetical flows which when subtracted one from the other would give the real flow. Would one be for all that justified in concluding that in the connecting tube there really are flowing, in opposite directions, two contrary currents which correspond to the absolute fluid heights in A and B? Assuredly not; such an idea has never entered the head of a hydraulician, given that one sees with one's eyes the single and unique current of water that flows as a result of the difference in levels. Why then should the question remain unanswered where electricity is concerned? Why? Because one does not see its imponderable fluids; because one willingly supposes that they have marvellous properties, and because one imagines that anything might be possible for them.

Thus, to return to the heart of the question, and to express it precisely, there are physicists who deny that two contrary currents can co-exist in the same solid body; and there are others, amongst the most famous, who attach a physical reality to these twin currents whose existence seems to be proved by calculation. If such currents do not act on the compass needle, it is not, they say, a reason to doubt their reality, because this absence of effect results necessarily from their contrary directions. In arguing thus, one would never dream that the heating effect of these twin currents would not persist; instead of debating, it would be better to experiment so as to find out whether or not there occurs this liberation of heat which accompanies electric currents whatever the circumstances, whatever the direction of their flow.

So, here are two distinguished physicists, MM. de la Provostaye and P. Desains who have felt that it would be opportune to perform an experiment of this type. They placed a platinum wire in the most favourable circumstances for it to be crossed, should it be possible, by two strong currents flowing in opposite directions. This wire when connected at its end to the poles of a powerful battery, first of all became brightly incandescent under the influence of a current crossing it, let us say from right to left; then while the phenomenon was at its height, the ends of the wire were connected to the poles of a second battery that was in every way similar to the first, but arranged to furnish a current from left to right. Immediately all incandescence ceased, and the platinum fell to its ordinary temperature.

We were quite ready to greet with acclaim the obvious inference from such a conclusive phenomenon; but by virtue of a state of mind whose cause escapes us, our authors continue to maintain the reality of two opposed currents inside the platinum wire, by attributing to them, for the needs of the cause and in this unique and special circumstance, the property of cancellation of their calorific power. We will not dwell on this interpretation which before long the authors will find as surprising as we do. What we want to select and emphasize is the fact that the absence of heat has confirmed the probable absence of current, because sooner or later this fact will engender the correct conclusion and it will consign to the emptiness of mathematical formulation a pure fiction which should not leave the realm of abstraction.

(1853 NOVEMBER 25)

On the Académie des Sciences

The effects of the holidays combining with those which have resulted from the political upheavals of the last several months, non-members were able to speak right from the beginning of yesterday's meeting of the *Académie*. M. de Quatrefages took advantage first, and told us about the mysteries of reproduction in animals with external fertilization. Then M. Decharte read an extract from a Memoir which attempted to place within the large class of the *di*cotyledons those plants which, at the instant of germination, have seemed, so far, to have *more* than two cotyledons or germinal leaves. Last came M. Audouard who, through discussion of the peregrinations of the cholera in Algeria during the years 1835 and 1836, favours classifying this terrible scourge, whose invasion we at present dread, among the contagious diseases. At the same time M. Audouard raises a question which is not without interest for a country or city like ours which have previously been visited by cholera: he wonders whether those persons who have endured the invasion of the disease for a first time without contracting it would be assured of resisting similarly a second time.

The order paper was light and the items threatened insufficient to fill the meeting suitably; but the honourable members found a way of passing the time by teasing each other on the subject of the rules; they took to discussing some of their most insignificant provisions.

M. Serres above all showed himself to be remarkably punctilious: he did not want the *Académie* to take this opportunity to consider the cholera. And why not? Because, according to the ancient and hallowed tradition, a Memoir cannot and must not be sent to commissioners unless it ends with a final paragraph expressed in the form of a conclusion. M. Audouard was unaware of this academic convention; instead of hazarding some conclusions, he posed a definite and simple question, and the *Académie* is not in the habit of replying to all those who dare to cross-examine it. Nevertheless, this time the company did not take offence, and the proposition put forward by M. Audouard – question or conclusion, it matters little – has been sent to a commission which, if it judges appropriate, will make it the object of close examination.

M. Geoffroy Saint-Hilaire also showed a truly excessive sensibility for the rules. M. de Quatrefages had just presented the most recent results of his studies in embryology, and, in asking him for certain clarifications, M. Lallemand offered him the opportunity to escape from the constraints which the *Académie* imposes on non-members who wish to speak before it. You are required to read seated, without intonation, without gesture, without doing anything to punctuate the delivery and maintain attention.

The kindly remark addressed to M. de Quatrefages by the learned surgeon was thus a mark of both interest and favour: a favour which the rules do not permit – as was made very clear. Thus M. Lallemand – either by an affected respect for these rules which were being invoked by the letter rather than by the spirit, or by a subtle mocking gesture – M. Lallemand left his place and went quite simply to sit on the public bench next to the naturalist, with whom he had a private conversation from which no one else could benefit. Thus M. Lallemand had the last laugh; and M. Biot was with him. 'When we were young,' he said with an air that befits the old, 'we were singularly flattered if Laplace or Lagrange deigned to ask us for some clarification, and in such cases the respect of some and the benevolence of others would never have permitted the discussion to become a source of controversy. Besides which, haven't we got a chairman [to control the debate]? What do you think he's there for?' We believe in your glowing recollections, M. Biot, and would wish that an irregularity which dates back to the *Académie*'s most glorious days is perpetuated as a tradition and is preserved among the customs of this company.

(1848 AUGUST 30)

On use of language

On the Anthropology of French Africa by M. Bory de Saint Vincent. Is it really at the *Académie* that the author was able to read such a Memoir as this in all its entirety? Is it really in the name of science that he wrote these long and empty phrases; where serious topics are mixed with comments whose sole failing is not to be useless? Has anthropology placed Arabs and Ethiopians sufficiently far from us to justify the gaiety and half-joking tone with which the author recounted the demise of the unhappy individuals whose skulls were displayed on the Chairman's desk? 'This chap was a member of a band of looters who having raided deep into the Mitidja, was run through with a sword, a circumstance for which the *Académie* can see the evidence, etc. This other fellow is the head of a bandit from the Sudan, killed in the Sahel, where one of the sword cuts which he was struck with reveals the thickness of the bones of the skull, etc.' What fortunate events, these sword blows and beheadings, helping to advance the progress of science! And what is the relevance of the beauty of Moorish women and the fullness of their charms? And these unhappy people *roasted* by the heat of the Sun, and these *grimy spectres* reminiscent of skeletons?...

(1845 JULY 16)

On astronomy

Two new planets have just revealed themselves as amongst the objects composing our solar system. One of them was even discovered from Paris, Rue de Seine, from the balcony of the house at No. 12, by means of a seaman's telescope, a simple spyglass, by a painter of historical scenes M. Hermann Goldschmidt. The story is singular. Nevertheless, it should not be supposed that pure chance and good luck delivered a real planet to an artist who was vaguely looking at the sky through a spy-glass in search of inspiration; M. Goldschmidt knows his constellations very well, in his house he has the Berlin maps which are a faithful and detailed representation of the stars in the sky. In addition, it seems M. Goldschmidt has, as one says, a head for comparisons, that his memory retains with surprising fidelity the arrangement of details, even when they are distributed arbitrarily. This is precisely the faculty which is most important for the discovery of these mobile objects which, for an attentive eye, disturb the figures of the constellations by their passage. So, examining the constellation of Aries with his telescope around 10.30 p.m. on November 15, M. Goldschmidt saw an unaccustomed little dot, a white spot of eighth or ninth magnitude. In addition this spot moved slowly in the ecliptic direction; it was obviously a planet.

But was it a new planet? In a sky where a score of these asteroids have already been discovered in the last few years it was quite possible to have encountered one

Extracts from the *Journal des Débats*

of the known ones. It was here that our artist showed himself much more of an astronomer than might have been expected. Taking the elliptical elements of the orbits of all these little planets, he calculated their current positions and assured himself that not one of them was in Aries at the time. Furthermore, he went quickly to the Observatory to make his happy encounter known, and soon the discovery was confirmed both as to its reality and its novelty.

Next one wondered what to call the new conquest, and at the risk of giving it a name which one day might be appropriate for another, M. Arago proposed *Lutetia...* (*The* Académie *awarded Goldschmidt a medal for his discovery.*[2] *He went on to discover thirteen more minor planets, including Harmonia, named to celebrate the end of the Crimean War, and Eugenia, the first asteroid named after a real person, the Empress Eugenie.*[3])

(1852 NOVEMBER 26)

On a crank

M. H. Carnot is still claiming that vaccination is a disastrous invention which weighs on the population, limits its growth and, by changing the distribution of mortality, produces an imbalance, as the author puts it, between *useless mouths and the hands that have to feed them.* In vain does M. Charles Dupin proffer irreproachable arguments based on discussion of official statistics to show the daily lengthening of the average length of life and improvement in well-being of the masses. M. Carnot persists in his error and manages to re-clothe it in new forms. But at the same time the difficulty he has in spreading his opinion amongst enlightened people becomes for him the occasion to compare himself to Columbus, to Galileo, once misunderstood, like him. On the appearance of this characteristic symptom all discussion must stop, and the time has come to leave the heretic statistician in his little corner to prophesy the epoch at which the population of France will start to fall; this event, he says, can be predicted with more certainty than the return of a comet after its first apparition. We accept the augury with great willingness because most comets, after a single apparition, recede never to return.

(1849 MAY 16)

On mathematics

Dr Broc spoke for three-quarters of an hour in an attempt to give us a new definition of a straight line. According to certain mathematicians, we have already had too much talk about the definition of the straight line, the shortest route from one point to another. In expressing one of the properties of the straight line thus, the mind does not picture the principal idea of what a straight line is. The phrase 'a straight line' arouses in us so simple an impression that we are unable to describe how it feels. Also, after so much effort wasted in trying to define everything and to prove everything, one has returned to the view that a certain number of terms are frankly self-evident, and that one must appeal to an inner sense to comprehend these ideas which are unique of their type and which are too simple to be defined. Thus one does not define life, nor time, nor space; one does not say what an angle is, and one is embarrassed to define force – or one spouts a lot of wasted words. Besides, what proves that activities of this sort are vain and superfluous is that the supposed deficiencies which they are trying to remedy have not hindered the march of science.

(1845 OCTOBER 15)

On cancer

Foucault reports the difference between what are now called squamous-cell cancers and erodent ulcers.

Cancer, this dreadful disease whose name alone inspires so much fright, attacks all organs and in its invasive progress, respects no tissue. Often it appears on the face, and in this case it is given the expressive name of *Noli me tangere* [Don't touch me]. In a way it is a warning that reminds the surgeon of the incurability of this affliction when it is well developed. Nevertheless Dr Lebert claims that true cancer of the skin has often been confused with an alteration of this organ which externally is fairly similar but which has the essential difference that it does not contain the strictly cancerous element, which can be recognized with certainty through globules that are easy to distinguish under the microscope. In contrast, this source of pseudo-cancer requires prompt and complete removal and is susceptible to complete cure. This should therefore be another of these happy applications of the microscope, where by increasing the power of vision, the instrument furnishes a diagnostic procedure which none other can supplant.

(1846 DECEMBER 23)

On a toad in a hole

A person who lives in the town of Blois, and whose name we have not been able to decipher, has just made the trip to Paris with the express intention of showing the *Académie* a voluminous flint broken into two pieces, and in which a living toad was found. After having read the report of this strange discovery, which was made on June 23 last by workmen who were digging a well, the representative from Blois lifted up one of the fragments and discovered a cavity which continues to be the usual retreat of the hermit animal. The *Académie* certainly saw the flint, the cavity and the toad; but the decisive moment was that when the stone was broken, and on this one must refer to the accounts of honest workers unused to making observations. However it must be said that tales of this sort are abundant in the history of science.

(The man from Blois, in the Loir-et-Cher, was one Dr Monins.[4] The Academy appointed a commission of four who travelled to Blois and wrote no less than an eleven-page report which enumerated a couple of dozen similar incidents over the two previous centuries.[5] The commissaires nevertheless remained 'perplexed'. After the reading of the report, the physiologist François Magendie sceptically noted that the toad's muscles were not atrophied after a supposed long captivity and suggested that one of the well-diggers might have popped an errant animal into the cavity.[6] 'The joke wouldn't have been bad,' he added. Anyone who has observed toads will know that when threatened they retreat backwards into cavities to hide. This behaviour is not shared by frogs.)

(1851 JULY 27)

Figure B.4. *Bufo viridis,* the green toad, with its characteristic dorsal stripe, recognised by the Academy's commissioners as the toad in the flint from Blois.

On a standard kilogramme

M. Deleuil presented a standard kilogramme calibrated according to a new method which permits the exact realization of the desired weight without damaging the gilding that normally covers all standard weights constructed from an oxidizable metal. In order to avoid breaking this thin skin of gold while effecting the progressive reduction of weight to the correct value, it had been thought that one could do no better than to make the object in two parts, and to connect the parts with a screw thread while at the same time excavating a small interior cavity in which to place the necessary adjustment slug. In reality, this was a standard in three parts, which could come apart if the screw joint became loose, which after a certain delay, indeed almost always happened. In order to give his standard kilogramme the solidity appropriate for such objects, M. Deleuil has made it out of copper and in one piece; he gilded it and burnished it, and he arranged that it was a little under-weight; then, to finish it off, he drove into its top a little dowel of solid gold which partly jutted out, from which successive abstractions were made in order to bring the standard to the correct weight.

(1852 FEBRUARY 28)

On lubricating oils

We are now going to discuss a very beautiful piece of work by M. Berthelot concerning the reaction of glycerine with acids, and the synthesis of the direct constituents of animal fats ...

Having reported M. Berthelot's work, which seems to us to speak so clearly for itself concerning its scientific merit, we will now present some thoughts concerning one of the most delicate applications of oily substances. When the detailed study of the composition of fats has led to an application as important as that of the invention of stearic candles, one no longer has any right to ask the eternal and impertinent *cui bono*? However, oily substances are not only used for lighting. Practical mechanics employs them to reduce friction, and in precision clockmaking in particular, oil plays an extremely important rôle.

How is it in the present state of science that we are unable to furnish craftsmen with an oil reliably endowed with properties which are found by chance in certain samples? A clockmaker is never sure of his oil until he has tried it, and the trial is only ever conclusive after a very long time. Because M. Berthelot knows how to make fatty substances, after having isolated and purified the components as he wishes, it would seem that he better than all others could apply himself to the regular production of a fine oil possessed of all the necessary properties. This application is not unworthy of a savant; the discovery of a smooth liquid which does not dry out and is able keep the rate of marine chronometers uniform is one of the most beautiful improvements that could be brought to clockmaking. To say it once more, the problem is not insoluble because chance sometimes provides an excellent oil; it is up to science to supplant this capricious adjunct.

(1853 SEPTEMBER 23)

On white bread

We willingly seize the opportunity to decry the overconsumption and obsession with white bread. If one has lived the bachelor life, if one has for some time suffered the diet of white rolls which are inflicted upon one with such insistence in public places, one ends up by holding in sovereign distaste this whiteness – or rather, this pallor – of dough, an invariable indication of absence of flavour. The repugnance for pure white bread is more widespread than one might think, and for business to continue to promote its use there must be a hidden advantage of which the consumer is unaware. In general, industry deploys far too much zeal in perfecting every sort of refining process, in effecting this excessive purification which strips the foodstuffs of their original character, removes every last trace of their true origin, and makes them converge to complete similarity. Who today could distinguish cane sugar from beet

Figure B.5. A mercury-in-glass barometer.

sugar now that the refiner's art has deprived the former of its aromatic elements and freed the latter of those repulsive constituents which originally gave it its notorious inferiority? It is true that sugar is very pretty, very white, but it is no longer a food, it is only a condiment. It is a chemical product represented by a fixed formula and on which one would very quickly die of hunger.

White bread has not arrived there, but it will not take much; if it delights the eye it no longer pleases the tongue; and while men still let themselves eat it, dogs, it is said, turn away ...

(1853 December 11)

On the barometer

... The barometer has even worked its way into apartments: it has become the finishing touch to a fully furnished home, and many people, confusing it with a thermometer because of the mercury, glass and graduated metal which both comprise, never fail to consult it when going out in order to decide whether to put on galoshes or dress shoes. For those who still have an unthinking taste for measurement, and who in the practice of science obstinately confuse the means and the end, the barometer has become an unending source of pleasure, because, whatever the weather, the harmless instrument always offers a reading which can be taken down; and once his notebook filled in, the man with a barometer is at least as satisfied with his day as was Titus[†] after having performed a good deed.

([†] *Titus Vespasianus, who despite debauched and violent beginnings, became a benevolent Roman emperor,* AD *79–81.*)

(1857 February 17)

On an album of flower photographs

These charming victims are always so truly captured, their forms so delicately accentuated, and the structure of their parts so exquisitely outlined that one does not have even the most fleeting thought of the missing colour. In this controlled gamut from white to deep brown, the rose seems to recover its pink blush, the foxglove its purple nuance, and the delicately striped iris appears to have kept its tender blue. One restores their variegated tints to the tulip, peony and poppy, and the coarse dalhia takes on colour at one's pleasure. The camelia remains white; but one perceives that it bruises at the slightest touch. As for the hollyhock, it is not necessary to have a real specimen to sense that it is velvety and wrinkled to the touch; the lupin thrusts up its serried stem, the heliotrope modestly unfurls its whorls, and the gladioli launch out obliquely like rockets at a fête. Amateurs, botanists, horticulturalists, artists, impossible that any of you can cast your eyes over this album without experiencing the greatest of pleasure.

(1854 December 7)

Appendix C

Photographs and instruments

Disappointingly few of Foucault's pictorial daguerreotypes have survived. Of the two earliest ones (signed 1842), one was recently sold (Fig. 3.15),[1] while the other is in the Cromer Collection at George Eastman House, Rochester (Fig. 3.16). A view of Parisian roofs is preserved in Florence (Museo Fratelli Alinari). The Société Française de Photographie (SFP) in Paris holds a number of unsigned items which it received in 1929 and were attributed to Foucault, probably reliably, by their original collector, Alfred Nachet of the famous Parisian microscope-making company (Figs. 4.22,[2] 4.23, 12.32 and Plate V). Eight daguerreomicrographs are held by the SFP and two by the Bibliothèque Nationale de France, while in Britain six are held by the Royal Microscopical Society (Fig. 4.8) and a further eight by the Wellcome Institute Library. Most were gifted by Nachet.[3] The Conservatoire National des Arts et Métiers (CNAM) in Paris holds a Foucault–Fizeau daguerreotype of the Sun (Fig. 4.21).

Even fewer of Foucault's original instruments survive. The photo-electric microscope (Figs 5.1 and 5.2) and a compound microscope owned by Foucault were catalogued in the Nachet microscope collection in 1929, but their current whereabouts is unknown.[4] Foucault bequeathed the pendulum and drive from the 1855 Exposition Universelle to CNAM, which also holds the original Panthéon bob from 1851 (Plate VI) and a globe marked by Foucault to illustrate the deviation of the swing plane.[5] As part of the same bequest, the 1862 speed-of-light apparatus was returned to the Observatoire de Paris (Plates XXII–XXIV), though the bellows are lost. The Paris Observatory also preserves original 10-, 20- and 40-cm reflecting telescopes (e.g. Plate XIV, Fig. 12.29), a 240-mm diameter objective lens (Plate XX), wooden models illustrating the knife-edge test (Plate XVII), and prototype parts for the 1855 pendulum (a drive, Plate X, and parachute parts similar to Fig. 9.37). Foucault's 80-cm telescope (Plate XVIII) is at the Marseilles Observatory. Foucault left the remainder of his apparatus and equipment to his friend Jules Regnauld,[6] and while most is lost, this legacy is the origin of the interference mirrors (Plate II) and spherometer (Fig. 12.27) held by the Paris Observatory and perhaps a sectioned Ruhmkorff coil held by CNAM.[7]

Though original equipment is rare, museums abound with copies which were sold by the instrument makers with whom Foucault worked. (It is not always made explicitly clear when an instrument is a copy.) When identifying objects, note that instrument makers often sold items through their catalogues that were actually made by others.[8] An example is the Ruhmkorff coil at the University of Urbino (Plate XIII) which carries a small brass plate inscribed 'Secrétan à Paris'.

The original gyroscope of 1852 was bequeathed to the Collège de France and has been lost, but copies are to be found at CNAM (Plate VIII), the Science Museum, London (Plate IX) and the Museu de Física of the University of Coimbra. The Smithsonian Institution, Washington holds an incomplete gyroscope. Italy is rich in physics museums, where photometers are to be found in the Museo per la Storia dell'Università in Pavia (Plate XI), at the University of Catania and at the Fondazione Scienza e Tecnica (Florence). Foucault current apparatus is to be found at CNAM (where there is also a polarizing prism), the École Polytechnique, Bologna (Museo di Fisica), Pavia (Plate XII) and the Smithsonian Institution. Mercury-cup circuit breakers that are either stand-alone or incorporated in a Ruhmkorff coil are particularly widespread, being held by at least CNAM, Florence, Pavia (Fig. 11.10), the Instituto Galileo Ferraris (Turin), the University of Urbino (Plate XIII), Vanderbilt and Toronto Universities, and the Science Museum. Vanderbilt also holds a possible Foucault prism. Small, silvered-glass reflecting telescopes are to be found at the Palais de la Découverte (Paris), the Deutsches Museum (Munich), the Science Museum (numbered 26) and the Smithsonian Institution (Plates XV & XVI), while there is a 40-cm one in Laval University's Physics Department (Quebec). (The Science Museum also holds silvered and platinized glass given by Foucault to John Herschel in 1857.) Copies of the 1862 spinning mirror are to be found at CNAM (the associated toothed-wheel remains at the École Polytechnique), Teyler's Museum (Haarlem), the Museum of the History of Science (Oxford) and the Museum of Scotland (Edinburgh). Foucault–Duboscq arc lights, which need to be distinguished carefully from Duboscq's earlier form, are held by Oxford, the Science Museum and the Národní Technické Muzeum (Prague). Since the example in Oxford is numbered 130, one can expect that many others exist elsewhere. Teyler's Museum holds an arc lamp of slightly different form, numbered 1 by Duboscq. A small Foucault heliostat is held by the Liceo Ennio Quirino Visconti in Rome, while the Paris Observatory possesses one with a presumably optical-quality circular mirror. Careful study is needed to recognize Foucault-type regulators, but the original and present governors on the posthumous siderostat at Paris Observatory are two. (The former is numbered 52.) CNAM holds a recording cylinder with a regulator like that illustrated in Fig. 15.19.[9]

This enumeration of instrument copies is certainly incomplete. The interested reader should consult museum catalogues, which are increasingly becoming available on-line, often with photographs. (Internet addresses can be found easily with a search engine.) The *Online Register of Scientific Instruments*, set up under the auspices of the International Union of the History and Philosophy of Science, will become increasingly valuable as it develops.

It is interesting to note the cost of commercialized instruments. Teyler's Museum paid 296 florins for its spinning mirror in 1863.[10] Duboscq's 1864 catalogue lists small and large Foucault heliostats for 800 and 1200 francs, respectively.[11] (Silbermann's heliostat had inflated to 500 fr.) Secrétan's catalogue for 1874 quotes the following prices:[12] pendulum 250 fr.; gyroscope 2500 fr.; photometer 45 fr.; Foucault current apparatus 350 fr.; 10-cm reflecting telescope mounted in metal on a tripod 500 fr.; *idem* 16-cm 1200 fr.; 16-cm reflecting telescope equatorially mounted in metal with drive 2500 fr; *idem* with wooden mounting 1650 fr.; *idem* with 50-cm mirror 29 000 fr.; *idem* with steel mounting 35 000 fr.; resilvering charge for a 16-cm mirror 7 fr. 50; *idem* for 50-cm 70 fr.; arc lamp 200 fr.; fan governor (Fig. 15.19) 250 fr.; *idem* with recording cylinder 500–600 fr.; siderostat 12 000 fr. To put these prices in context, see Table 2.1.

Appendix D

Building a Foucault pendulum

Many have been deceived by the apparent simplicity of the Foucault pendulum, for as Chapter 9 has made clear, building one is not without pitfalls. There is no unique design or universal recipe since the available height, accessibility of the suspension, and desired demonstration duration vary from case to case. This appendix provides guidance and references to the published literature for readers who wish to design and construct a pendulum. *It is the only part of this book which assumes prior knowledge of physics.* Foucault pendulums are also sometimes available commercially.[1]

The physics of the ideal, isotropic but anisochronic (anharmonic) spherical pendulum can be found in many mechanics texts, and has been outlined at different levels of sophistication by Olsson.[2] The period, P, of plane oscillations for a pendulum of length l from pivot to point mass m, and swinging through angular amplitude α (2α peak-to-peak) is given by $P = P_0(1 + \alpha^2/16)$, where g is the acceleration due to gravity and $P_0 = 2\pi/\omega_0 = 2\pi\sqrt{l/g}$. Orthogonal oscillations of linear amplitudes a and b produce an elliptical bob path of period $P_0(1 + (a^2 + b^2)/16l^2)$, which veers during one cycle by an angle $\Delta = 3\pi ab/4l^2$ in the same direction as the bob motion (Fig. 9.32), or at a rate $d\theta/dt = 3abg^{1/2}/8l^{5/2}$. These equations can be used to estimate the consequences of any ellipsing that develops relative to the terrestrial-rotation veering rate at latitude λ of $\Omega_\oplus \sin \lambda$, where $\Omega_\oplus$ is the angular velocity of the Earth. The effect of launching the pendulum from rest with respect to the Earth is to slow the veering rate by $3a_0^2 \Omega_\oplus \sin \lambda /8l^2$, where a_0 is the initial swing amplitude.[3] This effect is negligible provided the angular amplitude of the swing is small.

Note that for a real pendulum of reasonable length, the centre of gravity may be located well above the bob, and the length of the equivalent simple pendulum, l, which must be determined from the physics of the compound (solid) pendulum, may differ considerably from the physical length of the pendulum. (For Foucault's 67-m pendulum, both differences were about 1.0 m, with a radius of gyration about the centre of mass of 6.5 m.)

Pendulum oscillations that die away make for a poor demonstration, so a modern Foucault pendulum should be driven. The most elegant method is via parametric amplification and was suggested only a few weeks after Foucault's pendulum began to swing in the Panthéon.[4] The suspension point oscillates vertically at twice the pendulum frequency. It is raised when the pendulum is at the lowest point in its swing, and the tension in the wire is greatest (because centripetal force adds to the weight of the bob), and lowered at each extremity of the swing, when the tension is least. More

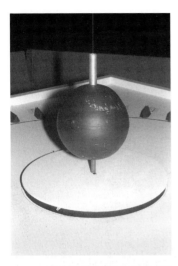

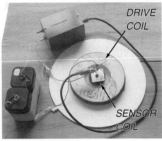

Fig. D.1. A simple electromagnetic pendulum drive. (*Upper*) The rod under the bob is a small bar magnet. (A high-flux neodymium–iron–boron magnet is ideal.) The superior aluminium tube hides the connection of wire to bob. At the top of the tube, a rubber grommet presses against the wire to provide defence against rocking. The drive and sensor coils are beneath the round, white perspex plate. (*Lower*) The coils and electronics. Passage of the bob magnet through the vertical induces a potential in the central detector coil. This triggers a delayed pulse in the drive coil, which repels the bob as it swings away.

energy is therefore supplied during the raising than is retrieved during the lowering, and this excess counteracts the damping loss; but what is more important is that over time energy is extracted from any component of the oscillation which is in antiphase with the motion of the drive. Parametric amplification therefore offers the significant advantage of damping out the minor axis of an ellipsing pendulum.

The pendulum in London's Science Museum is driven by parametric amplification and has been described by its author, Sir Brian Pippard.[5] The suspension is relatively complex mechanically, since it must oscillate, and the pendulum is counterbalanced so that the drive stepper motor does not have to move the full weight of the pendulum. It also contains adjustments to minimize anisotropy. Installation and adjustment of complicated and delicate apparatus of this sort require a suspension point which is easily accessible. A parametric drive will increase the amplitude without limit if the damping losses are proportional to the square of the velocity. For finite oscillations, it is necessary to adopt a cylindrical bob, which spawns eddies and is subject to enhanced losses.

An electromagnetic drive is easier to make. With modern electronics no moving parts are necessary and unlike Foucault's 1855 drive, the bob can be pushed outwards to reduce ellipsing. Figure D.1 show views of a simple, battery-powered, electromagnetic drive used with a 10.5-m, 18.5-kg Foucault pendulum set up in the King Edward Barracks, Christchurch, New Zealand, for a week-long *Science-Technology Extravaganza* in 1987. A cylindrical bar magnet was attached under the bob. The magnet functioned as a stylus indicating the swing-plane azimuth, but as the bob swung through the vertical, the magnet induced a potential in a 100-turn detector coil placed centrally under the bob's rest position. The potential was amplified; the positive-going output provided a trigger pulse to a one-shot, whose output pulse on its negative-going edge triggered a second one-shot. The output of this second one-shot switched on a power transistor and hence the current in a 400-turn drive coil, whose large diameter of about 160 mm was chosen to minimize centring errors and produce an accurately radial field which pushed against the retreating bob magnet. The one-shot time constants controlling the delay and duration of the drive pulse were adjusted empirically with the help of an oscilloscope until the pendulum bob maintained the required oscillation amplitude of ± 0.6 m ($\pm 3.3°$).

Pippard gives coupled differential equations describing the evolution of the swing plane angle, θ, and $\varepsilon = b/a$ in a parametrically driven pendulum.[6] If the fundamental frequencies of an anisotropic pendulum are $\omega - \delta$ and $\omega + \delta$, the anisotropy causes a modulation of the veering rate $d\theta/dt = 2\delta\varepsilon \cos 2\theta$. One plans that 2ε will be small, but setting this factor to unity gives an order-of-magnitude limit $\delta/\Omega_\oplus \sin\lambda < 0.1$ for less than 10% modulation in the veering rate and less than 1% reduction in its mean. (A modulated rotation speed $\Omega_1 + \Omega_2 \sin n\theta$ produces a mean speed $(\Omega_1^2 - \Omega_2^2)^{1/2}$.) If the anisotropy $\pm\delta$ is converted to an anisotropy in length, Δl, one finds $\Delta l/l < 0.4(\Omega_\oplus/\omega) \sin\lambda$, or $\Delta l < 9.3 \times 10^{-6} \sin\lambda\, l^{3/2}$ m, or a third of a millimetre for a 10-m pendulum at middle latitudes. This indicates the importance of precise metalwork and a sturdy suspension. The isotropy of the flexing wire is of course less important than that of the bob on account of the bob's large moment. A pendulum's isotropy should be investigated through timing small oscillations at various azimuths.

In 1931, F. Charron devised a useful trick to attenuate ellipsing in a driven pendulum.[7] An accurately centred, horizontal ring is set just below the suspension point. The oscillation amplitude and ring diameter are chosen so that the wire touches the ring at the extremity of each swing. Friction between the wire and the ring damps out the tangential motion inherent in ellipsing. In a sense, the Charron ring starts the pendulum afresh on each swing. To mimimize fatigue of the wire, it is desirable to curve the inner side of the ring. Alternatively, the Charron ring may be displaced to the ground, to be rubbed against by the bob or stylus. Some 'give' is then necessary in the ring.[8] In a Foucault pendulum installed at the Brown Boveri Research Center in Baden, Switzerland, tangential braking is elegantly provided by a high-intensity, samarium–cobalt stylus magnet which induces eddy currents (*courants de Foucault*) in a copper ring set into the floor.[9] The Charron ring or equivalent is so effective that a pendulum can be set going by hand without recourse to the traditional flammable loop (Plate VII).

Dick Crane presents a way of eliminating not ellipticity but its effect.[10] A bar magnet is placed along the axis of the drive coil with the poles oriented to repel the bob magnet. The additional repulsion severely reduces the anharmonicity of the pendulum, and hence the associated veering, irrespective of a and b, provided the magnet is at an appropriate depth, which can be found empirically without knowledge of the Foucault veering rate.

With a driven pendulum and a Charron ring, the importance of the pendulum's length, anisotropy and wire diameter is much reduced. Charron's pendulum was only 1.7 m long, and several designs have been published for short, portable Foucault pendulums which run continuously for weeks (e.g. Fig. D.2).[11] The shortest pendulum reported so far had a length of only 15 cm. Crane gives extensive details of a 70-cm pendulum which can be built in a home workshop.[12] With a Charron ring it is possible to achieve very acceptable performance even with a rapidly built pendulum (Fig. D.3).

The potential for fatigue fracture of the wire and consequent injury to spectators makes the suspension one of the most troublesome aspects of the Foucault pendulum. To minimize the effect of air currents, which can open up the orbit and hence cause anharmonic veering, it is desirable to adopt as thin a wire as is compatible with supporting the weight of the bob, but wires loaded to near their elastic limit and subject to cyclic stresses are highly susceptible to failure. Surface micro-cracks develop at the point of greatest stress and propagate into the wire, which breaks.

High-tensile, cold-drawn steel is appropriate for the suspension wire because the drawing process produces an isotropic internal structure and, more importantly, increases its ultimate tensile strength, S_{max}. Further, steel exhibits a distinct fatigue limit, $\sigma_{FL,0}$. This is characterized in the 'S–N' diagram, which indicates how many cycles of a particular stress amplitude, σ_c, are required to produce failure. Below the fatigue limit, which the S–N curve reaches by $\sim 10^7$ cycles, the migration of carbon and nitrogen atoms locks the dislocations and the wire can flex indefinitely. (This dynamic strain hardening then raises the fatigue limit.) However, the fatigue limit is reduced when the wire is subject to a static stress, σ_s, as well as a cyclic one. The reduction is represented in the so-called Goodman diagram, in which the cyclic stress amplitude for which failure will occur is plotted as a function of the mean stress. For conservative design, a linear interpolation should be used between the two points $(\sigma_{FL,\sigma_s}, \sigma_s) = (\sigma_{FL,0}, 0)$ and $(0, S_{max})$.[13] The design challenge is to ensure that the suspension operates below the fatigue limit.

A simple way to grip the suspension wire is in a pin vice whose jaws have been lapped to provide an accurate fit to the diameter of the wire. However, as the pendulum

Fig. D.2. A short Foucault pendulum ($l \sim 17$ cm) made by H. R. Crane.

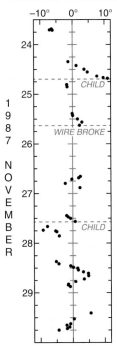

Fig. D.3. Performance of a 10.5-m Foucault pendulum with Charron ring set up in the King Edward Barracks, Christchurch. Dashed lines indicate various mishaps: children who transgressed the barrier to interfere with the bob, and wire failure.

swings, a highly loaded wire bends and strains quite sharply where it emerges from the vice jaws.[14] The displacement of a laterally displaced, vertical rod or wire of radius, ρ, Young's modulus, E, and under tension, T, is given by $C(\nu\xi + e^{-\nu\xi} - 1)$, where $\nu^2 = 4T/\pi E\rho^4$, ξ is the distance from the suspension point, and C is set by the requirement that at $\xi = l$, the displacement is a ($\approx C\nu l$). The maximum curvature is $C\nu^2$ (at $\xi = 0$) and the maximum stress of any filament $\sigma_{max} = C\nu^2\rho E \approx (2a/l\rho)(ET/\pi)^{1/2}$.

Fatigue is notoriously variable even for samples taken from the same wire. A *very* crude rule-of-thumb for the S–N curves for steel is that stresses of $0.9 S_{max}$ will cause failure after 10^3 cycles, while failure occurs at 10^6 cycles for $0.5 S_{max}$. It is probably conservative to estimate the fatigue limit at $0.3 S_{max}$. For steel, $E \approx 200$ GPa, while the ultimate tensile strength of steel wires ranges from $S_{max} \approx 1$ GPa for wires such as Foucault may have used to ~ 2.5 GPa for high-tensile, cold-drawn steel. The static and maximum cyclic stress amplitudes for the Christchurch pendulum ($\rho = 0.35$ mm) were 0.5 and 1.1 GPa respectively; and 0.2 and 0.5 GPa for Foucault's Panthéon pendulum. We can therefore see that the cyclic stress amplitudes exceeded the corresponding fatigue limits σ_{FL,σ_s} of 0.6 and 0.3 GPa, and it is no surprise that both wires failed. Because of the stochastic and uncertain nature of fatigue, a parachute remains an essential safety precaution (Fig. 9.37). The wire must not be damaged at its moorings either mechanically or by corrosion. Although wire is traditional for the suspension, it is reported that good results can be obtained with a cord provided it is run in for several weeks.[15]

To reduce cyclic stresses the wire can be made thicker ($\sigma_{max} \propto \rho^{-1}$) or carefully wrapped at its upper extremity with a helix of wire, though the use of a different metal, such as brass in the current Panthéon pendulum, is to invite corrosion since humidity is inevitable. In the Science Museum pendulum, it is a long rod rather than a wire which flexes. Flexure can be eliminated entirely with a knife-edge suspension, as in Baden, or a universal joint, as in the pendulum presented by The Netherlands for the United Nations building in New York, which was particularly designed to run for years without intervention.[16] As Kammerlingh Onnes knew, a suspension that lacks rotational symmetry needs adjustment masses so that vibrational isotropy can be achieved through equalization of the oscillation periods in its principal planes.

There is no need for an exceptionally massive bob in a driven pendulum, easing loading of the wire. A spherical bob is traditional, but an ellipsoidal shape has been used where external air currents were a concern.[9] Electromagnetic drives give a sideways kick to the bob, which may cause it to wobble. A palliative is to fix a tube rigidly to the top of the bob, and at a level above the bob–wire connection, provide a lossy link to the wire with a rubber grommet. If transverse travelling or standing waves are excited in the wire, the bob can be shielded from them or they can be damped with a small weight placed above the bob, or at a nodal point, respectively.

Showmanship aspects should not be neglected. An azimuth scale is essential. For small pendulums, strategically placed mirrors can make the suspension visible. The veering is slow. Experience teaches that greater impact is achieved when the bob knocks over skittles or cuts breaches in sand piles.

In a confident moment, Foucault wrote that his pendulum is 'of easy execution, of assured success'.[17] With care and intelligence, this can indeed be the case.

Notes

References are mostly only made to specialist sources. Titles are given only for selected articles. Background material has been taken without citation from reference works that should be available in any good library.

To minimize confusion, journal names are not abbreviated, with the exception of *C.R.A.S.*, *Comptes rendus hebdomadaires de l'Académie des sciences* and *JdD*, *Journal des Débats*. *Recueil* refers to Gariel C.-M. (1878) *Recueil des travaux scientifiques de Léon Foucault*, Gauthier-Villars, Paris. Various archives and their call numbers are indicated as follows: A.A.d.S., *Archives de l'Académie des Sciences*; A.d.P., *Archives de la Ville de Paris*; A.N., *Archives nationales*; I.d.F., *Bibliothèque de l'Institut de France*.

Foreword

[1] Dunkin E., *The Midnight Sky: Familiar Notes on the Stars and Planets.* Religious Tract Society, London (1869); Putnam, New York (1872) pp. 197–8.

[2] Young C. A. (1888, revised 1895) *A Text-book of General Astronomy for Colleges and Scientific Schools.* Ginn, Boston and London. p. 3.

Preface

[1] *Blackwood's Magazine,* **54**, 524 (1843); Ross S. (1962) *Annals of Science*, **18**, 65–85.

[2] Foucault L. (2001) *Recueil des travaux scientifiques.* Librairie Scientifique et Technique Albert Blanchard, Paris. ISBN 2 85367 214 X.

[3] Tobin W., Perfecting the Modern Reflector. *Sky & Telescope*, **74**, 358–9 (1987); Foucault's invention of the silvered-glass reflecting telescope and the history of his 80-cm reflector at the Observatoire de Marseille. *Vistas in Astronomy*, **30**, 153–84 (1987); Toothed wheels and rotating mirrors: Parisian astronomy and mid-nineteenth century experimental measurements of the speed of light. *Vistas in Astronomy*, **36**, 253–94 (1993); Tobin W., Pippard B. (1994) Foucault, his pendulum and the rotation of the Earth. *Interdisciplinary Science Reviews*, **19**, 326–37; Tobin W., Chevalier G. (1995) Le grand art des pièges à lumière. *Les Cahiers de Science & Vie*, No. 25, 50–67; Tobin W., Foucault, son pendule, et la rotation de la Terre. *L'Astronomie*, **110**, 50–60 (1996); Léon Foucault, *Scientific American*, **279**, No. 1, 52–9 or 70–7 (pagination varies) (1998).

[4] This inventory is probably the one prepared by E. Rolland and J. Lissajous as members of the committee set up by Napoléon III to publish Foucault's work. A.N. F^{17}3247.

[5] *JdD*, 1852 April 9.

Chapter 1 Introduction

[1] The *Nouvelles météorologiques, publiées sous les auspices de la Société Météorologique de France, Année 1868*, Libraire agricole de la maison rustique, Paris, pp. 130–40 reports that on 1868 March 2 the sky was covered, the average temperature was 7.0 °C (with a miniumum of 3.2 °C) and that the wind was moderate from the west.

[2] *Les Mondes*, **16**, (2ème sér.) 384 (1868).

[3] McKie D. (1966) *Endeavour*, **25**, 100–3.

[4] Grillot S. (1980) *Revue d'histoire du quatorzième arrondissement de Paris*, **25**, 7–24; Wahiche J-D. (1989) *L'Astronomie*, **103**, 249–57.

[5] Cassini is often described as the first director of the Observatory, but this is apparently incorrect. See *Nature*, **19**, 122–3 (1878).

[6] *L'Astronomie*, **103**, 291 (1989).

[7] Bigourdan G. (n.d.) *Annuaire pour l'an 1928 publié par le Bureau des Longitudes.* Gauthier Villars, Paris. A1–A72.

[8] Quoted by Guedj D. (1988) *La Révolution des savants.* p. 92. Gaillimard, Paris.

[9] *Correspondance de Napoléon Ier.* Tome VI. No. 4384. Imprimerie impériale, Paris (1860).

[10] Jamin (1885) Éloge historique de M. François Arago. Institut de France, Paris.

[11] Heilbron J. L. (1989) *American Journal of Physics*, **57**, 988–92.

[12] These adventures are described in Arago's charming *Histoire de ma jeunesse.* Kiessling Schnée, Brussels. (1854, reprinted 1985 Bourgois, Paris.)

[13] At the time called the First Class of the Institut National.

[14] Marcou J. (1869) *De la science en France*, fasc. 2, p. 141.

[15] *JdD*, 1853 October 7.

16 See Arago's introductory remarks to the 1841 course in his *Astronomie populaire*, Vol. I, Barral J.-A. (ed.) Gide, Paris (1857).

17 Tisserand F. F., quoted in *Nature*, **48**, 223–4 (1893).

18 E.g. *Revue scientifique et industrielle*, **13** (2ème sér.) 84 (1847).

19 The *arrondissements* were rearranged in 1860. The old 12th *arrondissement* corresponds to the south-east of the present 5th *arrondissement* and the north of the 14th.

20 Quoted in Solé R. (1998) *Les savants de Bonaparte*. Seuil, Paris.

21 Bonaparte L.-N. (1843) *C.R.A.S.*, **16**, 1180–1.

22 On his banishment to Elba in 1814, Napoléon I abdicated in favour of his infant son. Although the child did not reign, the Bonapartist view was that he had been a lawful French sovereign and the second Emperor Napoléon.

23 Farmer P. (1960) in *The New Cambridge Modern History*. Bury J. P. T. (ed.) Vol. X. Cambridge University Press, Cambridge.

Chapter 2 Early years

1 Dauzat A. (rev. Morlet M.-T.) (1988) *Les noms de famille en France* (Guenégaud, Paris) gives the roots FULC: people or falcon, and WALD: forest. The earlier notion that Foucault means 'governor of the people' is now discredited.

2 For a description of pre-revolutionary San Domingo, see McClellan J. E. (1992) *Colonialism and Science: Saint Domingue in the Old Regime*. Johns Hopkins University Press, Baltimore.

3 Bouillet M.-N. (1847) *Dictionnaire universel d'histoire et de géographie*. Hachette, Paris.

4 Most of the details concerning Foucault's antecedents come from the inventory established after his father's death. A.N. ET/XLIX/1248.

5 Compiled from documents in the A.N. and the A.d.P. One must wonder if Julienne Minée was related to the consitutional priest Julien Minée (1739 Nantes – 1808 Paris). *Index biographique français*, **743**, 127.

6 A.N. F[18]1764.

7 Some of which in fact appeared before he obtained his *brevet*, e.g. *Les Ligueurs de 1814, par M*** (de Nantes)*, Foucault, Paris (1814). This pamphlet was probably published soon after Foucault's marriage (it appeared in the *Bibliographie de la France* on 1814 October 29). The address given for the publisher, 17 Quai des Augustins, confirms that it is the correct Foucault.

8 [Foucault J. L. F.] (1816) *Avertissement du Libraire-Éditeur*. In Petitot (1817) *Répertoire du théâtre françois*. Tome I. Foucault, Paris.

9 At the time of his marriage in 1814, his father lived at 17 Quai des Augustins (A.d.P. *Fichiers des mariages Parisiens*); these may not have been suitable quarters for a married couple. In 1817 the bookshop was at 37 Rue des Noyers; it later moved to 9 Rue de Sorbonne.

10 The *Reconstitution de l'état civil – Baptêmes* at the Archives de Paris reports that Jean Aimé Barthélémy Foucault was born on 1822 November 7 and baptised at St Etienne du Mont the following day – as if he was not expected to live. The choice of names is suggestive. Sister Aimée was born on 1823 August 7, precisely 273 days later. Since a woman can ovulate within a fortnight of giving birth, it would only be necessary for Aimée's gestation to have been slightly shorter than average for Jean Aimé to be a brother.

11 Bertrand J. (1883) *Mémoires de l'Académie des sciences de l'Institut de France*, **42**, lxxvii–cciv. However, Bertrand's assertion that the family returned to Paris when Foucault's father died is clearly incorrect. The houses in Nantes were at 6 Rue Bonsecours (on the corner with Rue Kervégan, on the opulent Île Feydeau) and 9 Petite Rue St Clément (Rue des Orphelins). By the end of the 1830s they were being rented out.

12 Trees can be seen in Marville's photograph. See note 4.70.

13 *JdD*, 1845 July 1.

14 *JdD*, 1848 November 1.

15 Martin H. J., Chavtier R. (eds.) (1985) *Histoire de l'Édition française*, Vol. 3. Promodis, Paris.

16 *JdD*, 1848 October 4.

17 Lissajous (1869) *Revue des cours scientifiques de la France et de l'étranger*, **6**, 484–9.

18 Gariel C.-M. (1869) In *Annuaire scientifique publié par P.-P. Dehérain, Huitième année, 1868*. Masson, Paris. pp. 82–97.

19 *Almanach Royal et National pour l'an MDCCCXXXIX*. Guyot, Paris (1839).

20 Corlieu A. (1896) *Centenaire de la Faculté de Médecine de Paris*. Imprimerie nationale, Paris.

21 A.N. AJ[16]5327.

22 The artist's named signed on both pictures is Finck.

23 A.N. ET/XLIX/1504.

24 Perhaps Charles Lecorbeiller, pianist and composer (Fetis F. J. [1881–9] *Biographie universelle des musiciens*. [2ème ed.]). The *Journal des Artistes* dedicated a poem to him in 1843 (17ème année, 1er vol., 9–11).

25 The letter, dated 'Paris 10 avril 1840' and starting 'Mon cher ami,' is in the family collection. Perhaps, therefore, Foucault found it too self-indulgent to send.

26 A.d.P. V2E12084.

27 A.N. ET/XLIV/1504. They were valued at 500 francs, which even in 1868 and for probate purposes would have been a very low estimate. See Hill W. H., Hill A. F., Hill A. E. (1963) *Antonio Stradivari: His Life and Work (1644–1737)*. Dover, New York. No Stradivarius is listed as

[28] *JdD*, 1848 December 1.
[29] de Parville H. (1869) *Causeries scientifiques, 8ème année 1868.* Rotschild, Paris.
[30] Family collection.
[31] As judged by a demonstration of tuning forks at the Academy in 1845 (*JdD*, 1845 April 30) when Foucault was able to hear much higher frequencies than most of the Academicians. (Quoted numerical values are incorrect; early estimates of frequencies through aural matching of scales invariably gave overestimates. See Graff K. F. (1981) *Physical Acoustics*, **15**, 1–97.)
[32] *Les Mondes*, **16**, 337–8 (1868).
[33] *JdD*, 1849 March 14.
[34] *JdD*, 1846 August 5.
[35] *JdD*, 1846 January 6.
[36] *The British Medical Journal*, **1**, 454 (1895).
[37] A.N. AJ[16]232; Foucault L. *JdD*, 1854 January 19; Cheymol J., Soubiran A. (1968) *La Presse Médicale*, **76**, 2366–8.
[38] Donné Al., *JdD*, 1868 March 1.
[39] Redondi P. (1988) *History and Technology*, **6**, 203–25.
[40] Collection of the family of G. Darrieus. Letter dated 1854 September 24.

Chapter 3 The metallic eye

[1] E.g. Norman P. R. (1967) L. J. M. Daguerre, 1789–1851 and W. H. Fox Talbot, 1800–77, in *Early Nineteenth Century Scientists*, Olby R. C. (ed.) Pergamon, Oxford.
[2] Bontemps (see Chapter 11) as quoted by Figuier L. (1873–77) *Les Merveilles de l'Industrie*, Furne Jouvet, Paris, gives the following recipes. Crown glass: white sand 120 kg, potash 33 kg, soda 20 kg, chalk 15 kg, white arsenic 1 kg. Flint glass: sand 100 parts by mass, lead oxide 100 parts (minimum), potassium carbonate 30 parts.
[3] Brewster drew attention to Canada balsam's useful optical properties. See Mills A. A. (1991) *Annals of Science*, **48**, 173–85.
[4] Gernsheim H. & A. (1952) *Photographic Journal*, May issue.
[5] Brenni P. (1993) *Bulletin of the Scientific Instrument Society*, No. 39, 11–14.
[6] Harmant P. G., Lefebvre B. (1974) *Charles-Louis Chevalier.* Pavillon de la Photographie, Rouen.
[7] *Journal des Artistes*, **2**, 203 (1835).
[8] (1839) Le Daguerréotype. *C.R.A.S.*, **9**, 250–67.
[9] Quoted in *JdD*, 1839 January 28. The article is signed 'J.J.'
[10] Arago (1839) *C.R.A.S.*, **8**, 4–6.
[11] Arago (1839) *C.R.A.S.*, **8**, 170–2.
[12] Discussed by Arago in his posthumous *Astronomie Populaire*, **2**, 36–44, Gide, Paris (1857), where he notes that *parallatique* is often miswritten *parallactique*. The report of his remark at the Academy makes this mistake!
[13] Donné Al., *JdD*, 1839 August 20.
[14] Duboscq J. (1853) *Règles pratiques de la photographie sur plaque, papier, albumine et collodion d'après les meilleurs procédés connus.* Chez l'auteur, Paris.
[15] Donné Al., *JdD*, 1839 October 16.
[16] Airy W. (ed.) (1896) *Autobiography of Sir George Biddel Airy...* Cambridge University Press, Cambridge.
[17] Quoted in Deedes-Vincke P. (1992) *Paris – The City and Its Photographers.* Little Brown, Boston.
[18] Barger M. S., White W. B. (1991) *The Daguerreotype: Nineteenth-Century Technology and Modern Science.* Smithsonian Institution Press, Washington. This book contains a fuller historical summary of the discovery and development of the daguerreotype.
[19] Fizeau *père* had been Professor of Medical Pathology, but had been dismissed following the July Revolution. Corlieu A. (1896) *Centenaire de la Faculté de Médecine de Paris.* Imprimerie nationale, Paris.
[20] Arago (1840) *C.R.A.S.*, **10**, 488.
[21] Donné notes (*JdD*, 1841 July 7) that the Academy had decided no longer to report on memoirs that were printed.
[22] The Academy moved to a new meeting room in the Aile Le Bas of the Institut building in 1846. Figure 3.12 is based on the plan of this room given by Maindron E. (1888) *L'Académie des Sciences*, Alcan, Paris. Arago's announcement (Fig. 3.8) took place in the old meeting room in the Aile Le Vau. However, both rooms were very similar.
[23] Fizeau H. (1840) *C.R.A.S.*, **11**, 237–8.
[24] Gilding strengthens the image spherules by replacing mercury atoms in them with gold ones. It also enlarges them slightly, shifting the scattering to nearer the centre of the visible spectrum, so resulting in whiter highlights. See ref. 18.
[25] Donné (1839) *C.R.A.S.*, **9**, 485–6; Robert Cornelius' portrait of P. E. Du Ponceau, Spring 1840, reproduced in ref. 18.
[26] *JdD*, 1850 March 27.
[27] Lerebours (1841) *C.R.A.S.*, **12**, 1059–60.
[28] Fizeau (1841) *C.R.A.S.*, **12**, 1189–90.
[29] The quoted exposure times suggest a five- or tenfold increase in speed, and might be a reference to a compound lens with several elements designed by Chevalier, which could be arranged for landscapes or portraiture as desired. Chevalier's lens is illustrated in his *Nouvelles Instructions...* (ref. 31); see also van Hasbroeck P.-H. (1989) *150 Classic Cameras from 1839 to the Present*, Sotheby's, London. The famous Petzval portrait lens was introduced at about the

same time and with apertures as great as f/3.5, was up to twenty times faster. The Petzval lens only gave good definition over a small field because of astigmatism and field curvature, limiting its use to portraiture. See Thomas D. B. (1969) *The Science Museum Photography Collection.* HMSO, London.

[30] The Br_3^- ion forms easily in water.

[31] Chevalier C. (1841) *Nouvelles Instructions sur l'usage du Daguerréotype*, Chez l'auteur, Paris.

[32] 'Léon F.' (1841) *Journal des Artistes*, XVème année, **2**, No. 20, 305–7.

[33] Becquerel Ed. (1840) *C.R.A.S.*, **11**, 702–3; Biot (1841) *C.R.A.S.*, **12**, 101–12.

[34] Gaudin (1841) *C.R.A.S.*, **12**, 862–3.

[35] Moigno F. (1863) *Les Mondes*, **1**, 34. Other disparaging comments appear pp. 393–4, 630.

[36] Gaudin, Lerebours N.P. (1842) *Derniers perfectionnements apportés au daguerréotype. Troisième édition...suivie d'une notice sur l'emploi de l'eau bromée, par M. H. Fizeau.* Lerebours, Bachelier, Paris.

[37] Lerebours N.-M. P., Secretan G. (1846) *Traité de photographie, Cinquième édition, entièrement refondue...* Lerebours Secretan, Paris.

[38] Unless Fig. 4.23 is an electrotype copy of one, in which case it may predate Fig. 3.15.

[39] The sitter could well be sixty-one, which was Monmerqué's age when the portrait was taken. An engraving by Girard of Monmerqué as a younger man is preserved in the portrait collection of the Bibliothèque nationale de France (No. D216038). There are many similarities: broad brow and nose, curly hair, forward right eye, crease of chin, mild eyebrows, and probably the lips and creases of the face; and there are no striking dissimilarities (a mole under the right eye is perhaps still present on the daguerreotype, but has flattened with age).

[40] *Recueil*, pp. 57–8.

[41] Baudelaire expressed similar sentiments. See *Œuvres complètes de Charles Baudelaire. Correspondance générale*, **5**, 189. Conard, Paris (1949).

[42] *Cosmos*, **2**, 685–6 (1853).

Chapter 4 The 'delicious pastime' applied to science

[1] *JdD*, 1848 February 24.

[2] Guinagh K. (1983) *Dictionary of Foreign Phrases and Abbreviations.* Wilson, New York; Belfield-Lefévre H. (1837) *Recherches sur la nature, la distribution et l'organe du sens tactile.* Rigoux, Paris.

[3] *The Monthly Journal of Medico-Chirurgical Knowledge.* London. (1833)

[4] Belfield Lefèvre, *Revue scientifique et industrielle*, **11**, 417–62 (1842); **12**, 5–34 (1843).

[5] *Cosmos*, **3**, 104–5 (though this report written a decade later – mistakenly? – inverts the order of deposition), 158 (1853). It was not until about 1850 that electrotyped plates succeeded commercially.

[6] It would be interesting (but in the interests of conservation, unwise) to unseal the cases of Foucault's daguerreotypes and examine them for any plate marks indicative of electro-manufacture by Belfield-Lefèvre and partners Colas and L. J. Deleuil.

[7] *Cosmos*, **3**, 158 (1853).

[8] Daguerre (1839) *Historique et description des procédés du daguerréotype et du diorama.* Giroux, Paris. (Reprinted 1982, Rumeur des Ages, La Rochelle.)

[9] Belfield Lefèvre, Foucault L. (1843) *Revue Scientifique et Industrielle*, **14**, 198–9; *C.R.A.S.*, **17**, 260–2; *Annales de Chimie et de Physique* (3ème sér.) **9**, 507–8; *Bibliothèque universelle de Genève, Archives des sciences physiques et naturelles*, **46**, 377–9.

[10] Daguerre (1843) *C.R.A.S.*, **17**, 356–61.

[11] Choiselat (1843) *C.R.A.S.*, **17**, 605–8.

[12] Belfield-Lefèvre (1843) *C.R.A.S.*, **17**, 914–6.

[13] The *pli* (A.a.d.S. No. 440, deposited 1843 October 9) was opened in 1980 and sent to an expert at Kodak for comment. His intitial reaction was severe, but finally concluded that little could be said without performing 'long and difficult' experiments.

[14] Belfield-Lefèvre, Foucault L., *C.R.A.S.*, **23**, 713–14 (1846); *Annales de chimie et de physique*, **19** (3ème sér.) 125–7 (1847).

[15] Moigno (1847) *Revue scientifique et industrielle*, **14** (2ème sér.) 231.

[16] Chaix-Parat J. (1966) *Essai sur le renoncement d'Hector Berlioz à la carrière médicale.* Thèse pour le Doctorat en Médecine, Université de Paris.

[17] Huard P., Imbault-Huart M. J. (1974) *Revue d'histoire des sciences*, **17**, 45–62.

[18] Bertrand A. (1997) *Bulletin de l'Académie des Sciences et Lettres de Montpellier*, **28** (Nlle sér.) 99–113.

[19] The photographs reproduced by A. Lennox Thorburn are very poor and have been retouched around the eyes, while the origin of the drawing with a hat is unstated (*British Journal of Venereal Diseases*, **50**, 377–80 (1974)). Donné is possibly caricatured in an engraving by Maurisset (*La Caricature*, 1839 December 8) but there is no reason to assume the likeness is accurate.

[20] La Berge A. (1994) *Clio Medica*, **25**, 296–326.

[21] Robin C. (1871) *Traité du microscope.* Ballière, Paris.

[22] The first, *giardia intestinalis*, had been discovered by van Leeuwenhoek some 160 years earlier.

[23] This discovery is often credited to J. H. Bennett (1812–75), who had studied with Donné (Degos L. (1995) *Histoire de la médecine et des sci-*

ences, **11**, 1478–81). Donné's work may have been overlooked because it was first published in a textbook rather than in front of a learned society. See ref. 20.

[24] *L'Illustration*, **2**, 103 (1844).

[25] Donne A. (1842) *Conseils aux mères sur l'allaitement et sur la manière d'élever les enfants nouveaux-nés...* Ballière, Paris. See also La Berge A. F. (1991) *Journal of the History of Medicine and Allied Sciences*, **46**, 20–43.

[26] Donné A. (1864) *Conseils aux familles sur la manière d'élever les enfants...* Ballière, Paris.

[27] Sachaile C. [pseudonym for Lachaise C.] (1845) *Les Médecins de Paris jugés par leurs œuvres...* Chez l'auteur, Paris. One must wonder if this legend is confused. A.N. AP/300(I)/2435 records that Donné was paid 200 fr in 1842 February for treating the infant Duc de Chartres (born 1840 November 9), younger brother of the Comte de Paris (born 1838 August 24).

[28] Corlieu A. (1896) *Centenaire de la Faculté de Médecine de Paris.* Imprimerie nationale, Paris.

[29] Donné Al. (1844, but actually available in late 1843) *Cours de microscopie complémentaire des études médicales. Anatomie microscopique et physiologie des fluides de l'économie.* Ballière, Paris.

[30] Ref. 2.38

[31] The self-pitying letter quoted in Chapter 2 was written when Foucault was approaching 21 and seems incompatible with a new interest in life as Donné's assistant. More probably they met in 1841 or 1842, though the first clear trace of their collaboration begins in the autumn of 1843.

[32] Ref. 2.18.

[33] Donné Al., *JdD*, 1839 August 20.

[34] Donné Al., *JdD*, 1839 August 28.

[35] Donné Al., *JdD*, 1839 September 11.

[36] Donné (1839) *C.R.A.S.*, **9**, 376–8; Donné Al., *JdD*, 1839 September 18.

[37] Donné, *C.R.A.S.*, **9**, 411, 485 (1839); **10**, 288, 339, 667, 933 (1840).

[38] Chevalier [V.] (1840) *C.R.A.S.*, **10**, 423, 583.

[39] Fizeau, *C.R.A.S.*, **12**, 401, 509, 957 (1841); **19**, 119 (1844).

[40] *Recueil*, pp. 12–13.

[41] Figuier L. (c.1885) *Les Nouvelles Conquêtes de la Science.* Lahure, Paris.

[42] Robin C. (1871) *Traité du Microscope.* Ballière, Paris. Robin implies that he provided many of the specimens for the *Atlas*, but the only provider of specimens mentioned by name in the *Atlas* is a M. Bourgogne.

[43] Donne Al., *JdD*, 1843 December 22.

[44] Donné A., Foucault L. (1845) *Cours de microscopie complémentaire des études médicales. Anatomie microscopique et physiologie des fluides de l'économie. Atlas exécuté d'après nature au microscope-daguerréotype.* Ballière, Paris.

[45] The Observatory's monthly weather summaries were widely published. Data have been taken from *L'Illustration*, **3**, 294, 378; **4**, 32, 91 (1844).

[46] Two of the figures were actually based on drawings rather than daguerreotypes.

[47] Advertisement in the *JdD*, 1845 February 27.

[48] The selection of daguerreotypes was revised compared to that foreseen when the *Cours de Microscopie* was published, which explains the eccentric numbering of the figures: 0, 2, 3...15, 16, 16*bis*, 17...77, 78, 82...86.

[49] Ref. 29, pp. 35–9.

[50] E.g. *Bulletin de l'Académie Royale de Médecine*, **6**, 541 (1841).

[51] Richet G. (1997) *Histoire de la Médecine et des Sciences*, **13**, 45–8.

[52] Gernsheim A. (1961) *Medical and Biological Illustration*, **11**, 85–92; Turner G. L'E. (1973) *Journal of Microscopy*, **100**, 3–20.

[53] Tyndall J. (1896) The Electric Light. In *Fragments of Science.* Longmans Green, London.

[54] Donné (1845) *Bulletin de la Société d'encouragement pour l'industrie nationale*, **44**, 388–92.

[55] Hurter F., Driffied V. C. (1890) *Journal of the Society of the Chemical Industry*, **3**, 455–69.

[56] Fizeau, Foucault L. (1844) *C.R.A.S.*, **18**, 746–54.

[57] Data have been taken from *Annales de Chimie et de Physique*, **8** (3ème sér.) 509; **9** (3ème sér.) 128 (1843) and *L'Illustration*, **3**, 83, 240 (1844).

[58] *Les Mondes*, **27**, 44 (1871). This result, presented in discussion, was not recorded in the *Comptes rendus*.

[59] E.g. Debray H. (1862) *Annales de Chimie et de Physique*, **65** (3ème sér.) 331–40.

[60] *JdD*, 1853 August 11.

[61] E.g. *JdD*, 1849 June 1.

[62] Fizeau H., Foucault L. (1844) *C.R.A.S.*, **18**, 860–862. This and the previous memoir were printed together in several other places, e.g. *Annales de Chimie et de Physique*, **11** (3ème sér.) 370–83 (1844).

[63] Which Foucault certainly attended (*JdD*, 1851 May 10).

[64] Arago F. (1858) Quatrième mémoire sur la photométrie. In *Œuvres complètes*, Vol. 10. Gide, Paris. pp. 231–50. This memoir was read at the Academy on 1850 April 29; the parts concerning the solar daguerreotypes were produced from notes provided by Fizeau a few days earlier, and which are preserved in the Fizeau boxes at the A.A.d.S. Although Arago reported a 1.38-m focal length, the lens was no doubt the one used for the intensity measurements.

[65] Contradictory claims are made concerning attempts by N.-M. P. Lerebours to daguerreotype the Sun in 1842. See de Vaucouleurs G.

(1958) *La photographie astronomique du Daguerréotype au télescope électronique*, Albin Michel, Paris, and *Cosmos*, **2**, 444 (1853).

[66] This experimental set-up is mentioned explicitly by Wolf C. (1874) *Revue scientifique de la France et de l'étranger*, **14**, 356–7.

[67] A number of authors, unaware of this point, have expressed surprise that the solar spectrum was daguerreotyped (by J. W. Draper in the USA on 1842 July 27) well before the Sun was imaged, e.g. Norman D. (1938) *Osiris*, **5**, 560–94; Darius J. (1984) *Beyond Vision*, Oxford University Press, Oxford. However, the exposure times for dispersed solar light would have been easy to obtain manually.

[68] *JdD*, 1850 May 29.

[69] The focal length makes one think that a telescope might have been used, but the Observatory possessed no telescope this long. Only a small aperture would have been required, so perhaps the lens was not a telescope objective.

[70] The number of windows is correct, but to match the ground plan (A.N. F^{31} 31 Pièce 8) from this viewpoint, the image would need to be reversed. The houses in the background would then be on the western side of the Rue d'Assas. If the daguerreotype is an electrotype copy, however, this would mean that the original was taken with a reversing prism such as Foucault used in 1841 September (p. 35). The pompous archway visible in the lower right of the photograph might then be the one seen opposite No. 5 in a photograph of the Rue d'Assas taken by Charles Marville in 1868, though the cornices differ and there is a structure to the north of the arch which is absent in the daguerreotype taken 20 years earlier. See Mellot P. (1993) *Paris sens dessus-dessous*, Trinckvel, Paris, p. 185.

[71] Baker R. (1977) *History of Photography*, **1**, 111–16.

[72] *C.R.A.S.*, **12**, 401–2 (1841). Fizeau then copied the copy. *C. R. A. S.*, **12**, 509, 597 (1841).

[73] *Cosmos*, **2**, 685–6 (1853).

[74] Quoted by Buerger J. E. (1988) *French Daguerreotypes*, University of Chicago Press, Chicago, p. 7.

Chapter 5 The beautiful science of optics

[1] The *Recueil* (p. 60) prints $0^m.25$ for the negative electrode, but this must be a typographical error.

[2] A.A.d.S. *pochette* for 1843 December 18.

[3] Donné, Foucault (1844) *C.R.A.S.*, **18**, 696.

[4] Donné (1845) *Bulletin de la Société d'encouragement pour l'industrie nationale*, **44**, 388–92, 578–86, 977.

[5] Butrica A. J. (1988) *History and Technology*, **6**, 325–34.

[6] *JdD*, 1847 July 1.

[7] Pouillet (1844) *Élémens de physique expérimentale et de météorologie*. Vol. 2. (4ème ed.) Béchet, Paris. pp. 746–9.

[8] Le Roux F. P. (1867) *Bulletin de la Société pour l'encouragement de l'industrie nationale*, **24** (2ème sér.) 748–90.

[9] Sluijter F., *Physics World*, 2002 June, p. 20.

[10] Laplace (1796) *Exposition du système du monde*. Livre II. Cercle-Social, Paris. p. 198.

[11] *JdD*, 1847 September 29.

[12] For fuller details of developments in the early nineteenth century see Buchwald J. Z. (1989) *The Rise of the Wave Theory of Light*, University of Chicago Press, Chicago.

[13] Fizeau, Foucault L. (1849) *Annales de chimie et de physique*, **26** (3ème sér.) 138–48.

[14] Stated explicitly by Cornu A. (1898) Notice sur l'œuvre scientifiques de H. Fizeau. In *Annuaire pour l'an 1898 publié par le Bureau des Longitudes*. Paris, Gauthier-Villars (n.d.).

[15] The separate designations of H and K for these lines were not introduced until 1864.

[16] Fizeau H., Foucault L. (1845) *C.R.A.S.*, **21**, 1155–8.

[17] *JdD*, 1845 November 28.

[18] Fizeau, Foucault (1846) *C.R.A.S.*, **22**, 422.

[19] Barnard F. A. P. (1863) *Lectures on the undulatory theory of light*. Annual Report of the Board of Regents of the Smithsonian Institution for the year 1862. Government Printing Office, Washington.

[20] *JdD*, 1854 April 8.

[21] Jenkins F. A., White H. E. (1975) *Fundamentals of Optics*. McGraw-Hill Kogakusha, Tokyo.

[22] Curiously, the nature of the 'lame cristallisée' is not reported in Fizeau and Foucault's memoir; this detail comes from Foucault's article in the *Journal des Débats* two days later (1846 March 11). No doubt selenite, a transparent form, was used.

[23] *JdD*, 1846 March 11.

[24] *JdD*, 1848 July 7.

[25] [Babinet] (1848) *C.R.A.S.*, **26**, 680–2.

[26] Ref. 13; Fizeau H., Foucault L. (1850) *Annales de chimie et de physique*, **30** (3ème sér.) 146–59.

[27] For Foucault's comments on the system of reports see *JdD*, 1846 September 16 (reproduced as Fig. 6.2 for readers with excellent eyesight, or a magnifying glass).

[28] Ref. 2.18; de Parville H. (1869) *Causeries scientifiques, 8ème année 1868*. Rotschild, Paris. pp. 152–70. On the subject of medicine, de Parville says that 'ses goûts étaient ailleurs' but this can be taken as a form of words and not contradicting the comment by Gariel, a family member, that Foucault 'manifestait un goût assez vif' for medicine.

[29] Figuier L. (1884) *Les nouvelles conquêtes de la science*. Librairie illustrée, Paris.

[30] Lissajous (1869) *Revue des cours scientifiques de la France et de l'étranger*, **6**, 484–9 (reprinted

in the *Recueil*); Bertrand J. (1883) *Mémoires de l'Académie des Sciences de l'Institut de France*, **42**, lxxvii–civ.

[31] *JdD*, 1850 September 15.

[32] *JdD*, 1853 November 11.

[33] Ackerknecht E. H. (1967) *Medicine at the Paris Hospital 1794–1848*, Johns Hopkins Press, Baltimore.

[34] *Revue scientifique et industrielle*, **7**, 443–51 (1841).

[35] *JdD*, 1847 January 26.

[36] *La lancette française*, **1** (2ème sér.) 540 (1839) reports 1310 doctors and 200 *officiers de santé* for a Parisian population of 900 000.

[37] Fizeau, Foucault (1878) *Annales de chimie et de physique*, **15** (5ème sér.) 363–94. Although the memoir was sent to the Academy on 1847 August 2, Arago did not place it on the agenda until September 27. See *L'Institut*, **15**, 314–15 (1847).

[38] *JdD*, 1847 September 29.

[39] A.A.d.S. *pochette* for 1846 May 4. The *pli* is in Fizeau's writing and was finally opened in 1981.

[40] Fizeau, *Société Philomathique de Paris. Extraits des procès-verbaux des séances pendant l'année 1847*, 108.

[41] *JdD*, 1847 September 29.

[42] Lerebours (1846) *C.R.A.S.*, **23**, 634.

[43] Becquerel E. (1843) *C.R.A.S.*, **23**, 800–4.

[44] Foucault (1846) *C.R.A.S.*, **23**, 856–7.

[45] Fizeau H., Foucault L. (1847) *L'Institut*, **15** (1ère sect.) 28–9.

[46] Claudet (1847) *C.R.A.S.*, **25**, 554–5.

[47] Becquerel E. (1847) *C.R.A.S.*, **25**, 594–6; Gaudin (1847) *C.R.A.S.*, **25**, 639–40.

[48] The first description of this apparatus (see the *Recueil*, p. 37, Fig. 2) omitted the cylindrical lens and the lens L', which will have reduced the spectral purity of the image. The ray diagrams in both the *Recueil* and especially Ref. 49 are confusing because they make no distinction between real rays and ones used in determination of the position of the final image.

[49] Foucault L. (1853) *Cosmos*, **2**, 232–3.

[50] By the physicist C. S. Pouillet (see Ref. 49). The claim by Sherman P. D. (1981) *Colour Vision in the Nineteenth Century*, Hilger, Bristol, that the demonstration was copied *from* Pouillet is a clear misreading.

[51] Helmholtz (1854) *Report of the Twenty-Third Meeting of the British Association for the Advancement of Science. Notes and Abstracts of Miscellaneous Communication to the Sections*. John Murray, London. p. 5.

[52] Ref. 3.18.

[53] Bibliothèque de la Sorbonne, Ms 2089, Procès verbaux 1842 VI 18–1849 XII 29.

[54] Berthelot (1888) Notice sur les origines et sur l'histoire de la Société philomathique. In *Mémoires publiés par la Société philomathique à l'occasion du centenaire de sa fondation 1788–1888*. Gauthier-Villars, Paris; Duveen D. I. (1954) *Annals of Science*, **10**, 339–41; Thomas E. (ed.) (1990) *La Société philomathique de Paris et deux siècles d'histoire de la Science en France*. Presses Universitaires de France, Paris.

Chapter 6 The *Journal des Débats*

[1] Bellanger C., Godechot J., Guiral P., Terrou F. (1969) *Histoire générale de la presse française*. Vol. II. Presses Universitaires de France, Paris. p. 146.

[2] de Parville H. (1889) La Critique Scientifique: Donné, Foucault, Babinet, Ch. Daremberg. In *Le livre du centenaire du Journal des Débats*, Plon Nourrit, Paris.

[3] Sachaile in note 4.27.

[4] In 1851 a 3 months' subscription cost 17 francs. Figures for *Le Siècle*, *Le Constitutionnel* and *La Presse* were 12, 12 and 13 francs, respectively.

[5] Ruju P. A. M., Mostert M. (1995) *The Life and Times of Guglielmo Libri (1802–1869): Scientist, Patriot, Scholar, Journalist and Thief*. Verloren, Hilversum.

[6] *JdD*, 1845 May 23.

[7] *JdD*, 1845 July 9.

[8] However *Revues scientifiques* by Libri appeared in the *Revue des Deux Mondes*.

[9] *JdD*, 1846 May 21. This led to further recriminations on June 1/2.

[10] Bibliothèque nationale de France fr.n.a.3269 (347), dated 1846 October 6.

[11] *JdD*, 1846 September 30.

[12] *JdD*, 1846 October 21. Libri also wrote about Le Verrier's planet in the *Revue des Deux Mondes*, **16**, 378 (1846) under the pseudonym V. de Mars.

[13] Bibliothèque nationale de France fr.n.a.3269 (348), dated 1846 October 7.

[14] *JdD*, 1854 January 19.

[15] Tolls across the Seine were abolished in 1848 by the February Revolution.

[16] *JdD*, 1855 January 13.

[17] *JdD*, 1850 January 25.

[18] *JdD*, 1852 September 1.

[19] *JdD*, 1853 June 16.

[20] *JdD*, 1857 April 8.

[21] *JdD*, 1846 January 6, 1848 January 12, June 3, July 7, 1853 May 19.

[22] *JdD*, 1851 July 2.

[23] *JdD*, 1847 January 26.

[24] *JdD*, 1847 February 3.

[25] Dawson P. M. (1908) *A Biography of François Magendie*. Huntington, Brooklyn.

[26] *JdD*, 1847 February 5.

[27] *JdD*, 1847 February 17.

[28] *JdD*, 1847 February 24.

[29] *JdD*, 1847 March 24.

[30] *JdD*, 1846 April 19.
[31] *JdD*, 1846 September 30.
[32] *JdD*, 1849 June 1, and again on 1850 August 16/17.
[33] *JdD*, 1847 September 8.
[34] *JdD*, 1851 September 13.
[35] *JdD*, 1852 October 1.
[36] *JdD*, 1847 December 29.
[37] Fraissinet A. (1873) *La Nature*, **1**, 358–60; *Les Mondes*, **32**, 101–2 (1873).
[38] E.g. *JdD*, 1847 March 24, May 21, 1854 August 5.
[39] *JdD*, 1858 June 4.
[40] *JdD*, 1848 June 2.
[41] *JdD*, 1853 July 1.
[42] *JdD*, 1853 April 15.
[43] *JdD*, 1847 February 17, 1853 April 1, July 28.
[44] E.g. *Le moniteur scientifique*, **7**, 140–1 (1865); ref. 2.18.
[45] *JdD*, 1854 June 22.
[46] *JdD*, 1849 March 14.
[47] *JdD*, 1852 July 2.
[48] *JdD*, 1851 December 27.
[49] *JdD*, 1848 December 16.
[50] *JdD*, 1856 December 20.
[51] *JdD*, 1846 April 1.
[52] *JdD*, 1848 May 14.
[53] *JdD*, 1850 July 3; a phrase borrowed from the poet N. J. L. Gilbert (1751–80).
[54] *JdD*, 1854 January 19.
[55] *JdD*, 1851 January 4.
[56] *JdD*, 1845 October 22.
[57] *JdD*, 1849 July 12.
[58] *JdD*, 1847 October 27.
[59] Alis H. (1889) Le Roman-Feuilleton. In *Le livre du centenaire du Journal des Débats*, Plon Nourrit, Paris.
[60] Citron P. (ed.) (1978) *H. Berlioz: Correspondance générale*. Flammarion, Paris. p.120.
[61] *JdD*, 1853 May 29.
[62] *Cosmos*, **2**, 659 (1853); Roger H. *Le Constitutionnel*, 1853 May 23.
[63] *JdD*, 1853 June 2.
[64] *JdD*, 1854 March 3.
[65] *JdD*, 1854 June 8.
[66] *JdD*, 1846 April 1.
[67] E.g. *JdD*, 1855 February 17, 1856 October 23.
[68] *JdD*, 1853 March 19.
[69] Bertrand J. (1864) *Revue des Deux Mondes*, **51**, 96–115.
[70] *JdD*, 1845 November 28.
[71] *JdD*, 1850 February 8.
[72] *JdD*, 1845 August 27, 1848 February 2, 1857 April 8.
[73] *JdD*, 1852 January 23.
[74] *JdD*, 1853 October 7.
[75] *JdD*, 1845 September 11.
[76] *JdD*, 1851 August 27.
[77] *JdD*, 1850 June 14.
[78] Ref. 2.38
[79] *JdD*, 1846 January 6.
[80] *JdD*, 1853 December 11.
[81] *JdD*, 1845 December 10.
[82] *JdD*, 1848 March 30.
[83] *JdD*, 1848 March 12.
[84] *JdD*, 1848 May 14.
[85] *JdD*, 1848 May 30.
[86] *JdD*, 1848 July 7.
[87] *JdD*, 1853 February 12.
[88] For completeness, it should be stated that Foucault had one other excursion into the world of the paid writer. In 1848 he wrote for a serial publication entitled *Instruction pour le peuple: Cent traités sur les connaissances les plus utiles* (Dubochet Lechevalier, Paris). His two treatises on general physics, which sold at 25 c. each, covered properties of matter, gravity, acoustics and optics. His friend Jules Regnauld provided a third treatise on electricity and magnetism. The article on Foucault in the *Dictionary of Scientific Biography* claims that he also wrote schoolbooks. This is incorrect, and results from confusion with another L. Foucault, a teacher of mathematics at Pons.

Chapter 7 Mixed luck

[1] *JdD*, 1845 November 14.
[2] *JdD*, 1847 March 9.
[3] *JdD*, 1847 August 19. Foucault makes the mistake discussed in note 3.12.
[4] In general, forcing and damping alter a pendulum's period, but in a clock the forces are small and the effect is negligible.
[5] Foucault L. (1847) *C.R.A.S.*, **25**, 154–60.
[6] Adapted from drawings in the original, handwritten memoir. A.A.d.S. *pochette* for 1847 July 26.
[7] Pecqueur (1847) *C.R.A.S.*, **25**, 251–3.
[8] Bigourdan G. (n.d.) Le Bureau des Longitudes. In *Annuaire pour l'an 1931 publié par Le Bureau des Longitudes*, A1-A151. Gauthier-Villars, Paris.
[9] *Notice sur les travaux de M. Léon Foucault*. Bachelier (1850, 1859), Mallet-Bachelier (1863, 1865), Paris.
[10] Brenni P. (1994) *Bulletin of the Scientific Instrument Society*, No. 40, 3–6.
[11] The University of Chicago's Bruce Telescope furnishes an example. See Barnard E. E. (1905) *Astrophysical Journal*, **21**, 35–48.
[12] Faye had been considering it since at least 1845. See *Revue scientifique et industrielle*, **5**, (2ème sér.) 337 (1845).
[13] Faye (1847) *C. R. A. S.*, **25**, 375–80; *Recueil*, pp. 305–9, 310–2 where the electrical engineer Théodore du Moncel (1821–84) describes a pair of contacts for the electrical distribution of time which he claims were made by Foucault in 1858 or 1859.
[14] *C.R.A.S.*, **21**, 923 (1845); *JdD*, 1845 October 15, 22.

[15] Vérité (1863) *C.R.A.S.*, **56**, 401–2; Foucault L. (1863) *C.R.A.S.*, **56**, 645–6.

[16] Laugier (1847) *C.R.A.S.*, **25**, 415–9, 480–1. An electric clock was placed in the catacombs beneath the Observatory by Liais in the 1850s. (Le Bureau International de l'Heure (1992) *75 ans au service de l'heure universelle.* Bureau des Longitudes, Paris.)

[17] E.g. Laboulaye *et al.* (1867) Éclairage. In *Dictionnaire des arts et manufactures, de l'agriculture, des mines, etc.* Lacroix, Paris; Figuier (1882) in note 18; Moigno (1867) *Les éclairages modernes.* Gauthier-Villars, Paris.

[18] Figuier L. (1882) *L'art de l'éclairage.* Jouvet, Paris; ref. 5.29. I could find no report of the purported illumination of the Place de la Concorde in newspapers, while repetitions of the claim postdate Figuier's two publications and presumably derive from them (e.g. Éclairage in *La grande encyclopédie*, Vol. 14, Lamirault, Paris, p. 342 (1885–1902); Drohojowska (1888) *Les savants modernes et leurs œuvres.* Lefort, Lille). Writers who knew Foucault make no mention of illumination of the Place de la Concorde but report his use of the electric light in *Le prophète*, all of which strongly suggests that Figuier's claim is wrong. Deleuil and Archereau illuminated the Place de la Concorde on 1843 October 20 (*L'Illustration*, **2**, 152 (1843 October 28)). This night was foggy, as claimed by Figuier for the trial by Foucault and Deleuil.

[19] *JdD*, 1845 September 11.

[20] *JdD*, 1849 February 9.

[21] A.A.d.S. Foucault box, letter dated 1849 January 19.

[22] Foucault L. (1849) *C. R. A. S.*, **28**, 68–9.

[23] Foucault was probably unaware that the flow of current through an arc becomes easier as the current increases (the arc is said to have *negative* resistance), because his currents will have been limited by the internal resistance of the Bunsen cells and the platinum–acid regulator. When an arc is driven from a modern, low-impedance power supply, it is essential to use a series resistance to limit the current.

[24] *JdD*, 1846 March 11.

[25] Laussedat A. (1865) *Notice biographique sur Gustave Froment.* Hetzel, Paris.

[26] *The Illustrated London News*, p. 317 (1848 November 18).

[27] *The Illustrated London News*, p. 368 (1848 December 9).

[28] Mackechnie Jarvis C. (1958) The distribution and utilization of electricity. In *A History of Technology*, Vol. 5. Singer C., Holmgard E. J., Hall A. R., Williams T. I. (eds) Oxford University Press, Oxford.

[29] Foucault L. (1849) *C.R.A.S.*, **28**, 68–9.

[30] A.A.d.S. Foucault box, letter dated 1847 September 1.

[31] Dumas (1849) *C. R. A. S.*, **28**, 120–1.

[32] A year previously – Brevet d'Invention No. 7033, application dated 1848 January 14; Gaigneau (1847) *C.R.A.S.*, **28**, 157.

[33] See Darnis, *JdD*, 1856 December 21 for a brief discussion about nullity in French patent law.

[34] de la Rive A. (1849) *Bibliothèque universelle de Genève. Archives des sciences physiques et naturelles*, **10**, 222–5.

[35] The English inventors were unable to find financial backing and their company went into receivership in 1851. Staite died three years later.

[36] A.N. AJ^{13}223.

[37] Becker H. & G. (1985) *Giacomo Meyerbeer: Briefwechsel und Tagebücher*, Band 4, 1846–1849. De Gruyter, Berlin.

[38] Adam A. *Le Constitutionnel*, 1847 April 18.

[39] A.N. AJ^{13}429 records three part-payments to Froment of 150, 100 and 100 francs. The total cost of the 'electrical apparatus' was 850 francs more, which was no doubt principally the cost of the Bunsen cells.

[40] See Dunsheath P. (1962) *A History of Electrical Engineering.* Faber & Faber, London; and Éclairage in *La grande encyclopédie* (note 18).

[41] Duboscq J. (1877) *Catalogue des Appareils employés pour la production des phénomènes physiques au théâtre.* Duboscq, Paris.

[42] Because of his later associations with Foucault and the Opera, many authors incorrectly assume that Duboscq made the arc lamp used in *Le prophète*.

[43] Ref. 5.29, p.14.

[44] A.N. AJ^{13}335.

[45] Becker H. & G. (1989) *Giacomo Meyerbeer: A Life in Letters.* Amadeus Press, Portland. pp. 122–5.

[46] Ref. 2.38.

[47] Wheatstone C. (1838) *Philosophical Transactions of the Royal Society of London*, **128**, 371–94.

[48] The experiment was first presented at the Société Philomathique in 1848 December, and then a month later at the Academy on the same day as the regulated arc lamp (Foucault L. Regnauld J. (1848) *L'Institut*, **17** (1ère sect.) 3–5; (1849) *C.R.A.S.*, **28**, 78–80). In his letter to Dumas (note 21) Foucault wrote, 'Monsieur Fizeau knows of the end I was seeking, but he only saw the machine work when I invited him to come and gauge the recomposition of complementary colours...'

[49] Howard I. P., Rogers B. J. (1995). *Binocular Vision and Stereopsis.* Oxford University Press, Oxford. pp. 325–7.

[50] Foucault L. (1849) *L'Institut*, **17** (1ère sect.) 44–6; *Société Philomathique de Paris. Extraits des procès-verbaux des séances pendant l'année 1849*, 16–20; *Bibliothèque universelle de Genève. Archives des sciences physiques et naturelles*, **10**, 222–5.

⁵¹For a fuller review of the origins of spectroscopy, see Hearnshaw J. B. (1986) *The Analysis of Starlight*. Cambridge University Press, Cambridge.

⁵²'Every person who studies optics has to a greater or lesser degree searched for a monochromatic source of light,' Foucault noted later. He himself had produced a feeble monochromatic yellow light by burning a hydrogen jet into which sodium had been evaporated (*Revue des sociétés savantes. Sciences mathématiques, physiques et naturelles*, **1**, 298–9 (1862)).

⁵³See Whitmell C. T. L. (1876) *Nature*, **13**, 188–9; and Wilson D. B. (ed.) (1990) *The Correspondence Between Sir George Gabriel Stokes and Sir William Thomson, Baron Kelvin of Largs*. Cambridge University Press, Cambridge, letter 277, 1871 July 5. Stokes' memory was unreliable, however, since he had already written to Thomson about Foucault's discovery (letter 104, 1854 March 28).

⁵⁴Reviewed by Sutton M. A. (1988) *Platinum Metals Review*, **32**, 28–30.

⁵⁵Reprinted by Figuier L. (1862) *L'année scientifique et industrielle, 6ème année*. Hachette, Paris.

⁵⁶*JdD*, 1860 July 16.

⁵⁷For a review, see James F. A. J. L. (1983) *Ambix*, **30**, 137–62.

⁵⁸Moigno F. (1861) *Cosmos*, **19** (2ème sér.) 136–40.

⁵⁹*Philosophical Magazine*, **19**, 193–7 (1860); *Annales de Chimie et de Physique* (3ème sér.) **58**, 476–8 (1860).

⁶⁰Eleven years later Kirchhoff was in a better position to interpret the reversal. The omnipresence of the D lines had been explained, and ideas about the conservation of energy had developed during the 1850s, which were essential for recognizing the link between absorption and emission.

⁶¹There had been an earlier 'communication by MM. Lefevre & Foucault on pathographic images' on 1843 August 12. The Sociéte's archives have been deposited in the Bibliothèque de la Sorbonne. See box 125 and Ms 2089–2092 for relevant minute books; 2099 F110 for the 1852-53 membership list; and box 134 for the 1851 attendance sheets.

⁶²*L'Institut*, **17** (1ère sect.) 11 (1849); *Société philomathique de Paris. Extraits des procès-verbaux des séances pendant l'année 1848*, 81–3.

⁶³On 1849 December 29.

⁶⁴I.d.F. Ms 3711 No. 39. The letter is dated 1856 August 13.

⁶⁵A.A.d.S. Fizeau box No. 2.

⁶⁶A.A.d.S. *pochette* for 1852 March 1.

⁶⁷Including the one to replace him when he moved to honorary membership following election to the Académie des Sciences.

Chapter 8 Demise of the corpuscular theory

¹*JdD*, 1850 May 15.

²Wheatstone C. (1834) *Philosophical Transactions of the Royal Society of London for the year MDCCCXXXIV*, 583–91.

³See (1835) *Report of the Fourth Meeting of the British Association for the Advancement of Science; held at Edinburgh in 1834*, John Murray, London; Arago (1837) *C.R.A.S.*, **7**, 954–65.

⁴*JdD*, 1883 November 2 & 3. The spellings *Breguet* and *Bréguet* were both used; the former is adopted here.

⁵[Breguet L. F. C.] (1873) *Notice sur les travaux de M.L. Bréguet, Artiste, Membre du Bureau des Longitudes*. Gauthier-Villars, Paris.

⁶Reported by Cornu A. (1900) Sur la vitesse de la lumière. In *Rapports présentés au Congrès international de physique réuni à Paris en 1900 sous les auspices de la Société française de physique*, Vol. 2. Gauthier-Villars, Paris.

⁷Arago (1850) *C.R.A.S.*, **30**, 489–95; ref. 14.

⁸The identity of the Montmartre station is revealed in *Revue scientifique et industrielle*, **5**, (3ème sér.) 393–7 (1849).

⁹*JdD*, 1849 December 20.

¹⁰Fizeau: A.N. LH 977/56 (reconstitution); Froment: Lami, E-O. (1885) *Dictionnaire encyclopédique et biographique de l'industrie et des arts industriels*. Librairie des dictionnaires, Paris.

¹¹Delaunay (n.d.) Notice sur la vitesse de la lumière. In *Annuaire pour l'an 1865 publié par le Bureau des Longitudes*. Gauthier-Villars, Paris. Translated into English as (1865) Essay on the velocity of light. *Annual Report of the Board of Regents of the Smithsonian Institution... for the year 1864*. Government Printing Office, Washington.

¹²*C.R.A.S.*, **30**, 668 (1850).

¹³More accurately, it is not the mass that needs to be evenly distributed for smooth spinning, but quantities called the *products of inertia* which need to be zero.

¹⁴Foucault L. (1850) *C.R.A.S.*, **30**, 551–60.

¹⁵*Recueil*, p. 518.

¹⁶*JdD*, 1850 April 30.

¹⁷*JdD*, 1850 May 4.

¹⁸Fizeau H., Bréguet L. (1850) *C.R.A.S.*, **30**, 562–3.

¹⁹Fizeau H., Bréguet L. (1850) *C.R.A.S.*, **30**, 771–4.

²⁰A.A.d.S. Foucault box, letters dated 1850 May 31, June 20.

²¹Foucault L. (1853) *Sur les vitesses de la lumière dans l'air et dans l'eau*. Bachelier, Paris. 36 pp., 1 plate.

²²Bournard F. (1899) *Un bienfaiteur de l'Humanité. Pasteur. Sa vie, son œuvre*. Toltra, Paris. p. 19.

²³*Almanac Impérial pour 1853 présenté à leurs Majestés*. Guyot Scribe, Paris. p. 105.

²⁴*Les Mondes*, **1**, 141–3 (1863).

²⁵A.A.d.S. Foucault box.

[26] The formula for the mean speed of propagation V over the space between the rotating mirror, m, and concave mirror, M, is $V = 8\pi n l^2 r/d(l + l')$, where the mirror speed is n r.p.s., d is the linear deviation measured at the eyepiece, l is the distance from m to M, l' is the distance from m to L and r is the distance from L to a. Foucault's thesis, however, prints δ instead of d at several points, where δ is the angular rather than linear deviation. One might think these typographical rather than logical errors, except for (i) Foucault's inability to derive correct formulae in his doctoral examination, (ii) the fact that though derived the formulae are not actually applied to observations in the thesis (although indicative values of d are calculated correctly), and (iii) the repetition of the errors in the *Annales de chimie et de physique* the following year. They are corrected in the *Recueil*.

[27] Moigno also had criticisms, but misunderstood why Foucault did not present a numerical value for the speed of light (*Cosmos*, **2**, 534–5 (1853)).

[28] *JdD*, 1850 January 25, 1852 May 21.

[29] Sarton G. (1936) *The Study of the History of Science*. Harvard University Press, Cambridge.

Chapter 9 The rotation of the Earth

[1] Kuhn T. S. (1957) *The Copernican Revolution*. Harvard University Press, Cambridge; Acloque P. (1982) *Cahiers d'histoire & de philosophie des sciences* (Nlle sér.) No. 4; Gapaillard J. (1993) *Et pourtant, elle tourne!* Seuil, Paris; *Cahiers d'histoire & de philosophie des sciences* (Nlle sér.) No. 25 (1998).

[2] E.g. '... it seems to me that it is much more natural to make the roast turn in front of the fire than to make the fire turn around the roast; and since the cooks of the Earth have always acted thus, why should the great architect of the world abstract himself alone from the laws of simple common sense?' Quoted in *La Presse*, 1851 March 14.

[3] The angle in radians is $(1/g)\Omega_\oplus^2 R \cos\lambda \sin\lambda$, where g, $\Omega_\oplus$, R and λ are the acceleration due to gravity, the angular velocity and radius of the Earth, and the latitude, respectively.

[4] The effect of air resistance is negligible. The eastwards deviaton is $(2/3)(2/g)^{1/2}\Omega_\oplus h^{3/2} \cos\lambda$ where h is the drop.

[5] *JdD*, 1849 June 30.

[6] Costabel P. (1984) *La Vie des Sciences*, **1**, 235–49.

[7] Costabel P. (1989) In *Studies in the History of Scientific Instruments*. Blondel C., Parot F., Turner A., Williams M. (eds). Rogers Turner, London. pp. 129–33.

[8] Fizeau went on the following year to measure a probable shift in water flows in accord with Fresnel's predicted partial drag. (Fizeau H. (1851) *C.R.A.S.*, **33**, 349–55.) The presence of Fresnel drag in laboratory experiments but its absence due to the terrestrial motion of the laboratory was fully explained some fifty years later by Einstein's Theory of Relativity.

[9] Ref. 5.29 p.20.

[10] Foucault L. (1851) *C.R.A.S.*, **32**, 135–8.

[11] *JdD*, 1849 December 29.

[12] *JdD*, 1851 March 31.

[13] *JdD*, 1852 October 19.

[14] *JdD*, 1852 December 16.

[15] Poisson (1837) *C.R.A.S.*, **5**, 660–7. This is an excerpt; the full text was published in the *Journal de l'École polytechnique*.

[16] Terrien, *Le National*, 1851 February 19. Terrien's words have frequently been misquoted, e.g. in ref. 108.

[17] The details given by Ph. Blanchard in *Le Siècle*, 1851 February 16, make it clear that the same 5 kg bob was used as in Foucault's cellar.

[18] Oprea J. (1996) *American Mathematical Monthly*, **102**, 515–22.

[19] Sher D. (1969) *Journal of the Royal Astronomical Society of Canada*, **63**, 227–8.

[20] Silvestre E. (1851) *C.R.A.S.*, **33**, 40–1.

[21] *JdD*, 1851 July 27.

[22] Centre canadien d'architecture/Caisse nationale des monuments historiques et des sites (1989) *Le Panthéon: Symbole des révolutions de la Nation au Temple des grands hommes*. Picard, Paris.

[23] A.N. F[17]3153, letter dated 1851 July 8.

[24] For a more accurate but later sketch of the experiment see Drohojowska (1888) in note 7.18.

[25] *Le National*, 1851 March 26.

[26] Roger H., *Le Constitutionnel*, 1851 March 15.

[27] *L'Univers*, 1851 March 4.

[28] *L'Univers*, 1851 March 31.

[29] A.N. F[21]845. The possible use of the Panthéon was also invoked on February 16 in *Le Siècle*.

[30] Guyot J. (1837) *C.R.A.S.*, **4**, 888; Babinet (1851) *C.R.A.S.*, **32**, 705; *Cosmos*, **2**, 447–9 (1852). Guyot's unbelievable result was ignored by the Academy. A.N. F[21]845; Mène Ch. (1851) *C.R.A.S.*, **33**, 39, 222–4.

[31] Foucault claimed 18 cm, but he was adopting round numbers. Recent measurements on the bob show its diameter to be 172 mm or $6\frac{1}{3}$ French inches.

[32] *Le Siècle*, 1851 March 12.

[33] 'Since 11 o'clock this morning, several hundred people have been in a circle under the Panthéon dome ...' *La Presse*, 1851 March 21; *Le National*, 1851 March 26.

[34] *Le Constitutionnel*, 1851 March 27.

[35] *Putnam's Monthly Magazine of American Literature, Science and Art*, **8**, 416–21 (1856).

[36] A.A.d.S. *pochette* for 1851 July 7.

[37] *The Times*, 1851 April 12.

[38] *The Times*, 1851 June 4.

[39] Conlin M. F. (1999) *Isis*, **90**, 181–204.

[40] Binet (1851) *C.R.A.S.*, **32**, 157–9, 197–205. The memoir was published over two weeks because of its length. For an attractive modern Coriolis treatment, see Opat G. I. (1991) *American Journal of Physics*, **59**, 822–3.
[41] Liouville (1851) *C.R.A.S.*, **32**, 159–60.
[42] Reprinted as Poinsot (1851) *Théorie nouvelle de la rotation des corps.* Bachelier, Paris.
[43] Bertrand J. in Poinsot L. (1872) *Eléments de statique.* (12ème ed.) Gauthier-Villars, Paris.
[44] Poinsot (1851) *C.R.A.S.*, **32**, 206–7.
[45] Ref. 2.11.
[46] Bertrand (1882) *C.R.A.S.*, **94**, 371–2.
[47] Powell B. (1851) *Notices of the Proceedings at the meetings of the members of the Royal Institution*, **1**, 70–7.
[48] Terquem (1851) *C.R.A.S.*, **32**, 244; Dubuat (1814) *Annales de mathématiques pures et appliquées*, **5**, 55–9, 216–20; Antinori (1851) *C.R.A.S.*, **32**, 635–6; Poleni J. (1742–3) *Philosophical Transactions*, No. 468, 299–306; Dehaut (1860) *C.R.A.S.*, **51**, 575.
[49] *Cosmos*, **17**, 508 (1860).
[50] Belfield-Lefebvre (1852) *Revue scientifique et industrielle*, **1** (4ème sér.) 19–21 & Pl. I; *The Times*, 1851 April 29.
[51] Foucault (1851) *L'Institut*, **19** (1ère sect.) 269; *Société Philomathique de Paris. Extraits des procès-verbaux des séances pendant l'année 1851*, 58–9; The full report is published in the *Recueil*, pp.392–400.
[52] E.g. Hart J. B., Miller R. E., Mills R. L. (1987) *American Journal of Physics*, **55**, 67–70.
[53] Even Donné made this misleading simplification (*JdD*, 1853 February 12).
[54] *C.R.A.S.*, **32**, 352 (1851).
[55] *C.R.A.S.*, **32**, 367–8 (1851).
[56] Crosland M. (1992) *Science under Control – The French Academy of Sciences 1795–1914.* Cambridge University Press, Cambridge.
[57] *La Presse*, 1851 April 4.
[58] *Le National*, 1851 March 20. Regnault and Liouville had been elected most recently, in 1840 and 1839, aged 29 and 30 respectively. Biot and Arago has been elected in 1803 and 1809 aged 28 and 23.
[59] *The Times*, 1851 April 11.
[60] Stokes G. G. (1850) *Transactions of the Cambridge Philosophical Society*, **9**, 8–106.
[61] E.g. Charron (1931) *C.R.A.S.*, **198**, 208–10.
[62] 300 francs for 'the actual pendulum' and 150 francs for 'the setting up of the divided circle'. A.N. F^{17}3153, letter received 1851 May 1.
[63] For a recent review, see Schulz-DuBois E. O. (1970) *American Journal of Physics*, **38**, 173–88.
[64] Dufour, Wartmann, Marignac (1851) *C.R.A.S.*, **33**, 13–15.
[65] Airy G. B. (1851) *The London, Edinburgh and Dublin Philosophical Magazine*, **2** (4th ser.) 147–9.
[66] For a discussion see the excellent Acloque P. (1981) *Oscillations et stabilité selon Foucault.* Éditions du CNRS, Paris.
[67] *The Times*, 1851 April 26.
[68] A.N. F^{21}845.
[69] Bertrand (ref. 2.11) says the Panthéon experiment ran 'pendant plusieurs mois', suggesting it was not re-established. However, the posthumous inventory of Foucault's papers lists five notebooks of observations on the *Pendule du Panthéon* dated from 1851 April 23 to 1852 October 24. The inventory was clearly compiled rapidly and contains some obvious errors. More reasonably, the Panthéon pendulum was not re-erected after its wire broke, and these lost notebooks contained investigations into looping with rods and pendulums located elsewhere.
[70] *Revue scientifique et industrielle*, **1** (4ème sér.) 220–1.
[71] *Les Mondes*, **16**, 467 (1868); **19**, 633 (1869).
[72] *The Times*, 1851 April 16.
[73] *JdD*, 1846 February 18.
[74] Indicated by Foucault's careful acknowledgments of Poinsot in his memoirs.
[75] E.g. Short J. (1752) *Philosophical Transactions of the Royal Society of London*, **47**, 353–3.
[76] von Bohnenberger J. G. F. (1819) *Annalen der Physik*, **60**, 60–71.
[77] Sang E. (1856) *Transactions of the Royal Scottish Society of Arts*, **4**, 413–20.
[78] *JdD*, 1852 October 1.
[79] Foucault L. (1852) *C.R.A.S.*, **35**, 421–4 gives no numerical value, which comes from Foucault L. (1855) *C.R.A.S.*, **41**, 450–2.
[80] The inventory of his papers titles one: 'Deviation observée au gyroscope le vendredi 30 avril 1852, entre 2^h et 3^h chez Froment. Vendredi 5 mai inclinaison, 10 mai déclinaison.'
[81] Ref. 2.38.
[82] *JdD*, 1852 September 22.
[83] Person (1852) *C.R.A.S.*, **35**, 417–20.
[84] *Cosmos*, **15**, 108–10 (1859).
[85] Sire G. (1852) *C.R.A.S.*, **35**, 431–2. Sire did not realize his letter had arrived, and wrote to the Academy again (A.A.d.S. *pochette* for 1852 October 11).
[86] Foucault L. (1852) *C.R.A.S.*, **35**, 421–4.
[87] Foucault L. (1852) *C.R.A.S.*, **35**, 424–7.
[88] The period for horizontal oscillations is $T_{horiz} = 2\pi(I/(I_S \Omega_\oplus \omega_S \cos\lambda))^{1/2}$ where I_S and I are the moments of inertia about the spindle and the other two principal axes, $\Omega_\oplus$ and ω_S are the angular velocities of the Earth and rotor, and λ is the latitude. Periods in the meridian are smaller by a factor $\cos\lambda$.
[89] Foucault L. (1852) *C.R.A.S.*, **35**, 602; the full text is published in the *Recueil*, pp. 413–15.
[90] A.A.d.S., *pochette* for 1852 October 11.
[91] *Cosmos*, **2**, 645–7 (1853).
[92] *JdD*, 1852 November 26.
[93] *Cosmos*, **3**, 362–4 (1853).

[94] Person C.-C. (1852) *C.R.A.S.*, **35**, 549–52, 753–4.
[95] Quet (1852) *C.R.A.S.*, **35**, 602–3, 669, 686–8.
[96] No doubt actually Childwall Hall, demolished in 1949.
[97] Ref. 2.40.
[98] *Transactions of the sections* in *Report of the twenty-fourth meeting of the British Association for the Advancement of Science* p.56, John Murray, London (1855).
[99] In the *Recueil* Lissajous told Gariel that 'it seemed to him' that this suspension was made for the Panthéon, but Lissajous was speaking almost thirty years after the event. It seems overwhelmingly probable that the safety parachute was constructed for the 1855 pendulum (see note 101).
[100] *Catalogue officiel.* (2ème éd.) Imprimerie nationale.
[101] The prototype motor (Plate X) and parachute parts (Inv. IA-19-61) in the Observatory's collection are the basis for this assertion.
[102] *Cosmos*, **7**, 67–73 (1855).
[103] Brisse L. (1857) *Album de l'Exposition universelle*, Tome 1. Abeille Impériale, Paris. p. 196.
[104] Govi G. (1855) *L'Illustration*, **26**, 198–9.
[105] *Visites et études de S.A.I. le Prince Napoléon au Palais de l'Industrie ou guide pratique et complet à l'Exposition universelle de 1855*, Perrotin, Paris (1855).
[106] Exposition universelle de 1855. (1856) *Rapports du jury mixte international*, Imprimerie impériale, Paris. p. 401. In addition, Froment was one of the judges for precision engineering and could hardly have participated in awarding a prize for an instrument he had himself constructed.
[107] A.N. F^{17}3726.
[108] E.g. de Fonville W. (1887) *La Nature*, 15ème année, 2ème sem., 409–11.
[109] *La Nature*, 15ème année, 2ème sem., 334 (1887); A.N. ET/XLIX/1504.
[110] Details from *L'Astronomie*, **14**, 465–80 (1902).
[111] *1851–1902–1995 Le pendule de Foucault au Panthéon.* Caisse nationale des monuments historiques et des sites/Musée du Conservatoire national des arts et métiers, Paris (1995).
[112] Inv. 12658.
[113] Theorists make further subdivisions. *Active* gravitational mass is the source of the gravitational field, which acts upon *passive* gravitational mass.
[114] An exposure of almost 3 hr was required with a 42 curie ^{60}Co source placed 3 m from the bob.
[115] Mach E. (1883) *Die Mechanik in ihrer Entwickelung historisch-kritisch dargestellt.* Brockhauss, Leipzig.
[116] See Barbour J. B., Pfister H. (eds) (1995) *Mach's Principle: From Newton's Bucket to Quantum Gravity.* Birkhäuser, Boston.
[117] This speculative numerical value makes assumptions about the mass of the universe and adopts an inverse distance law for the Machian interaction.
[118] Poincaré H. (reprinted 1968) *La Science et l'hypothèse.* Flammarion, Paris.
[119] Einstein A. (1918) *Annalen der Physik*, **55**, 240–4.
[120] Li L.-X. (1998) *General Relativity and Gravitation*, **30**, 497–507.
[121] Braginsky V. B., Polnarev A. G., Thorne K. S. (1984) *Physical Review Letters*, **53**, 863–6; Ciufolini I., Pavlis E., Chieppa F., Fernandes-Viera E., Pérez-Mercader J. (1998) *Science*, **279**, 2100–3.
[122] Wheatstone C. (1851) *Proceedings of the Royal Society of London*, **6**, 65–8.
[123] Figuier L. (1852) *Exposition et histoire des principales découvertes scientifiques modernes.* Vol. 2. Masson Langlois Leclercq, Paris.
[124] *JdD*, 1852 June 20.

Chapter 10 Biding time

[1] Blondel C. (1994) in *Les professeurs du Conservatoire national des arts et métiers*, Tome 1. Fontanon C., Grelon A. (eds.) Institut national de recherche pédagogique, Conservatoire national des arts et métiers, Paris. pp.168–82.
[2] *C.R.A.S.*, **35**, 803, 823 (1852).
[3] A.N. F^{17}20758, letters dated 1853 February 24, March 10.
[4] A.A.d.S. Archives A. Lacroix; A.N. F^{17}4758.
[5] The *pli cacheté* in which Foucault describes his regulator is vague (ref. 7.66) but presumably the necessary control was achieved by wiring the electromagnet in parallel with the arc while the plates and acid were in series.
[6] See e.g. Faraday M. (1839) *Experimental Researches in Electricity.* Vol. I. Taylor, London. §783.
[7] E.g. *Experimental Researches* (ref. 6) §968–73, 984, 1017 and especially 1032.
[8] Foucault L. (1853) *C.R.A.S.*, **37**, 580–3.
[9] *Cosmos*, **4**, 242 (1854).
[10] *JdD*, 1854 February 16.
[11] A.d.l.R. (1849) *Bibliothèque universelle de Genève. Archives des sciences physiques et naturelles*, **10**, 222–5.
[12] A.d.l.R. (1853) *Bibliothèque universelle de Genève. Archives des sciences physiques et naturelles*, **24**, 268–70.
[13] A.d.l.R. (1854) *Bibliothèque universelle de Genève. Archives des sciences physiques et naturelles*, **25**, 65–7.
[14] Foucault L. (1854) *Cosmos*, **4**, 248–50, *Bibliothèque universelle de Genève. Archives des sciences physiques et naturelles*, **25**, 180–3.
[15] Meidinger H. (1854) *Bibliothèque universelle de Genève. Archives des sciences physiques et naturelles*, **24**, 268–70; the actual reaction, which passes by the producion of peroxodisulphuric

16. Matteucci (1854) *Cosmos*, **4**, 390.
17. Jamin J. (1854) *C.R.A.S.*, **38**, 390–2, 443–4; *Cosmos*, **4**, 274–5.
18. *Cosmos*, **4**, 289–91 (1854). See also e.g. Faraday M. (1855) *The London, Edinburgh and Dublin Philosophical Magazine and Journal of Science*, **10**, 98–107.
19. Despretz C. (1856) *C.R.A.S.*, **42**, 707–10.
20. A final configuration appears in Foucault L. (1854) *Bibliothèque universelle de Genève, Archives des sciences physiques et naturelles*, **26**, 126–7 and figure.
21. *JdD*, 1856 May 17.
22. de Parville in note 5.28.
23. *Proceedings of the Royal Society of London*, **7**, 571–4 (1855).
24. *Minutes of Council of the Royal Society from December 10th 1846 to November 30th 1858*. Taylor Francis, London (1858).
25. Ref. 2.40
26. Powell B. (1854) *Notices of the Proceedings at the meetings of the members of the Royal Institution*, **1**, 393–9.
27. Powell B. (1855) *Monthly Notices of the Royal Astronomical Society*, **15**, 182–5.
28. Moigno F. (1855) *Cosmos*, **6**, 145–6.
29. Boase F. (1892–1921, reprinted 1965) *Modern English Biography*. Cass, London.
30. Observatoire de Paris, Ms 1037. Letter dated 1854 October 15.
31. James F. A. J. L. (ed.) (1999) *The Correspondence of Michael Faraday*. Vol. 4. Institution of Electrical Engineers, London.
32. *The Times*, 1854 October 27.
33. *JdD*, 1856 December 20.
34. Boas Hall M. (1976) *History of Technology*, **1**, 143–8.
35. Dériberé M. & P. (1979) *Préhistoire et histoire de la lumière*, France-Empire, Paris.
36. Moigno F. (1855) *Cosmos*, **6**, 593–7.
37. Rumford's original photometer was actually of somewhat different form. See Walsh J. W. T. (1965) *Photometry*. Dover, New York.
38. Babinet J. (1853) *C. R. A. S.*, **37**, 774–5.
39. Foucault L. (1855) *Rapport sur le pouvoir éclairant des produits gazeux fournis par la distillation de la tourbe*. Wiesener, Paris. 28 pp.
40. Audouin P., Bérard P. (1862) *Annales de chimie et de physique*, **65** (3ème sér.) 423–88; Regnault V., Dumas J. *idem* 486–95 (note pages are misnumbered).
41. Ed. Becquerel in his *La lumière, ses causes et ses effets*, Vol. I, Firmin Didot, Paris (1867) p. 101 shows a Foucault photometer where the starch film is protected between two glass sheets and the movable vane is terminated at the screen end with a wider, orthogonal blade. A Foucault photometer preserved at the University of Catania incorporates a filter to limit intensity comparisons to red wavelengths.

Chapter 11 The Observatory physicist

1. Quoted by Anon. (actually Brewster D.) (1854) *North British Review*, **20**, 459–500.
2. *JdD*, 1853 October 7.
3. Bigourdan G. (n.d.) Le Bureau des Longitudes (4e partie). In *Annuaire pour l'an 1931 publié par Le Bureau des Longitudes*. Gauthier-Villars, Paris. Funds for the telescope had been voted in 1848.
4. Quoted in Le Verrier (1870) *Extrait du Discours prononcé au Sénat dans la Question de l'Observatoire, le 8 février 1870*.
5. See Institut de France (1911) *Centenaire de la naissance de U.-J.-J. Le Verrier*. Gauthier-Villars, Paris; *Moniteur Universel*, 1854 February 3.
6. A detailed account of Neptune's discovery can be found in Baum R., Sheehan W. (1997) *In Search of Planet Vulcan: The Ghost in Newton's Clockwork Universe*. Plenum Trade, New York.
7. *JdD*, 1846 November 12.
8. Le Verrier (1868) *C.R.A.S.*, **66**, 380.
9. Observatoire de Paris Ms 1037, letter dated 1854 October 15; I.d.F. Ms 3711 No. 40.
10. Family collection, letter dated 1854 September 22.
11. Le Verrier, (1854) *Mémoire sur l'état actuel de l'Observatoire impérial de Paris et projet d'organisation scientifique*. The report was reprinted in abbreviated form in *Annales de l'Observatoire impérial de Paris (Mémoires)*, **1** (1856).
12. Quoted in Bigourdan G. (n.d.) Le Bureau des Longitudes (5e partie). In *Annuaire pour l'an 1932 publié par Le Bureau des Longitudes*. Gauthier-Villars, Paris.
13. A.N. F^{17}20758, 1854 December 28.
14. A.N. F^{17}20758.
15. *Cosmos*, **6**, 255–6 (1855).
16. I.d.F. Ms 3711 Nos 44, 45.
17. *JdD*, 1853 May 5, 1854 June 8, September 14, though Foucault had already mentioned the emerging idea of the conservation of energy, such as on 1848 September 15.
18. *JdD*, 1854 September 14.
19. Brenni P. (1994) *Bulletin of the Scientific Instrument Society*, No. 41, 4–8.
20. *JdD*, 1855 October 18.
21. Foucault L. (1855) *C.R.A.S.*, **41**, 450–2.
22. *JdD*, 1855 October 18.
23. Comment following Foucault L. (1855) *Cosmos*, **7**, 309–10.
24. Instrument at the Musée des arts et métiers, Paris, Inv. 7403.

[25] E.g. Ganot A. (1856) *Traité élémentaire de physique expérimentale et appliquée et de météorologie.* (6ème ed.) Chez l'auteur-éditeur, Paris; *La Nature*, (1878) Sem. II, p. 221.
[26] Dimensions are quoted in *Notice sur les travaux de M. Léon Foucault.* Mallet-Bachelier, Paris. 34 pp. (1863)
[27] *Les Mondes, Science Pure*, **1**, 134–9 (1862). Moigno was printing a summary written by him in 1856 May, which he says 'we did not dare publish' at the time.
[28] Bertin M. (1867) *Rapport sur le progrès de la thermodynamique en France.* Imprimerie impériale, Paris.
[29] Family collection, letter dated 1855 September 29.
[30] Wilson (1990) in note 7.53.
[31] Joule J. P. (1843) *The London, Edinburgh and Dublin Philosophical Magazine and Journal of Science*, **23** (3rd ser.) 263–76, 347–55, 435–43.
[32] See e.g. Murray R. H. (1925) *Science and Scientists in the Nineteenth Century.* Ch. IV. Sheldon Press, London.
[33] For the history of the induction coil see Shiers G. (1971) *Scientific American*, **244**, May issue; Hackmann W. D. (1989) In *Studies in the History of Scientific Instruments.* Blondel C., Parot F., Turner A., Williams M. (eds.) pp. 235–50. Rogers Turner, London.
[34] Foucault L. (1856) *C.R.A.S.*, **42**, 215–7.
[35] Foucault L. (1856) *C.R.A.S.*, **43**, 44–7.
[36] Foucault L. (1856) *L'Institut*, **24** (1ère sect.) 151.
[37] Foucault L. (1856) *L'Institut*, **24** (1ère sect.) 221.
[38] Illustrated in the *Recueil*, p. 355.
[39] Foucault L. (1857) *L'Institut*, **25** (1ère sect.) 265–6.
[40] Foucault L. (1857) *Société Philomathique de Paris. Extraits des procès-verbaux des séances pendant l'année 1857*, 133–4.
[41] Figuier L. (1858) Rapport sur l'Exposition universelle de 1855. In *L'année scientifique et industrielle*, 2ème année, pp. 463–74. Hachette, Paris.
[42] *Cosmos*, **5**, 1 (1854).
[43] Brisse L., Jubinal A., Gage P. (1856) *Album de l'Exposition Universelle*, Vol. 1. Abeille impériale, Paris.
[44] Foucault (1856) Appareils d'éclairage. In *Exposition Universelle de 1855. Rapports du Jury Mixte International publiés sous la direction du S.A.I. le Prince Napoléon, Président de la Commission impériale.* Imprimerie impériale, Paris. pp. 452–5.
[45] Gordon (1870) *The Home Life of Sir David Brewster.* Edmonston Douglas, Edinburgh.
[46] Foucault, Babinet (1856) Phares lenticulaires. In *Exposition Universelle de 1855. Rapports du Jury Mixte International publiés sous la direction de S.A.I. le Prince Napoléon, Président de la Commission impériale.* Imprimerie impériale, Paris. pp. 455–6.
[47] In 1844, while still director at Choisy-le-Roi, Bontemps had offered 55-cm flint and crown discs to the Bureau des Longitudes. (*L'Illustration*, 1844 December p. 242.)
[48] Ref. 3.2.
[49] Observatoire de Paris. Archives 1850–1942. Carton 6. Dossier: objectifs. Report dated 1856 November 21.
[50] The flint was slightly oval so Foucault also checked the refractive indices with slivers cut from each end of the long diameter.
[51] Le Verrier (1868) *C.R.A.S.*, **66**, 380–9; A.N. F^{17} 3730.

Chapter 12 Perfecting the telescope

[1] The spellings *Secretan* and *Secrétan* were both used; the latter is adopted here.
[2] Ref. 7.10.
[3] Foucault L. (1858) *Cosmos*, **13**, 162–8.
[4] Newton I. (1704) *Optice*. Lib. I, Prop. VII, VIII.
[5] Ref. 3.16.
[6] Liebig J. (1835) *Annalen der Pharmacie*, **14**, 131–67.
[7] British patent No. 9968 (1843). A later specification by Drayton, who had become a 'practical chemist' in Regent Street, specified heating to 160° Fahrenheit (No. 12358, 1848).
[8] *L'Institut*, **25** (1ère sect.) 38; *Société Philomathique de Paris. Extraits des procès-verbaux des séances pendant l'année 1857*, 15; Foucault L. (1857) *C.R.A.S.*, **44**, 339–42.
[9] Foucault's accounts are inconsistent, presumably because he stopped the mirror down by varying amounts to improve image quality. Cf. Foucault L. (1858) *L'Institut*, **26** (1ère sect.) 151.
[10] Bibliothèque de la Sorbonne, Ms 2044, minute for 1858 May 1.
[11] *JdD*, 1862 May 11.
[12] *JdD*, 1857 February 26.
[13] Steinheil (1857) *C.R.A.S.*, **45**, 968–9.
[14] *Cosmos*, **11**, 652 (1857).
[15] Steinheil (1858) *Monthly Notices of the Royal Astronomical Society*, **19**, 56–60.
[16] He did not understand how image defects could indicate errors of his mirrors' cross section (Steinheil (1858) *Astronomische Nachrichten*, No. 1138, 145–50; *Cosmos*, **14**, 32–3 (1859)), but nevertheless succeeded in making at least a 12-inch mirror (*Monthly Notices of the Royal Astronomical Society*, **20**, 26 (1859)). His 1860 price list made no mention of silvered-glass mirrors (Steinhel C.A. (1860) *Astronomische Nachrichten*, No. 1235, 171–6).
[17] Foucault L. (1857) *L'Institut*, **25** (1ère sect.) 265; *Société Philomathique de Paris. Extraits des procès-verbaux des séances pendant l'année 1857*, 104; *C.R.A.S.*, **45**, 238–41; *Cosmos*, **11**, 217–20. There can be no doubt that Foucault

18. Foucault L. (1858) *Report of the 27th meeting of the British Association for the Advancement of Science, Dublin August–September 1857. Notices and Abstracts of Miscellaneous Communications to the Sections.* John Murray, London. pp. 5–6, 6–8.
19. Royal Society manuscripts HS.7.377, HS.12.354, 355.
20. Observatoire de Paris, Ms 1037, envelope franked 29 SEPT 57.
21. His entry in Lord Rosse's visitors' book is difficult to decipher from microfilm. Unfamiliar with British customs, it is possible he signed 'L.L.D.'
22. *L'Institut*, **25** (1ère sect.) 343 (1857).
23. The mirrors are now in the Science Museum, London, Inv. 1943-55. One is a concave silvered mirror that is too fast for astronomical use ($\sim$f/1). The other is a sector cut from a disc of thin platinized glass, indicating Foucault investigated platinization as well as silvering; but platinization involves heating to $\sim$320 °C. Foucault's platinized glass has visibly softened and slumped, which is quite unacceptable for optical use. The quotation comes from a letter dated 1910 June 17 in the associated file in the Museum's Documentation Centre.
24. *Cosmos*, **11**, 368 (1857).
25. A.N. F^{17}3719A, letter dated 1857 October 16.
26. A.N. F^{17}3719A, letter dated 1858 January 9. He was also unwell. (I.d.F. Ms 3716.)
27. A.N. F^{17}3719A, letter dated 1857 November 10.
28. A.N. F^{17}20758, letter dated 1858 January 4.
29. I.d.F. Ms 3711 No. 41, letter dated 1858 January 4.
30. A.N. F^{17}20758, letter dated 1858 January 6.
31. A.N. F^{17}20758, letter dated 1859 February 2.
32. A.N. F^{17}20758, letter dated 1858 January 9.
33. A.N. F^{17}20758, letter dated 1858 January 13.
34. A.N. F^{17}3719A, letter dated 1858 February 23. I.d.F. Ms 3716 suggests the *décret* was signed on March 3.
35. *L'Institut*, **26** (1ère sect.) 151, 161, 186, 221 (1858); *Société Philomathique de Paris. Extraits des procès-verbaux des séances pendant l'année 1858*, 47–8, 41 (incorrectly dated), 49–50, 51–2.
36. Foucault L. (1859) *Annales de l'Observatoire impérial de Paris (Mémoires)*, **5**, 197–237 and plate.
37. *JdD*, 1847 June 9.
38. *JdD*, 1853 August 11.
39. *Cosmos*, **12**, 518–19 (1858).
40. Ref. 3. Remarks about the opticians' scepticism are repeated in ref. 36.
41. Flammarion C. (1911) *Mémoires biographiques et philosophiques d'un astronome.* Flammarion, Paris. p. 139.
42. A.N. F^{17}20758, letter dated 1858 July 3.
43. A.N. F^{17}20758, letter dated 1858 July 1.
44. A.N. F^{17}20758, letter dated 1858 July 23.
45. Struve was a member of another scientific dynasty, like the Cassinis and the Becquerels. Five astronomer Struves spanned four generations.
46. Apparently Foucault was not aware that the orbit in γ And is rapid and that the separation had increased to 0.6 arcsecond by 1858. Cf. ref. 36.
47. *Cosmos*, **13**, 328 (1858).
48. *Cosmos*, **15**, 138–40 (1859).
49. *Cosmos*, **13**, 411 (1858).
50. Faye (1858) *C.R.A.S.*, **47**, 619–21. Faye's comment earned a rebuke from Le Verrier, who was keen to point out that Faye had not used the Paris Observatory's new instruments. Le Verrier (1858) *C.R.A.S.*, **47**, 673–4.
51. Olson R. J. M., Pasachoff J. M. (1998) *Fire in the Sky.* Cambridge University Press, Cambridge.
52. *Monthly Notices of the Royal Astronomical Society*, **20**, 24 (1859). For photographs of Donati's Comet, see Pasachoff J. M., Olson R. J. M., Hazen M. L., (1996) *Journal for the History of Astronomy*, **27**, 129–45.
53. Draper H. (1864, reprinted 1904) *Smithsonian Contributions to Knowledge*, **14**, Art. 4.
54. The report published in Foucault L. (1858) *C.R.A.S.*, **47**, 205–7 is abbreviated. For the full memoir, see ref. 3.
55. In the autumn of 1865 Secrétan began selling 10-cm Foucault telescopes enclosed in a metal tube and mounted on a tripod. (*Les Mondes*, **9**, 272–3 (1865).)
56. *JdD*, 1858 June 23.
57. Instruction sheet accompanying the telescope in the Smithsonian Institution (Plate XV).
58. Foucault L. (1859) *L'Institut*, **27** (1ère sect.) 62; *Société Philomathique de Paris. Extraits des procès-verbaux des séances pendant l'année 1859*, 16; a fuller text is given in the *Recueil*, pp. 285–6.
59. *Intellectual Observer*, **1**, 380–1 (1862). The engravings date from 1858 (*Magasin Pittoresque*, **26**, 312).
60. A.N. F^{17}20758, letter dated 1859 January 31; Babinet, *JdD*, 1859 February 9.
61. Foucault L. (1859) *C.R.A.S.*, **49**, 85–7; and refs. 36, 48.
62. The reasoning does not apply to Cassegrain designs, where the central hole is normally cut following the optical polishing.
63. Sautter L. (1915) *Louis Sautter (1825–1912). D'après son journal intime et sa correspondance.* Fischbacher, Paris.
64. Lissajous (1869) *Revue des Cours scientifiques de la France et de l'Étranger*, **6**, 484–9.

[65] *Cosmos*, **15**, 138 (1859).
[66] *Monthly Notices of the Royal Astronomical Society*, **20**, 147 (1860).
[67] *Cosmos*, **15**, 397–401 (1859).
[68] A manuscript sold by Charavay (3 rue de Furstenberg, Paris) dated 1860 April outlined the planned construction procedures.
[69] From Algiers Bulard reported that he had the wooden mount of the 33-cm telescope, but a different mirror. It is not clear whether the 50-cm telescope is this instrument or an additional one. Bulard C. (1861) *C.R.A.S.*, **53**, 509–12; (1862) **55**, 879–81. The wooden mount was later replaced with metal: Lagrula J-L. (1932) *Journal des Observateurs*, **15**, 167–70; Baillaud B. (1880) *Annales de l'Observatoire de Toulouse*, **1**, v–vii.
[70] A.N. F17 3726 details the Observatory expedition.
[71] Georgelin Y., Arzano S. (1999) *L'Astronomie*, **113**, 12–17 (to Siam). The 40-cm was also used for an eclipse in Algeria in 1905.
[72] *Moniteur Universel*, 1860 July 29 p. 906, August 6 p. 942; Yvon Villarceau, *Mémoires et compte rendu des travaux de la Société des ingénieurs civils. Année 1860*, 344–7.
[73] Hind J. R. (1859) *Total Solar Eclipse. 1860, July 18. Revised Path of the Shadow, &c.* H.M.S.O., London.
[74] *JdD*, 1860 September 1.
[75] Family collection, letter dated 'Angers le 7 mars 1852'. Perhaps the amphitype was taken by or with the Abbé Moigno, who the following year was making albumen positives on glass by contact printing from glass negatives, and who will almost certainly have experimented with amphitypes and ambrotypes (Moigno F. (1853) *Cosmos*, **2**, 265–6).
[76] *Magasin Pittoresque*, **31**, 31–2 (1863).
[77] See Yvon Villarceau (1868) *C.R.A.S.*, **67**, 270–8; Rothermel H. (1993) *British Journal for the History of Science*, **26**, 137–69; Hingley P. D. (2001) *Astronomy & Geophysics*, **42**, 1.18–1.22. E. Mouchez does not even mention Foucault's work in his *La photographie astronomique à l'Observatoire de Paris*, in *Annuaire pour l'an 1887 publié par le Bureau des Longitudes*. Gauthier-Villars, Paris (n.d.).
[78] Foucault L. (1862) *C.R.A.S.*, **54**, 859–61.
[79] Fuentès P. (1997) *L'Astronomie*, **111**, 270–2.
[80] *Le Temps*, 1862 May 7.
[81] *C.R.A.S.*, **54**, 626–8, 888–9, 1012 (1862); *Monthly Notices of the Royal Astronomical Society*, **22**, 277 (1862); *Cosmos*, **20**, 391–3 (1862).
[82] A.N. F17 20758.
[83] Fizeau (1868) *C.R.A.S.*, **66**, 932–4; Stéphan (1874) *C.R.A.S.*, **78**, 1008–12.
[84] Michelson A. A., Pease F. G., (1921) *Astrophysical Journal*, **53**, 249–59.
[85] Family collection, letter dated 1857 February 27, indicating that Foucault was on the path to *retouches locales* very quickly.

Chapter 13 The size of the solar system

[1] *Cosmos*, **19**, Report of *Académie des Sciences* meeting on 1861 July 22.
[2] The Astronomer Royal (1857) *Monthly Notices of the Royal Astronomical Society*, **17**, 208–21.
[3] Reich F. (1837) *C.R.A.S.*, **5**, 687–700; Baily F. (1843) *Memoirs of the Astronomical Society of London*, **14**, 1–120; Cavendish H. (1798) *Philosophical Transactions of the Royal Society of London for the year MDCCXCVIII*, 469–526.
[4] Arago F. (1857) *Astronomie populaire*. Vol. 4. Gide, Paris. p. 418.
[5] Shuster-Fournier C. (1998) *L'Orgue, Cahiers et Mémoires*, Nos. 57–8.
[6] *Cosmos*, **18**, 536 (1861).
[7] I.d.F. Ms 3711 Nos. 5, 7.
[8] French patent No. 53432, 1862 March 18.
[9] Illustrated in the *Recueil* and described in Foucault L. (1862) *C.R.A.S.*, **55**, 792–6.
[10] Le Verrier (1867) *C.R.A.S.*, **65**, 878–84.
[11] Foucault L. (1862) *C.R.A.S.*, **55**, 501–3.
[12] Babinet (1862) *C.R.A.S.*, **55**, 537–40.
[13] Wolf C. (1885) *C.R.A.S.*, **100**, 301–9.
[14] Cornu A. (1900) Sur la vitesse de la lumière. In *Rapports présentés au Congrès international de physique réuni à Paris en 1900 sous les auspices de la Société française de physique*. Vol. 2. Gauthier-Villars, Paris.
[15] A.A.d.S. Fizeau box No. 3.
[16] *Les Mondes*, **15**, 511 (1867). Le Verrier became emotional at the Academy meeting (*Les Mondes*, **15**, 509–10 (1867)) and may have said things he did not quite mean. His later printed text highlights the dangers of experimenter bias (ref. 10).
[17] Draft in the family's possession.
[18] Cornu A. (1874) *Journal de l'École polytechnique*, **27**, 133–80.
[19] Cornu A. (1876) *Annales de l'Observatoire de Paris (Mémoires)*, **13**, A1–A315.
[20] Taylor I. D. (1970) *School Science Review*, **52**, 28–38; see also Domkowski A. J., Richardson C. B., Rowbotham N. (1972) *American Journal of Physics*, **40**, 910–12.

Chapter 14 Recognition

[1] At one time they met on Thursday afternoons (*Cosmos*, **7**, 598 (1855)) but later it was in the morning (ref. 2.38).
[2] *C.R.A.S.*, **144**, 668 (1907).
[3] *Bulletin de la Société française de photographie*, **1**, 62 (1855).
[4] *Bulletin de la Société française de photographie*, **12**, 80 (1865); *Cosmos*, 3 (2ème sér.) 398–400 (1866).
[5] *Bulletin de la Société française de photographie*, **1**, 62, 133 (1855); **2**, 117, 170–1 (1856); **3**, 62, 337 (1857); **4**, 67–70 (1858); **5**, 153, 156 (1859); **6**, 63, 223 (1860); **7**, 13-16, 150–1 (1861); **8**, 36, 119–22, 286 (1862); **9**, 35, 101

(1863); **10**, 38 (1864); **11**, 32, 33, 205 (1865); **12**, 57 (1866); **13**, (1867).

[6] *Bulletin de la Société française de photographie*, **14**, 57 (1868).

[7] In truth, so was every other retiring Committee member.

[8] Quoted in Aubenas S. (1994) *D'encre et de charbon*. Bibliothèque nationale de France, Paris.

[9] The Committee chosen for the first prize was reappointed for the second. *Bulletin de la Société française de photographie*, **6**, 113 (1860) **7**, 186 (1861).

[10] *Bulletin de la Société française de photographie*, **5**, 121 (1859).

[11] *Bulletin de la Société française de photographie*, **13**, 89–112.

[12] A letter from E. Chevreul dated 1865 April 12 informs Foucault that the Emperor had confirmed his appointment to the *Comité consultatif* that morning. (Family collection.)

[13] *JdD*, 1857 February 5.

[14] *Cosmos*, **2**, 131 (1853).

[15] *Cosmos*, **14**, 293–4 (1859).

[16] *Les Mondes*, **1**, 337–9 (1863); **6**, 138–52 (1864).

[17] When Foucault had first seen Lenoir's motor in 1860, he had predicted that with the rapidly developing network of gas pipes in Paris, it would easily provide medium amounts of power to small workshops, which it soon did. (*JdD*, 1860 June 7.)

[18] Ref. 11.12.

[19] *Discours de M. Yvon Villarceau*. Institut impérial de France, Paris (1868). The minutes of the Bureau's weekly Wednesday meetings which would elucidate Foucault's contribution are unfortunately lost.

[20] Chantrel J. (1868) Causerie scientifique. *Revue du Monde Catholique*, **17**, 849.

[21] *Revue scientifique et industrielle*, **41**, 135 (1851).

[22] *Review scientifique et industrielle*, **41**, 141 (1851).

[23] *C.R.A.S.*, **46**, 408; and A.A.d.S. Foucault box.

[24] *Cosmos*, **12**, 249 (1858).

[25] *Cosmos*, **12**, 330–1 (1858).

[26] *C.R.A.S.*, **46**, 564 (1858).

[27] Moigno F. (1859) *Cosmos*, **15**, 315–16.

[28] *C.R.A.S.*, **52**, 668 (1861).

[29] *C.R.A.S.*, **52**, 715 (1861).

[30] *Cosmos*, **18**, 435–6 (1861).

[31] *Notice sur les travaux de M. Léon Foucault*. Mallet-Bachelier, Paris (1863), (1865).

[32] Draft of a letter belonging to the Foucault family.

[33] A.A.d.S. *pochette* for 1863 May 11. Foucault's name has of course been deleted from the *Comptes rendus* report (*C.R.A.S.*, **56**, 919 (1863)).

[34] *Cosmos*, **22**, 626 (1863). This transcription seems more trustworthy than the one proffered by Moigno (note 35).

[35] *Les Mondes*, **1**, 390–1 (1863).

[36] *C.R.A.S.*, **56**, 945 (1863).

[37] Family collection, letter dated '27 Mai'. The letter is in obvious reply to Foucault's draft, but the initials with which it is signed are only perhaps M.A. de J. (M.A.V. Donné (née de Joantho)), but in a correspondence full of funny names, the initials may have some other meaning.

[38] *Les Mondes*, **6**, 523–4 (1864); *C.R.A.S.*, **59**, 725, 1060 (1864).

[39] *Notice des travaux scientifiques de M. Favé*. Gauthier-Villars, Paris (1864).

[40] A.A.d.S., Foucault box, letter dated 1864 April 3.

[41] *Les Mondes*, **5**, 276–81.

[42] A paraphrase of John 3:18.

[43] Ref. 6.69.

[44] Family collection, letter dated 1864 May 19.

[45] *Réfutation par M. E. Chevreul des allégations contre l'administration du Muséum d'histoire naturelle proférées à la tribune du corps législatif dans la séance du 19 juin 1862; suivie d'une lettre du Colonel Favé, aide de camp de l'Empereur*. Mallet-Bachelier, Paris (1862).

[46] *Les Mondes*, **6**, 766 (1864).

[47] *Cosmos*, **1**, (2ème ser.) 55–6 (1865).

[48] *Les Mondes*, **7**, 84 (1865).

[49] *Les Mondes*, **7**, 123–4 (1865).

[50] *Le Moniteur scientifique*, **7**, 140–1 (1865).

[51] *Cosmos*, **1** (2ème sér.) 111 (1865); Bryan G.H. (1900) *Nature*, **61**, 614–16.

[52] *Les Mondes*, **7**, 225 (1865).

[53] *Les Mondes*, **7**, 280 (1865).

[54] *Les Mondes*, **9**, 167 (1865), **11**, 132, **12**, 306 (1866); *C.R.A.S.*, **61**, 516 (1865); *Les Mondes*, **2** (2ème sér.), 177–8 (1866).

[55] *C.R.A.S.*, **60**, 518, 545 (1865); **62**, 478, 777 (1866); **64**, 450–4 (1867).

[56] Ref. 2.11.

Chapter 15 Control: the quest for fortune

[1] *JdD*, 1855 April 13. In 1858–60 Foucault was appointed by the court in Tours as one of three expert assessors in a case brought by Girard for patent infringement. The other experts found there had been no infringement, but Foucault disagreed (Archives Départementales d'Indre-et-Loire).

[2] French patents: heliostat, 53377 (1862 March 17, *Certificat d'addition* 1862 October 18); regulators, 55346 (1862 August 23 with 12 *Certificats d'addition* to 1866 January 9); electric light regulator, 60460 (1863 October 15, *Certificat d'addition* 1867 November 4); regulators, 69585 (1865 December 6 with 3 *Certificats d'addition* to 1867 October 15). English patents: 1862 December 30, No. 3479; 1864 June 25, No. 1597; 1864 December 21, No. 3169; 1866 July 4, No. 1777. Belgian import licences: 13686 (1863 January 5); 15502 (1863 December 24); 16532 (1864 April 28); 19246 (1866 March 5).

[3] Ref. 2.17.
[4] Ref. 2.23, which notes commercial arrangements between Duboscq and Foucault beginning in 1857.
[5] *Bulletin de la Société française de photographie*, **7**, 58–9 (1861).
[6] Mills A. A., *Journal of the British Astronomical Association*, **95**, 89–95 (1985); *Annals of Science*, **43**, 369–406 (1986).
[7] Duboscq (1862) *C.R.A.S.*, **54**, 618–20.
[8] Duboscq (1862) *C.R.A.S.*, **55**, 644–5.
[9] *Cosmos*, **21**, 470–2 (1862).
[10] Clavel (1852) *C.R.A.S.*, **35**, 604–5.
[11] Duboscq J. (1862) *C.R.A.S.*, **54**, 741.
[12] He does not appear to have been any close relation of Hector Berlioz.
[13] *Cosmos*, **18**, 646–7 (1861).
[14] The Hungarian-American escape artist Eric Weiss (1874–1926) adopted the stage name Houdini in homage to Robert-Houdin.
[15] Robert-Houdin (1855) *C.R.A.S.*, **40**, 1141–3.
[16] The operation of the gearing is described by Gariel C. M. (1886) *Traité pratique d'électricité*, Vol. II. Doin, Paris, pp. 94–8 which is usefully read in conjunction with the (sometimes confused) description given in *Les Mondes*, **2** (2ème sér.) (=**11**) 620–4 (1866).
[17] Illustrated in Foucault (1867) Appareil régulateur de la lumière électrique. In *Dictionnaire générale des sciences théoriques et appliquées*. Privat-Deschanel, Focillon Ad. (eds). Delagrave, Paris. pp. 1581–3. The illustration is reproduced by Mackechnie Jarvis (ref. 7.28), though misdated.
[18] Foucault L. (1863) *Brevet d'invention No. 60460*. The original form illustrated in the patent and by Saint-Edme E. (1864) *Cosmos*, **24**, 121–6 differs slightly.
[19] Observatoire de Paris Ms 1037. Letter dated 1857 October 15.
[20] Reynaud L., *Le Moniteur Universel*, 1866 May 21 609–10; Figuier L. (1867) *L'année scientifique et industrielle, Onzième année [1866]*. Hachette, Paris, pp. 48–56; Quinette de Rochemont (1870) *Annales des ponts et chaussées. Mémoires*, **19** (4ème sér.) 309–46.
[21] *Les Mondes*, **18**, 43–4 (1868); **19**, 327–8, 549–50; **20**, 89–90, 105 (1869).
[22] Battesti M. (1997) *La marine de Napoléon III*. Service Historique de la Marine, Paris.
[23] *Les Mondes*, **13**, 405–6 (1867).
[24] *Les Mondes*, **13**, 493 (1867).
[25] *Les Mondes*, **16**, 700–2 (1868).
[26] *JdD*, 1859 July 8.
[27] This may seem to contradict the earlier claim that the period of a pendulum is *more* when the amplitude is greater, but in one cycle a conical pendulum also veers by an amount Δ (Fig. 7.6). When only a 360° motion of the bob is considered, there is no contradiction.
[28] *Cosmos*, **21**, 103–4 (1862).
[29] Foucault L. (1862) *C.R.A.S.*, **55**, 135–6.
[30] *Les Mondes*, **6**, 487–92 (1864).
[31] *Les Mondes*, **1**, 253 (1863).
[32] A.N. F^{17}3730, letter dated 1863 August 1.
[33] Sainte-Claire Deville H. (1868) *C.R.A.S.*, **66**, 389–96.
[34] E.g. only four on governors: *C.R.A.S.*, **57**, 738–40 (1863); **61**, 278–9, 430–1, 515–16 (1865).
[35] Sautter L. (1864) *Mémoires et compte rendu des travaux de la Société des ingénieurs civils. Année 1864*, 65–8.
[36] Sautter L. (1880) *Notice sur les phares, fanaux, bouées et signaux sonores*. Chaix, Paris.
[37] Ref. 42; the governor and clock face are illustrated in the *Certificat d'addition* dated 1863 December 8 to patent 55346.
[38] Service Historique de la Marine, Vincennes. 5 DD^1 73:2461; 7 DD^1 108, 127, 168, 255, 294.
[39] Normand (1867) *Mémoires et compte rendu des travaux de la Société des ingénieurs civils. Année 1867*, 639–42.
[40] A.A.d.S. Foucault box, letter dated 1865 August 6.
[41] Smaller, as judged by the lesser price of 430 francs.
[42] Foucault L. (1865) *Notice sur les travaux de M. Léon Foucault*. Mallet-Bachelier, Paris. 37 pp.
[43] *Mémoires et compte rendu des travaux de la Société des ingénieurs civils. Année 1864*, 73 (1864).
[44] Ref. 6.69.
[45] In the patents, and e.g. *Cosmos*, **23**, 545–8 (1863), retracted and corrected by **24**, 37–9 (1864).
[46] Mayr O. (1971) *Notes and Records of the Royal Society of London*, **26**, 205–28.
[47] *Les Mondes*, **2**, 207–11 (1866).
[48] Vogel H. C. quoted in ref. 49.
[49] Wilterdink J.-H. (1918) *Annalen van de Sterrwacht te Leiden*, **11**, A1–A89.

Chapter 16 Unfinished projects

[1] *Recueil*, pp. 296–300.
[2] *Cosmos*, **18**, 536 (1861).
[3] A.N. F^{17}3730, letter dated 1862 February 27.
[4] A.N. F^{17}3730, letter dated 1864 January 6.
[5] A.N. F^{17}3730, report dated 1864 January 20.
[6] Family collection, letter dated 1864 May 19.
[7] *Les Mondes*, **5**, 133–5 (1864).
[8] Sainte-Claire Deville H. (1868) *C.R.A.S.*, **66**, 338–42.
[9] Martin A. (1852) *C.R.A.S.*, **35**, 29–30. The following year he extended the process to metal plates (**36**, 703–6) with the idea that the plates could then be engraved by hand for printing, but in fact this development gave rise to tintype photography which was much favoured by beach and itinerant photographers until well into the twentieth century.

[10] Martin A. (1867) *Thèses présentées à la Faculté des sciences de Paris.*

[11] The events of this section are disentangled from the accounts given by Sainte-Claire Deville H. (1868) *C.R.A.S.*, **66**, 338–42, 389–92; Le Verrier (1868) *C.R.A.S.*, **66**, 380–9, 393–6.

[12] de Parville in ref. 5.28.

[13] *Les Mondes*, **14**, 242–7 (1867); **15**, 164 (1867); D'Abbadie A. (1868) *C.R.A.S.*, **66**, 589–90.

[14] André C., Angot A. (1881) *L'astronomie pratique. 4e partie. Observatoires de l'Amérique du Sud.* Gauthier-Villars, Paris.

[15] Laussedat A., *Revue des Cours scientifiques de la France et de l'étranger*, **5**, 259–64 (1868); **14**, 155–61 (1874).

[16] A.N., ET/XLIX/1504.

[17] Foucault L. (1866) *C.R.A.S.*, **63**, 413–15. Unlike a filter, the thin silver does not degrade the image quality.

[18] Ref. 2.38.

[19] Letter to Turgenev dated 1867 May 7.

[20] *Cosmos* **5** (2ème sér.) 664–71 (1867); *Les Mondes*, **14**, 242–7, 344–6 (1867); Rayet G. In *Annuaire scientifique publié par P-P. Dehérain. Septième année 1868.* Masson, Paris, pp. 232–46.

[21] I.d.F. Ms 3711 No. 7, draft by Foucault.

[22] *Le Temps*, 1867 July 10.

[23] *Les Mondes*, **14**, 717–26 (1867).

[24] *Exposition universelle de 1867 à Paris. Catalogue Général publié par la Commission Impériale. 2ème livraison. Matériel et application des arts libéraux (Groupe II – Classes 6 à 13).* Dentu, Paris (1867); Wolf C. (1869) *C.R.A.S.*, **69**, 1222–6.

[25] Porter C. T. (1908) *Engineering Reminiscences.* Ch. 13. Wiley, New York; Pickering T. B. (1883–84) *Transactions of the American Society of Mechanical Engineers*, **5**, 113–21.

[26] *Mémoires et compte rendu des travaux de la Société des ingénieurs civils*, Année 1867. 332–4. The governor and engine are illustrated in Armengaud Aîné & E., fils (1869) *Les progrès de l'industrie à l'Exposition universelle*, Tome 2, Pl. 1-2. Morel, Paris.

[27] Worms de Romilly (1872) *Annales des Mines* (7ème sér.) **1**, 36–64.

[28] Ref. 15.46.

[29] *Les Mondes*, **14**, 598.

[30] Family collection, letter dated 1867 August 11.

[31] Quoted by Moigno in *Les Mondes*, **16**, 343 (1868).

[32] Pasteur L. (1868) quoted in *Les Mondes*, **16**, 221–3.

[33] A.N. ET/XLIX/1502.

[34] A.d.P. DQ^8 1531 *Table des Successions et Absences.*

[35] *Les Mondes*, **16**, 343–4 (1868).

[36] *Les Mondes*, **16**, 337 (1868).

[37] A.N. ET/XLIX/1504 and *acte de décès*.

[38] E.g. *JdD* and *Le Siècle*, 1868 February 14.

[39] A.A.d.S. 1B17 *Feuilles de présence 1867 à 1872.*

[40] Flammarion C., *Le Siècle*, 1868 February 17.

[41] Grandeau L., *Le Temps*, 1868 March 13.

[42] Sainte-Claire Deville H. (1868) *C.R.A.S.*, **66**, 338–42.

[43] de Parville H., *Le Constitutionnel*, 1868 February 16.

[44] Le Verrier (1868) *C.R.A.S.*, **66**, 380–9.

[45] *Cosmos*, **2** (3ème sér.) 20–5 (1868).

[46] Ref. 45. Le Verrier in fact discussed the siderostat in the printed version of his remarks *C.R.A.S.*, **66**, 393–6 (1868).

[47] *Le Temps*, 1868 March 13.

[48] de Parville H., *Le Constitutionnel*, 1868 March 4.

[49] Reported by Flammarion in *Le Constitutionnel*, 1868 March 14.

[50] *Le Constitutionnel*, 1868 March 10.

[51] Le Verrier (1868) *C.R.A.S.*, **66**, 442.

[52] Duruy V. (1868) *C.R.A.S.*, **66**, 441–2.

[53] A.N. F^{17} 3242.

[54] de Parville H., *Le Constitutionnel*, 1868 March 10; *Cosmos*, **2** (3ème sér.) 22 (1868).

[55] *Bulletin administratif*, No. 179, 70–4 (1868 August).

[56] *Le Petit Parisien*, 1902 October 21.

[57] Sainte-Claire Deville (1869) *C.R.A.S.*, **69**, 1221–2; Wolf C. (1869) *C.R.A.S.*, **69**, 1222–6; Wolf C. (1872) *Annales scientifiques de l'École normale supérieure*, **1** (2ème sér.) 51–84.

[58] *Rapport annuel sur l'état de l'Observatoire de Paris pour l'an 1880*, p. 11.

[59] Ashbrook J. (1958) *Sky & Telescope*, **17**, 509.

[60] The mirror is on display in the cinema at the Observatoire de Haute-Provence. A wide bevel hides the damage.

[61] Tobin W. (1987) *Vistas in Astronomy*, **30**, 153–84.

[62] Family collection, codicil by Aimée Foucault dated 1880 November 13; ref. 56.

[63] Family collection, will dated 1874 February 19.

[64] Family collection, undated draft.

[65] This book is now in the Paris Observatory library.

[66] Riche A. (1891) *Frédéric Le Play*. Poussielgue, Paris; Brun F. (1876) *Conférence Léon Foucault: Étude sur la chanson de Roland.* Plon, Paris.

[67] A.N. F^{21} 219.

[68] Veyrat G. (1988) *Les sentinelles de l'Hotel de Ville.* Saurat Paris.

[69] Aird C. (1981) *His Burial Too.* Bantam Books, New York; Eco U. (1989) *Foucault's Pendulum.* Harcourt, Brace, Jovanovich, New York.

[70] The Institute ordered a further marble bust of Foucault from a sculptor called Mengue and placed it in the garden of its country house at Prunay, west of Paris.

Chapter 17 Commentary

[1] The childrens' ages are concordant, cf. Fig. 2.1.

[2] Dubois L., *Le Courrier Français*, 1868 February 27.

[3] Ring F., *La Situation*, 1867 November 24.
[4] de Parville H. (1869) *Causeries scientifiques 8ème année 1868*. Rotschild, Paris, pp. 152–70.
[5] A.N. $F^{17}20758$.
[6] Lissajous (1869) *Revue des Cours Scientifiques de la France et de l'Étranger*, **6**, 484–9.
[7] Observatoire de Paris, Ms 1037, letter dated 1861 September 20.
[8] Grandeau H., *Le Temps*, 1868 March 2.
[9] Family collection, draft of a letter.
[10] Ref. 2.38.
[11] Observatoire de Paris, Ms 1037, letter dated 1854 October 15.
[12] In *Mémoire sur l'État actuel de l'Observatoire impérial présenté par les astronomes à Son Exec. le Ministre de l'Instruction publique*. Imprimerie Lahure, Paris (1870).
[13] A.N. $F^{17}3719A$.
[14] Klosterman L.J. (1985) *Annals of Science*, **42**, 1–40.
[15] *Les Mondes*, **7**, 225 (1865); ref. 2.38.
[16] Morin (1868) *Discours de M. Morin prononcé aux funérailles de M. Foucault*. Institut impérial de France, Paris.
[17] Bertrand, in the *Avertissement* to the *Recueil*.
[18] Grandeau L., *Le Temps*, 1868 March 2.

Colour plates

[1] Gautier H. (1867) *Les Curiosités de l'Exposition universelle de 1867*. Delagrave, Paris.
[2] In the 1970s A. Baranne of the Marseilles Observatory measured the shape of the front on the 80-cm mirror and found it to be hyperboloidal, as expected when *retouches locales* have corrected the spherical aberration of the eyepiece and relay lenses.

Appendix B Extracts from the *Journal des Débats*

[1] For the decimetre-sized vessels and centimetre-sized tube that Foucault's language evokes, the flow would be limited by the water's inertia and would indeed depend on the square root of the pressure difference. This is Torricelli's Law, but unfortunately breaks the strict analogy that Foucault is attempting to establish with electrical currents. Had he connected his reservoirs with a capillary, however, viscosity would have limited the flow, which would then have depended linearly on the pressure difference, and the analogy would have been complete. Poiseuille's work on the flow of water through thin tubes had been presented to the Academy in the previous decade.
[2] *C.R.A.S.*, **35**, 873 (1852).
[3] Schmadel L.D. (1992) *Dictionary of Minor Planet Names*, Springer Verlag, Berlin.
[4] Monins (1851) *C.R.A.S.*, **33**, 60–1.
[5] Duméril (1851) *C.R.A.S.*, **33**, 105–15.
[6] Magendie (1851) *C.R.A.S.*, **33**, 115–16.

Appendix C Photographs and instruments

[1] By Sotheby's on 1999 October 27.
[2] Which is catalogued, apparently incorrectly, as signed and dated (Marbot B. (1976) *Une invention du XIXe siècle. Expression et technique. La photographie. Collections de la Société française de photographie*. Bibliothèque nationale, Paris).
[3] In 1905 for the RMS (*Journal of the Royal Microscopical Society for the year 1906* 122) and in 1929 to the SFP and the Wellcome Institute.
[4] Nachet A. (1929) *Collection Nachet. Instruments Scientifiques et Livres Anciens*. Petit, Paris.
[5] Inventory 8045.
[6] Inventoried in A.N. ET/XLIX/1504.
[7] Inventory 13993.
[8] Brenni P. (1989) in *Studies in the History of Scientific Instruments*. Blondel Ch., Parot F., Turner A., Williams M. (eds) Rogers Turner, London. pp. 169–78.
[9] Inventory 8985.
[10] Turner G. L'E. (1983) *Nineteenth-century Scientific Instruments*. Sotheby, London.
[11] *Catalogue systématique des appareils d'optique construits dans les ateliers de J. Duboscq*. Hennuyer, Paris (1864).
[12] *Catalogue et prix des instruments de science qui se trouvent ou s'exécutent dans les magasins et ateliers de Secrétan. Deuxième partie, Géodésie, Astronomie, Météorologie, Marine*. Paris (1874).

Appendix D Building a Foucault pendulum

[1] Supply of architectural pendulums by the California Academy of Sciences is currently interrupted. The clockmaker Marcel Betrisey (www.betrisey.ch) advertises clocks incorporating Foucault pendulums.
[2] E.g. Synge J. L., Griffith B. A. (1949) *Principles of Mechanics*. McGraw-Hill, New York; Olsson, M.G., The precessing spherical pendulum. *American Journal of Physics*, **46**, 1118–19 (1978); Spherical pendulum revisited. **49**, 531–4 (1981). The term 'precession' in Olsson's first title is a common misnomer in the Foucault pendulum literature. See remark on page 139.
[3] Ref. 9.66.
[4] Franchot (1851) *C.R.A.S.*, **32**, 505, where Faye says that Foucault had the same idea during trials at the Observatory. Franchot's first idea was to move a mass within the bob (*Le Siècle*, 1851 April 18); later he suggested the more-practical expedient of moving the suspension point (*C.R.A.S.*, **32**, 768–70).
[5] Pippard A. B. (1989) Foucault's pendulum. *Proceedings of the Royal Institution of Great Britain*, **63**, 87–100.
[6] Pippard A. B. (1988) The parametrically maintained Foucault pendulum and its perturbations.

Proceedings of the Royal Society of London, A, **420**, 81–91.

[7] Charron F. (1931) *C.R.A.S.*, **192**, 208–10; *Bulletin de la Société astronomique de France*, **45**, 457–62. A cycloidal cone had earlier been suggested by Whittle G. (1887) *English Mechanic & World of Science*, **1185**, 346.

[8] Moppert C. F., Bonwick W. J. (1980) The New Foucault Pendulum at Monash University. *Quarterly Journal of the Royal Astronomical Society*, **21**, 108–18.

[9] Masner G., Vokura V., Maschek M., Vogt E., Kaufmann H. P. (1984) Foucault pendulum with eddy-current damping of the elliptical motion. *Review of Scientific Instruments*, **55**, 1533–8. This paper provides a graphic illustration of the importance of drive *push* over *pull* to minimize ellipsing.

[10] Crane H. R. (1981) Short Foucault pendulum: A way to eliminate the precession due to ellipticity. *American Journal of Physics*, **49**, 1004–6. Crane's magnet has been analysed by Hecht K. T. (1983) The Crane Foucault pendulum: An exercise in action-angle variable perturbation theory. *American Journal of Physics*, **51**, 110–14.

[11] Kruglak H., Oppliger L., Pitter R., Steele S. (1978) A short Foucault pendulum for a hallway exhibit. *American Journal of Physics*, **46**, 438–40; Kruglak H., Pitter R. (1980) Portable, continuously operating Foucault pendulum. *American Journal of Physics*, **48**, 419–20; Kruglak H., Pitter R., Steele S. (1980) A short, movable Foucault Pendulum. *Sky & Telescope*, **60**, 330–2; Kruglak H., Steele S. (1984) A 25 cm continuously operating Foucault pendulum. *Physics Education*, **19**, 294–6.

[12] Crane H. R. (1995) Foucault pendulum 'wall clock'. *American Journal of Physics*, **63**, 33–9.

[13] Dieter G. E. (1976) *Mechanical Metallurgy*. McGraw-Hill, New York.

[14] The strain could be limited by fixing the wire inside a trumpet-shaped hole that tapers down to the wire diameter. The trumpet would also act as a Charron ring.

[15] Marillier A. (1998) L'expérience du pendule de Foucault au Palais de la découverte. *Revue du Palais de la découverte*, **26**, No. 258, 31–45. This article provides an excellent introduction to pendulum construction.

[16] Haringx J. A., van Suchtelen H. (1957/8) The Foucault pendulum in the United Nations Building in New York. *Philips Technical Review*, **19**, 236–41.

[17] Ref. 9.12.

Index

For persons, the commonly used forename and life dates are given when known.

aberration
  chromatic, 21, 200
  positive and negative, 208
  spherical, 21, 35–36, 210
  stellar ('of starlight'), 118–119, 137, 227–228
Académie des Beaux-Arts, 4, 27
Académie des Sciences, 2, 4, 8, 29–30, 92–93, 267, 272, 275, 278
  Comptes rendus, 8, 30, 89
  elections, 30, 94, 100, 131, 154–155, 223, 238–246, 274
  journalists' access, 8, 80, 89
  meeting procedure, 1, 8, 29–30, 299–300
  Mémoires des savants étrangers, 68
  origins, 3, 4, 7
  pli cacheté, 29
  prizes, 237, 246
Académie des Sciences et Lettres (Montpellier), 224, 238
Academy of Sciences, *see* Académie des Sciences
acceleration
  centrifugal, 150
  centripetal, 149–150
  Coriolis, 150
achromatic doublet, *see* lens
air bag (for telescope mirror), 213–214, 217
Airy disc, 206–207
Airy, George B. (1801–92), 28, 52, 158, 184, 201, 207, 218, 228, 262
Albert [Prince Consort] (1819–61), 179
d'Alembert, J. Le R. (1717–83), 137
Algeria, 214, 218–219, 267, 299, 300
Algiers Observatory, 214, 215, 218
aluminium, 180, 195, 216, 262, 292
ambrotype, 220–221, 265
amphitype, 220, 305
anaesthesia, 84–86
Ancien régime, 2
Ångström A. A. (1814–74), 112
angular diameter (of stars), 224–225
anholonomy, 143
Annales de chimie et de physique, 68, 105

apportioner, 252
Arago, D. François J. (1786–1853), 6–9, 24–25, 27–29, 30, 47, 50, 51–54, 61, 64–65, 67, 68, 76, 78, 80, 82–83, 87, 96, 120–121, 124, 128–129, 132, 138, 139–140, 147, 149, 154, 162, 178, 183–184, 185, 189, 195, 230, 234, 257, 265, 301
arc, *see* electric arc
Archereau, H. A. (1819–93), 102
Archives des sciences physiques et naturelles, 176
Aristarchos (c. 310–230 BC), 133
Aristotle (384–322 BC), 133–134, 136
asteroids, 84, 274, 300–301
astronomical unit, 227, 234
Astronomie populaire (Arago's), 8, 55, 80, 100, 119, 123, 140
Atilla, Ethele ['the Hun'] (c. 406–53), 145
Audouard, 299
autocollimation, 264–265
aviso, 255, 259
Babbage, Charles (1792–1871), 280, 282
Babinet, Jacques (1794–1872), 67–68, 78, 93, 137, 149, 181, 190, 194–195, 202, 203, 209, 213, 214, 232, 240, 282
Bailly, J.S. (1736–93), 4
balance
  dynamic, 126–127, 162, 256
  static, 127
Balard, Antoine-Jérôme (1802–76), 31, 113, 115, 131–132, 180, 236
Balzac, Honoré de (1799–1850), 81
barometer, 304
battery
  of Bunsen cells, 107–108
  Volta, 48
  without metals, 175–176
Becquerel, A. Edmond (1820–91), 33, 74–75, 78, 87, 113, 155, 173, 195, 235–236, 240
Becquerel, A. Henri (1852–1908), 33
Becquerel, Antoine C. (1788–1878), 33, 178, 240

Belfield-Lefèvre, Henry (d. 1853), 37–39, 56, 152–153, 230
Bellarmine, R. F. R. (1542–1621), 135
Benzenberg, J. F. (1777–1846), 137
Berchtold, Charles von, 236
Berget, Alphonse (1860–1933), 287
Berlioz, Auguste, 250
Berlioz, Hector (1803–69), 88, 130
Bernard, Claude (1813–78), 69, 87, 90
Berthelot, P. E. Marcelin (1827–1907), 303
Bertin, L. F. ['Bertin l'Aîné'] (1766–1841), 79
Bertin, L. F. ['Bertin de Veaux'] (1771–1842), 79
Bertin, L. M. Armand (1801–1854), 82–83, 93
Bertrand, Joseph L. F. (1822–1900), 90, 94, 235, 243–244, 246, 262, 272, 283
Berzelius, J. J. (1779–1848), 22
Bessel, Friedrich W. (1784–1846), 119, 121
Bibliothèque universelle de Genève, 176
Binet, Jacques P. M. (1786–1856), 140, 149–150
binocular colour mixture, 108–109
Biot, Jean-Baptiste (1774–1862), 7, 24, 61, 154, 300
birefringence, *see* light, double refraction
Bismarck, Otto E. L. von (1815–98), 9, 269
blanc d'Espagne, 217
Bochart de Saron, J. B. G. (1730–94), 4
Bohnenberger, J. G. F. von (1765–1813), 162, 165
Bonaparte, C. Louis-Napoléon, *see* Napoléon III
Bonaparte, N. J. C. P. ['Jérôme-Napoléon'] (1822–91), 168, 255
Bonaparte, Napoléon, *see* Napoléon I
Bontemps, Georges (1801–after 1882), 196
Bory de Saint Vincent, 300
Bouguer, Pierre (1698–1758), 52, 87, 137
Boutigny, P. H. (d. 1884), 113
Bradley, James (1693–1762), 118–119, 137, 161
bread, 14, 88, 92, 303–304
Breguet, Abraham L. (1747–1823), 121
Breguet, Louis F. C. (1804–83), 121–122, 124–125, 128–129, 239
Brewster, David (1781–1868), 15, 61, 112, 119, 128, 168, 195, 248

British Association for the Advancement of Science, 238
  Dublin meeting (1857), 204
  Edinburgh meeting (1834), 120
  Liverpool meeting (1854), 20, 166–167, 177–178, 185
Broc, 301
bromination, *see* daguerreotype
Brougham, Henry (1778–1868), 119, 128
Buff, Heinrich (1805–78), 176
Bulard, C., 214, 218, 282
Bunsen
  battery, 107–108
  cell, 48, 50–51, 107
Bunsen, Robert W. E. (1811–99), 48, 111–112
Burdin, Claude (1788–1873), 242
Bureau des Longitudes, 4–5, 8, 184, 238, 244, 257, 267, 272, 282

Cagniard-Latour, Charles (1777–1859), 115, 126–127, 154–155, 239
Callan, Nicholas V. (1799–1864), 192
calorific rays, 70–72
camera obscura, 21–22, 220
Canada balsam, 22
cancer, 302
Carcel lamp, 14, 73
Carnot, H., 301
Carnot, N. L. Sadi (1796–1832), 188
Cassini, Jean-Dominique (1625–1712), 3, 118, 124, 135
Cauchy, Augustin L. (1789–1857), 80, 87, 90, 149, 239
Cavaillé-Coll, Aristide (1811–99), 230
Cavendish, Henry (1731–1810), 229
celestial sphere, pole, 25
Chacornac, Jean (1823–73), 87, 206, 218–220, 223, 282
Chance Brothers, 196, 269
Chance discs, 196–198, 263–267, 273–274, 276
Charcot, Jean M. (1825–93), 271
Charles X (1757–1836), 6, 79
Charron ring, 308–310
Chasles, Michel (1793–1880), 131
Chateaubriand, F. René de (1768–1848), 79
Chaudé, 105
chemistry, 87, 90, 303
Chenavard, Paul M. J. (1807–95), 145
chestnut, 13
Chevalier, Charles L. (1804–59), 23–24, 28, 32, 34–35, 40, 50, 55, 58, 105, 189, 235, 265
Chevalier, J. V. Vincent (1770–1841), 23, 42
Chevreul, M. Eugène (1786–1889), 87, 89
cholera, 188, 299
chopping, 71
chronology (summary), 295
Clapeyron, B. P. Émile (1799–1864), 239, 242
Clark, Alvan (1804–87), 223
Claudet, Antoine F. J. (1797–1867), 31, 75
Clerk Maxwell, James (1831–79), 234, 262

clock
  astronomical, 100–101
  escapement, 97
  synchronized, 101
coffee, 19
Colbert, J.-B. (1619–83), 2, 215
collaboration, 95, 124, 282
Collected works, see Recueil
Collège de France, 6, 271, 306
Collège Stanislas, 15, 29
collimator, 195, 200, 279
Combes, Charles P. M. (1801–72), 272
comet, 84, 88, 301
  Donati's, 213–214
  of 1842, 91
  Swift-Tuttle, 223
Comité Consultatif des Arts et Manufactures, 237
Comptes rendus, see Académie des Sciences
Comte, Auguste (1798–1857), 92, 111
conduction (electrical, by fluids), 174–177, 283
Conservatoire (national) des Arts et Métiers, 4, 168, 173, 272, 305–306
Cook, James (1728–79), 228
Cook, Thomas (1808–92), 269
Copernicus, Nicolas (1473–1543), ix, 134–136
Copley Medal, 111, 178, 189, 238
Coriolis, G. G. de (1792–1843), 150
Corliss, George H. (1817–88), 270
Cornu, M. Alfred (1841–1902), 124, 129, 233–234
Cosmos, xii, 20, 75, 177
coup d'état
  of 1799, 5
  of 1851, 9, 93, 160, 173, 183
Cours de microscopie (Donné's, *Atlas*), 42, 44–47, 276
Couvent des Carmes, 34, 55, 220, 280, 305
cranks, 89, 91, 161, 301
Crimean war, 167, 186, 243, 301
Curie, Marie (1867–1934), xi
Cuvier, J. L. N. F. ['Georges'] (1769–1832), 44, 78

D lines (reversal of), 109–111, 191, 283
Daguerre, L. J. M. (1789–1851), 23–24, 26–28, 31, 38, 42
daguerreotype, 10, 14, 27–28, 37, 42, 217
  bromination, 31–32, 34, 39, 77
  continuing rays, 33, 73–74, 77
  electrotype copy, 55–56
  gilding, 29, 30
  image formation, 38, 72–73, 77
  image structure, 28, 42
  micrographs, 44–47, 305
  origins, 24–25, 27–28
  physical mechanism, 76–77
  protecting rays, 73
  response and sensitivity, 30–31, 38–39, 72–76, 286
  solarization, 38, 73

taken by Foucault, 34–36, 45, 54–56, 73, 305
d'Alembert, J. Le R. (1717–83), 137
Davy's egg, 48
Davy, Humphry (1778–1829), 48
day (sidereal and solar), 26
de la Rive, Auguste A. (1801–73), 106, 128, 148, 158, 176, 244, 264
de la Rue, Warren (1815–89), 37, 221
Decharte, 299
Delacroix, Eugène (1798–1863), 195
Delaroche, Paul (1797–1856), 24, 56, 83
Delaunay, Charles E. (1816–72), 238, 272, 274
Deleuil, L. J. (1795–1862), 50–51, 102, 108, 174, 269, 302–303
Deloge, 104
Delorme, 104, 106
Denonvilliers, Charles P. (1808–72), 69
Desains, Q. Paul (1817–85), 115, 206, 212, 240, 282, 298
Descartes, René (1596–1650), x, 59–60, 119, 169, 243
Despretz, César M. (1789–1863), 16, 112, 131, 173, 177, 240
dew, 88
dichoptic colour mixture, 108–109
differential solids, 210, 291, 305
diffraction, *see* light
dispersion, *see* light
Doctorat ès sciences, 5, 37, 130–132, 173, 266
Dolland J. (1706–61), 21
Dom Pedro II, (1825–91), 168, 237, 252
Donati's comet, 213–214
Donati, G. B. (1826–73), 213
Donné, M. F. Alfred (1801–78), 19, 36, 39–46, 50, 57–59, 65, 78, 80, 82, 92, 94, 104, 113–114, 148, 164, 198, 214, 224–225, 241–242, 246, 268, 271, 278, 281–282
Donné, M. A. V. [*née* de Joantho] (1816–79), 241–242, 281
Doppler–Fizeau effect, 112
Draper, Henry (1837–82), 214
Drayton, Thomas, 201–202, 217
dreams, erotic, 86
Drummond light, *see* limelight
Drummond, Thomas (1797–1840), 41
Duboscq, Jules L. (1817–86), 106, 247–250, 252–254, 267, 269, 272
Duhamel, J. M. C. (1797–1872), 240
Dumas, Alexandre D. de La P. (1802–70), 81, 88
Dumas, Jean-Baptiste A. (1800–84), 28, 105, 112, 129–131, 174, 178, 182, 235, 237, 282
Dumoulin-Froment, P., 269, 288
Dupin, P. Charles F. (1784–1873), 301
Duponchel, C. Edmond (1795–1868), 106
Dutrochet, R. J. Henri (1776–1847), 86
dynamics, 133, 135–136, 149–150, 170, 172

Earth
  flattening, 7, 137

Index

mass, 230
principal motions, 133
rotation, ix, 7, 133–172
eclipse
   1860 solar, 218–221, 267
   Jupiter's moons, 118, 227
   lunar, 133
École de Médecine, 17, 39
École Normale Supérieure, 4, 277
École Polytechnique, 4, 7, 28, 218, 219, 246, 267, 306
Eichens, F. Wilhelm (1818–84), 213, 222, 230, 267–270, 275–276, 291
Eiffel, Gustave (1832–1923), 278
Einstein, Albert (1879–1955), 170, 220
electric arc, 43, 48–51, 59, 102, 109–111, 256
   regulated lamp
      at sea, 254–255
      at the Opera, 106–107
      auxiliary regulator, 114, 174
      Deleuil's, 106
      Duboscq's, 106, 250, 306
      Foucault's, 101–106, 109, 114, 176, 248, 252–255, 269, 272, 306
      lighthouse trials, 253–254
      Serrin's, 251, 254
      Staite's, 104
electricity
   conduction by fluids, 174–177, 283
   heating effect, 252, 296–299
   speed of, 120, 187
electrolysis, 174–177
electroplating, *see* electrotyping
electrotome, 193
electrotyping, 10, 37, 55–56
elephants, 134
Empire, Second, 9–10, 245, 276
Encke, J. F. (1791–1865), 228
energy, conservation of, 188–191, 289, 306
England, 111, 166, 177–179, 282
escapement, 97–98
état civil, xii
ether (anaesthetic), 84–86
ether (luminiferous), 138, 169–170, 186
Euler, Leonhard (1707–83), 151
Exposition Universelle
   of 1855, 10, 167, 194–196, 305
   of 1867, 10, 260, 269–270, 273, 288
   of 1900, 167, 275

Faraday's Law, 174–177
Faraday, Michael (1791–1867), 87, 174–179, 189, 191
Farcot, 256
Favé, Ildephonse (1812–94), 94, 190, 212, 243–246
Faye, Hervé E. A. A. (1814–1902), 100–101, 213, 218, 221, 238, 282
Figuier, Louis G. (1819–94), xii, 43, 69, 84, 102, 107, 138, 172, 244, 280
Fizeau, A. Hippolyte L. (1819–96), 28–29, 30–32, 34, 36, 42, 44, 47–55, 59, 62–68, 70–76, 78, 95, 112, 114, 115, 121–125, 128–129, 137–138, 155, 186, 192, 224, 230, 233, 240–241, 266, 285
Flammarion, Camille (1842–1925), 168, 211, 282, 287
Flaubert, Gustave (1821–1880), 201
Flaud, H., 270
floaters, 15
Flourens, M. J. Pierre (1794–1867), 87, 241
flux (luminous), 48–49, 181
Foiret, Jacques (d. 2001), 168
foot (forces in), 69
force, *see* acceleration
Foucault, A. A. F. [sister] (1823–1904), 12, 16, 67, 179, 272, 277, 280, 282, 291
Foucault, Aimée N. [mother, *née* Lepetit] (1793–1880), 12, 14, 38, 55, 67–68, 103, 271–272, 276–278, 282
Foucault, J. B. Léon (1819–1868)
   appearance, 18, 41, 161, 248
   beliefs, outlook, 14, 90–93, 145, 247, 272, 281
   birth, xii, 12
   education, 15–16, 283
   emotions, 16–18, 179, 246, 304
   friends, 2, 16, 19–20, 38, 42, 56, 68, 90, 93, 94, 104, 113–114, 123, 124, 126, 129, 152, 157, 164, 180, 220, 221, 230, 235, 243, 247, 267, 271, 273, 275, 276, 281, 283, 305
   illness and death, 1–2, 268, 270–272, 275, 280
   job hunting and employment, 88, 173–174, 185–187, 204–206, 211–212, 215, 257, 267, 272, 280–281
   medical studies, 16–17, 69–70
   mental state, 113–114, 198, 206, 211, 279, 280–281
   personality, 15, 19–20, 41, 107, 280–281, 283
   portraits, ii, 16, 248, 278, 285, 291
   posthumous inventory, xii, 16, 124, 267
   relationship with women, 17–20, 91, 114–115, 225
   tastes, 16, 18–19, 268
Foucault, J. B. M. [grandfather] (d. 1839), 11–12
Foucault, J. L. F. [father] (c. 1784–1839), 11–14, 17, 67
Fox Talbot, W. H. (1800-77), 26, 111–112
France (map of), 293
Fraunhofer lines, 63–64, 71, 73, 110–111, 198, 285, 286
Fraunhofer, Joseph von (1787–1826), 63, 285
Fresnel, Augustin J. (1788–1827), 61, 64, 78, 119, 138, 195, 215
Froment, P. Gustave (1815–65), 104–105, 107, 114, 123, 124, 139, 156, 160, 162, 167, 187, 195, 207, 216, 230, 232, 239, 246, 269, 288

Gaigneau, 105

Galilei, Galileo (1564–1642), ix, 52, 97, 118, 135, 145, 284
Gambey, H. P. (1787–1847), 44, 78, 96, 98, 104, 184, 189, 265
Gariel, C. M. (1841–1924), 12, 151, 276–277
Garnier, Gustave (1834–92), 277–278
gas carbon, 50–51, 57, 59, 109
gaslight, 181–182
Gaudin, Marc-Antoine A. (1804–80), 32–34, 55, 75
Gauss, J. K. F. (1777–1855), 136
Gay-Lussac, L. Joseph (1778–1850), 154
gearing
   helicoidal, 121, 123
   planetary, 252, 275
Geneviève [Saint] (c. 422–502), 145
Geoffroy Saint-Hilaire, Isidore (1805–1861), 299
Georgette Du Buisson, 255
Germain, Sophie (1776–1831), xi
Girard, Aimé C. A. (1830–98), 93, 220, 235–236
Girard, L. D. (1815–71), 104, 126, 247
Giroux, Alphonse, 28, 31
Goethe, J. W. von (1749–1832), 40
Goldschmidt, Hermann (1802–66), 300–301
governor, 261–262, 264, 269–270, 275, 283
   fan, 257–258, 261–262, 269–270, 306
   for lighthouses, 257–258
   for steam engines, 257–260
   for telescope drives, 98, 257, 275
   Foucault's first, 256–257
   naval, 258–261
   oscillations, 260–261, 270
   reponsive to acceleration, 260–261
   Watt's, 256
grandes écoles, 4, 246
Grandidier, A. L. Marguerite (1846–1928), 12, 276, 280
Grassot, Eulalie (c. 1805–39), 12, 17–18
's Gravesande, W. J. (1688–1742), 250
Great Britain (steamship), 88
Greenwich Observatory, 187
Gros, Antoine (1771–1835), 147, 235
Guéneau de Mussy, H. (1814–92), 179
Guérard, J. Alphonse (1796–1874), 271
Guérin, 33
Guglielmini, G. B. (1763–1817), 137
Guizot, F. P. G. (1787–1874), 82–83
gyrocompass, 165, 240, 284
gyroscope, 161–167, 168, 178, 179, 189, 195, 269, 271, 284, 288, 306
   interpretation, 169–172

Hadley, John (1682–1744), 207
Hall, C. M. (1703–71), 21
Halley, Edmond (1656–1742), 119
Hamann, E.-F., 165
Harvard College Observatory, 94, 100, 221
Haussmann, G. E. (1809–91), 294
heat, mechanical equivalent of, 188–191, 289, 306

heliostat, 42–44, 249–250, 269, 306
Helmholtz, Hermann L. F. von (1821–94), 76
Henry, Joseph (1797–1878), 191
Hering, Ewald (1834–1918), 109
Hero (fl. c.1st century AD), 117
Herschel, F. William (1738–1822), 70, 185, 268
Herschel, John F. W. (1792–1871), 28, 41, 52, 70, 72–73, 204
La Hève, 253–254
Hipparchus (fl. 160–125 BC), 134
Hockes, 178
Holmes, F. H. (fl. 1840–75), 250, 253
Hooke, Robert (1635–1703), 60, 136, 137
Houdin, *see* Robert-Houdin
'house of precision', 68, 139, 268, 276, 278
Hugo, Victor (1802–85), 9
Humboldt, F. H. Alexander von (1769–1859), 24
Huygens, Christiaan (1629–95), 52, 60–61, 97–98, 118–119, 256

Iceland spar, 60–61
imagination, ii
Imperial Academy of Sciences (St Petersbourg), 238–239
induction coil, 191–194, 237–238, 269, 290, 305–306
 contact breaker, 191–194, 205, 269, 290, 306
induction currents ('Foucault currents'), 188–191, 289, 306
inertia, 138, 153, 158, 169
Ingres, J. A. D. (1780–1867), 79, 195
Institut National, 2, 4
intensity (luminous), 48–49
interference, *see* light
inventory (posthumous), xii, 16, 124, 267
Ireland, 204
isochronism, *see also* governor, pendulum, 97, 100, 256–257, 260–261

Jamin, Jules C. (1818–86), 90, 113, 155, 177, 240
Janin, Jules (1804–74), 24
Jeanneton, 114–115
Joule, James P. (1818–89), 190–191
Journal des Artistes, 23–24, 33
Journal des Débats, xii, 14, 39, 79–94, 104, 202, 246, 296–304
 letters to, 128, 165
 readership, 79, 82, 84, 94
June Days (1848), 9, 68, 93, 145
Jupiter, 135, 204, 263
 moons' eclipses, 118, 227

Kammerling Onnes, Heike (1853–1926), 157
Kelvin, *see* Thomson W.
Kepler's Laws (planetary motion), 118, 134–135, 229
Kepler, Johannes (1571–1630), 117–118, 134, 227, 229
Kirchhoff, Gustav R. (1824–87), 8, 111–112, 191

knife-edge test, *see* optical fabrication

La Condamine, C. M. de (1701–74), 137
Lagrange, J. Louis de (1736–1813), 151, 243, 300
Lalande, J. J. L. de (1732–1807), 80
Lallemand, C. François (1790–1854), 299
Lamarle, A. H. E. (1806–75), 165–166
Lambert, J. H. (1728–77), 52
Laplace, P. S. (1749–1827), 5, 53, 60, 78, 136–137, 150, 188, 300
Lassell, William (1799–1880), 178–179, 200–201, 264
latitude, 25
Laugier, Lucie (1820–1900), 8, 238
Laugier, P. A. Ernest (1812–72), 100–101, 238
Lavoisier, A. L. de (1743–94), 4, 78
Lebert, 302
Lecorbeiller, Charles (d. 1842?), 17
Leeuwenhoek, Antony van (1632–1723), 40
Légion d'honneur, 40, 123, 130, 223, 230, 238
Lenoir, J. J. E. (1822–1900), 238, 251
lens, *see also* optical fabrication
 19-cm (7-inch), 266
 24-cm (9-inch), 266, 291, 305
 achromatic doublet, 22, 40, 263
 at 1867 Exposition Universelle, 269
 autocollimation, 264–265
 silvered, 268
Lense–Thirring effect, 170–171
Le Play, Frédéric (1806–82), 194–195, 269–270, 276
Lerebours, N. J. (1761–1840), 96, 100
Lerebours, N.-M. P. (1807–73), 32, 34, 55, 74, 100, 199, 209
Le Verrier, Urbain J. J. (1811–77), x, 1–2, 78, 83, 113, 178, 184–187, 196, 198, 201, 204–206, 211–212, 215, 218–221, 223, 224, 227–230, 232–234, 238, 257, 263–267, 272–274, 281–282
Leviathan (telescope), 200, 204, 223
Liais, Emmanuel (1826–1900), 187, 205–206, 282
Libri-Carucci Dalla Sommaia, Guglielmo B. I. T. (c. 1802–69), 82–83, 282
Lieberkühn, J. N. (1711–56), 41
Liebig, Justus von (1803–73), 201
light
 diffraction, 60, 62, 72
 dispersion, 21
 double refraction, 60, 64
 interference, 60–61, 70–72, 224
  with long path differences, 62–64, 285, 305
 nature of, 52–53, 59–61, 119–121
 polarization, 52–53, 61, 65–66
  chromatic, 64–68, 71, 104
  circular and elliptical, 66–67
 refraction, 59

refractive index, 59, 114, 197–198
speed of, 61, 117–132, 187
 absolute value, 119, 122–124, 227–234, 257, 271, 275, 292, 305–306
 air–water experiment, 124–130, 195
 Arago's experiment, 120–121, 128, 132
 Fizeau's toothed wheel, 122–124, 230, 233
 uniformly tinted, 75, 286
lighthouse
 electric light, 253–254
 optics, 195–196, 215, 257
Lima Observatory, 266
limelight, 41, 42–43, 48, 50–51
Lindsay, A. W. C. (1812–80), 275
Liouville, J. (1809–82), 150–152, 154
Lissajous, Jules A. (1822–80), 157, 159, 217, 247–248, 261, 275–276, 281
lizard, 13
Lormier *père*, 107
Louis XIV (1638–1715), 2
Louis XVI (1754–93), 3, 4, 6
Louis XVIII (1755–1824), 6
Louis-Napoléon, *see* Napoléon III
Louis-Philippe (1773–50), 6, 27, 83, 145, 173
Luynes, H. d'A., Duc de (1802–67), 213–214, 236–237

Mach, Ernst (1838–1916), 169–170
Madamet, 261
Magendie, François (1783–1855), 85–86, 280, 282, 302
magneto (Alliance), 250–253
Malus, E. L. (1775–1812), 61, 78
'man of science', x
map
 1860 solar eclipse, 219
 France, 293
 Paris, 294
 pendulums in USA, 149
Marié-Davy, E. H. (1820–93), 86–87
Mars, 135, 228, 230, 232
Marseilles, *see* Observatoire de Marseille
Martin, Adolphe (1824–96), 265–266, 273, 275–276, 283
Masson, A. P. (1806–60), 155, 192
mathematics, 90, 131, 262, 283, 301
Matteucci, Carlo (1811–68), 87, 177
Maupertuis, P. L. M. de (1698–1759), 137
Mauvais, F. Victor (1809–1854), 100
Maxwell, James Clerk (1831–79), 234, 262
Mayer, J. R. (1814–78), 188–189
measures, 5, 7, 104, 119, 283, 302–303
Mécanique céleste (Laplace's), 5, 53, 137, 150
Medical School, 17, 39
Meidinger, J. Heinrich (1831–1905), 177
Mémoires des savants étrangers, *see* Académie des Sciences
Mercury, 219, 220
meridian, 25

Index

Mérimée, Prosper (1803–70), 269
Mersenne, Marin (1588–1648), 136
metric system, *see* measures
Meyerbeer, Giacomo (1791–1864), 106
Michelson, Albert A. (1852–1931), 225
microscope, 305
    achromatic, 40
    catadioptric, 214–215, 279
    microscope-daguerréotype, 42–44
    photo-electric, 57–59, 305
    solar, 41
microscopy, 39–40, 42–47
minor planets, 84, 274, 300–301
mirror, magic, 209
Moigno, François N. M. (1804–84), 20, 75, 89, 102, 112, 176–178, 204, 210, 213, 235, 254, 271
    comments by, 34, 38, 39, 131, 152, 154, 166, 168, 175, 181, 187, 190, 202, 204, 210, 213, 214, 218, 230, 233, 236, 237, 239, 240, 243, 245, 246, 250, 255, 256, 257, 269, 270, 271, 272, 282
Monins, 302
Monmerqué, L. J. N. (1780–1860), 14, 35
de Montijo, Eugénie (1826–1920), 271, 301
Moon, 124, 186, 218, 220–221, 248, 275, 278
Morin, Arthur J. (1795–1880), 272, 283
Morse, Samuel B. (1791–1872), 28
music, 17–18

Nachet, J. Alfred (1831–1908), 55, 305
Nantes, 11, 13
Napoléon I ['Napoléon Bonaparte'] (1769–1821), 5, 7, 9, 11, 79, 93, 145, 147
Napoléon III ['Louis-Napoléon'] (1808–73), 9, 90, 94, 145, 160, 173–174, 180, 183, 185–187, 194–195, 198, 205–206, 211, 212, 214, 235, 245–246, 259, 274, 281–282, 294
    financial support by, 93–94, 124, 145, 160–161, 190, 212, 237, 243, 275
Neptune, 82–83, 88, 179, 185
Newcomen, Thomas (1663–1729), 280
Newton's Laws, 136, 149, 243
Newton, Isaac (1642–1727), x, 60, 118, 136, 149, 155, 161, 169, 172, 188, 197, 200–201, 229, 283
Nicol prism, 65–66
    Foucault's improvement, 203, 279, 306
Nicol, W. (1768–1851), 65
Niepce, J. N. (1765–1833), 22–23, 27
Nollet, F. (1794–1853), 250
nutation, 161–162

Observatoire de Marseille, 224–225, 274, 291, 305
Observatoire de Paris, 3, 6–7, 94, 121, 125, 139, 183–187, 196, 198, 199, 204–206, 211, 230, 235–236, 257, 263, 267, 271, 273, 278, 281–282, 291, 305
    Arago's dome, 96, 100, 184, 263

Œrsted, H. Christian (1777–1851), 148
oil (lubricating), 303
opthalmia, 57, 193
optical fabrication, 225–226, 282, 283
    ellipsoidization, 211
    first test, 207–208
    flats, 265
    hyperboloidization, 217
    lenses, 264–265, 273
    mirrors, 199–200, 206, 215–217, 221–222, 264
    parabolization, 212–213
    retouches locales, 210, 226, 265
    second test, 208–209
    third (knife-edge) test, 208–210, 226, 291
optical power, 207
optics institute, 265
Orsini, Felice (1819–58), 206
Oudet, 44
Owen, Richard (1804–92), 178
oysters, 19, 88

Page, C. G. (1812–68), 192
Palais de l'Industrie, 167
Panthéon, 144–147, 160, 168, 225
Papin, Denis (1647–c. 1712), 280
parallax
    solar, 227–228, 230, 232, 234
    stellar, 117–119, 135, 137
Paris
    Commune, xii
    map of, 294
    Observatory, *see* Observatoire de Paris
Pascal, Blaise (1623–62), 13, 280
Passot, 242
Pasteur, Louis (1822–95), 9, 69, 87, 161, 178, 243, 282
patents, 105, 237, 247, 252, 257, 262, 272
Pavé, 104
Pease, Francis G. (1881–1938), 225
Pecqueur, 256
pendulum, 306
    Charron ring, 308–310
    conical, 97–100, 256
    ellipsing, 100, 158–160
    interpretation, 149–154, 169–172
    isochronism, 97, 100
    mania, 148–149
    Meridian Room, 125, 139–141, 156
    Palais de l'Industrie, 158, 167–168, 271–272, 289, 305
    Panthéon, 144–148, 156, 160, 168–170, 278, 286–287, 305, 307, 310
    practical difficulties, 155–160, 283, 307–310
    reveals Earth's rotation, 138–139
    Rue d'Assas, 68, 139, 156, 276, 278
    simple, 97
    sine factor, 140–144, 161
    suspension, 140, 147, 158, 160, 167, 305

    table-top, 141–142
Person, C. C. (b. 1801), 165–166
Petrie, William (1821–1904), 104–106, 252
Peytier, 238, 241
Phillips, Édouard (1821–89), 244–245
photography, *see also* daguerreotype, 220, 249–250, 304
    astronomical, 186, 220–221, 235–236, 267, 275
    origins, 22–23, 26
photometer, 181–182, 289, 306
photometry, 47–51, 181–182
physicist (sibilant consonants), x
physiology, 90, 108–109, 262
Picard, Jean (1620–82), 118
Place de la Concorde, 102, 167
Plantamour, Émile (1815–82), 187, 198
Pliny the Elder (23–79 AD), 152
Poincaré, J. H. (1854–1912), 170
Poinsot, Louis (1777–1859), 150–151, 161, 171–172, 238, 243–244
Poisson, S. Denis (1781–1840), 141, 150
Poitevin, Alphonse (1819–1882), 236
polarization, *see* light
Pontifical Academy (Rome), 238
Porro, Ignace (1795–1875), 221
Pouillet, C. S. M. M. R. (1790–1868), 140, 149, 173, 183, 240, 245
pouvoir optique, 207
Powell, Baden (1796–1860), 152, 178
precession, 161–162, 164
    misnomer, 139
    of equinoxes, 134, 161
principal axes, 126
prism
    Nicol, 65–66
    Foucault's improvement, 203, 279, 306
Le prophète, 106–107
de la Provostaye, F. H. (1812–63), 155, 240, 298
Prussian Royal Academy of Sciences (Berlin), 238
Ptolemy (c. 90–c. 168 AD), 134
Pulkova Observatory, 100, 101, 213, 221

Quatrefages, J. L. Armand de (1810–92), 299
Quet, J. A. (1810–84), 166

Radiguet, 127
Raymond, Xavier (b. 1812), 104
Recueil des travaux scientifiques, ix, 274–275, 276–277, 279
    reprint, xi
reference frame, 136, 143, 149–150, 169–170, 186
refraction, refractive index, *see* light
Regnauld, Jules A. (1820–95), 19, 86, 104, 108–109, 178–179, 186, 204, 253, 272, 275, 276, 281, 305
Regnault, H. Victor (1810–78), 67, 89, 105, 154, 182, 185, 188, 235–236
regulator, *see* governor

Reich, Ferdinand (1799–1882), 137
relativity, 153, 170–171, 220, 227
retouches locales, see optical fabrication
Revolution
    February (1848), 6, 8, 83, 92, 103
    French (1789), 3–5
    July (1830), 6, 20, 79, 82
Riche, Auguste (1801–73), 276
Robert-Houdin, J. E. (1805–71), 252
Rolland, Eugène (1812–85), 244–245, 275
Rømer, Ole (1644–1710), 118
Rosse, William Parsons, Earl of (1800–67), 200–201, 204, 218, 222–223, 264–265
rotascope, 162
Rouargue, Émile (c. 1795–1865), 167
Rouargue, Adolphe (1810–84), 12, 167, 280
rouge, 199, 202, 217
Royal Astronomical Society (London), 238
Royal Society of Edinburgh, 238–239
Royal Society of London, 178, 238
Rue d'Assas, 13–14, 35, 55, 67–68, 205, 211, 268
Rue de Fleurus, 268
Ruhmkorff coil, 191–194, 237–238, 269, 290, 305–306
Ruhmkorff, H. D. (1803–77), 189–198, 237, 269, 290

Saint Etienne du Mont (church), 12–13, 23
Saint Gobain (glassworks), 215, 264
Saint Sulpice (church), 15, 230, 272
Saint-Venant, Adhémar J. C. B. de (1797–1886), 239, 242, 244
Sainte-Claire Deville, Charles J. (1814–76), 180
Sainte-Claire Deville, Henri E. (1818–81), 1–2, 161, 180, 216, 235, 257, 266, 271–275, 277, 282
San Domingo, 11–12
Sand, George [pseudonym] (1804–76), 81
Sarton, George A. L. (1884–1956), 132
Saturn, 223
Sautter, Louis (1825–1912), 196, 215–216, 222, 255, 257–259, 262, 265, 270
Schiaparelli, Giovanni (1835–1910), 223
Secchi, Angelo (1818–78), 148, 219, 221
Second Empire, 9–10, 245, 276
Second Republic, *see* Revolution, February
Secrétan, Auguste (1833–74), 269
Secrétan, Marc F. L. (1804–67), 199, 206, 210, 213–216, 218, 235, 247, 257, 261–262, 263, 267, 269, 290, 291, 305
Serres, E. R. Augustin (1786–1868), 299
Serrin, Victor, 251–254, 283
sidérostat, 266–267, 274–276, 306
Silbermann, J. T. (1806–65), 42, 44, 249–250, 267
silvering (of glass), 127, 200–202, 217, 268, 306
Silvestre, A. F. Édouard de, 144–145
Sire, G. E. (1826–1906), 165
siren, 126–127

Sirius, 94, 223
slavery, 8, 11, 91–92
Snel, Willebrord van Royen (c. 1580–1626), 59–60
Société d'Encouragement pour l'Industrie nationale, 58
Société de Physique et d'Histoire Naturelle (Geneva), 238–239
Société Française de Photographie, 235–236, 265, 305
Société Philomathique, 78, 112–113, 115–116, 235
Société Royale des Sciences (Liège), 238–239
solarization, *see* daguerreotype
Soleil, J. B. F. (1798–1878), 44, 105–106, 195, 248
Sorbonne, 5, 75, 82, 173, 281
space, *see* reference frame
Spain, 218–219
speculum metal, 200
spherometer, 216, 305
Staite, W. Edward (1809–54), 104–105, 109, 252
Steinheil, K. A. (1801–70), 202, 225, 269
Stéphan, J. M. Édouard (1837–1923), 224–225
stereoscope, 109, 247–248
Stokes, G. G. (1819–1903), 111–112, 155, 190
stonewort, 86
Struve, Otto W. (1819–1905), 213
Sun, 124, 186, 268
    1860 eclipse, 218–221, 267
    calorific rays, 70–72
    chemical composition, 111–112
    corona, 219–221
    daguerreotype, 52–55
    distance, 112, 227–234
    Fraunhofer lines, 63–64, 71, 73, 110–111, 198, 285, 286
    intensity, 49–51
    limb darkening, 52–55
    parallax, 227–228, 230, 232, 234
    photosphere, 50, 52, 54, 111
    prominences, 219–220
    spectral flux, 71, 73, 286
    spots, 52–55
Swan, J. W. (1828–1914), 256
Sylvester, J. J. (1814–79), 159, 161

Tardieu, Joseph E., 152
telegraphy, 10, 15, 101, 120–121, 133
telescope, *see also* optical fabrication
    equatorial drive, 25–26, 98–99, 164, 179, 213, 257, 261–262, 269–270
    optical principle, 200
    performance evaluation, 206–207, 218, 263
    reflecting, 200–201, 263, 306
        1.2-m, 263–264, 267, 273, 276
        18/20-cm, 204, 219, 305
        32- & 36-cm, 207
        33-cm, 213–215, 263

        40-cm, 215–218, 219, 305
        80-cm (Marseilles), 218, 221–225, 257, 291, 305
        commercial exploitation, 214–215, 290–291, 306
        Foucault's first, 201–202, 237, 279, 290, 305
        Lassell's, 178–179
        Lord Rosse's, 200, 204, 223
        mirror air bag, 213–214, 217
        paraboloidal, 211, 213
        Porro's, 221
    refracting, *see also* lens, 187, 221, 238, 263, 275
        Lerebours' 38-cm, 96, 100, 263
        Secrétan's 33-cm, 257, 263
Terrien, A., 141, 144–145, 147, 154
thermometers (miniature), 70
Thomson, William (1824–1907), 112, 178, 189–191
time (electrical distribution), 101, 187
Titan, 223
toad, 302
tobacco, 19
Tour Saint Jacques, 168
transit (of Venus), 228
Trinity College Dublin, 204, 238
tripoli, Venetian, 217
Triton, 179
Tyndall, John (1820–93), 178

units, *see* measures

Vaillant, J. B. P. (1790–1872), 242, 257
valve (steam), 258
Venus, 135, 202, 228, 230
Verne, Jules (1828–1905), 81
vibrating rod, 138, 156–157
Victoria [Queen] (1819–1901), 179, 248
Volta, Alessandro (1745–1827), 48

Wallot, J. W. (1743–94), 4
Warner & Swasey Co., 100
Watt, James (1736–1819), 256
wave motion, 52–53
wave–particle duality, 132
weights, *see* measures
Wheatstone, Charles (1802–75), 24, 33–34, 108–109, 111, 120, 171, 179, 195, 248
Whirpool nebula, 223
Winterhalter, F. X. (1805–73), 195
Wolf, Charles J. E. (1827–1918), 274–275
Woodward, Joseph J. (1833–84), 46
Worms de Romilly, Paul, 270
Wurtz, C. Adolphe (1817–84), 180

Young, Thomas (1773–1829), 60–61
Yvon Villarceau, Antoine J. F. (1813–83), 218–220, 238, 272

zenith, 25

DATE DUE

```
SCI QC 16 .F626 T63 2003

Tobin, W.

The life and science of
 L eon Foucault
```